A Course in Statistical Mechanics

Harold L. Friedman

Department of Chemistry
State University of New York
at Stony Brook

Prentice-Hall, Inc., Englewood Cliffs, New Jersey 07632

Library of Congress Cataloging in Publication Data

Friedman, Harold L. (date)
A course in statistical mechanics.

Bibliography: p.
Includes index.
1. Statistical mechanics. I. Title.
QC174.8.F75 1985 530.1'3 84-24939
ISBN 0-13-184565-9

Editorial/production supervision
and interior design: Nancy Velthaus
Cover design: Edsal Enterprises
Manufacturing buyer: John Hall

Printed in the United States of America

10 9 8 7 6 5 4 3 2 1

ISBN 0-13-184565-9 01

Prentice-Hall International, Inc., *London*
Prentice-Hall of Australia Pty. Limited, *Sydney*
Editora Prentice-Hall do Brasil, Ltda., *Rio de Janeiro*
Prentice-Hall Hispanoamericana, S.A., *Mexico*
Prentice-Hall Canada Inc., *Toronto*
Prentice-Hall of India Private Limited, *New Delhi*
Prentice-Hall of Japan, Inc., *Tokyo*
Prentice-Hall of Southeast Asia Pte. Ltd., *Singapore*
Whitehall Books Limited, *Wellington, New Zealand*

Contents

Preface *ix*

1 Fundamentals

1.1 Introduction *1*
1.2 Basic Concepts *2*
1.3 Axioms *5*
1.4 Canonical Ensemble and Distribution Function *6*
1.5 Thermodynamics *9*
1.6 The Canonical Ensemble and Thermodynamics *12*
1.7 Grand Ensemble and Distribution Function *14*
1.8 The Grand Ensemble and Thermodynamics *16*
1.9 Fluctuations. General Aspects *19*
1.10 Fluctuations in Energy *20*
1.11 Fluctuations in Density *21*
1.12 Large Fluctuations *22*
1.13 Fluctuations in Composition *22*
1.14 Equivalence of Ensembles *24*
Problems *25*

2 Independent Subsystems

2.1 Introduction *29*
2.2 Distinguishable Subsystems *30*
2.3 A Collection of Simple Harmonic Oscillators *30*

2.4 Einstein's Model of a Crystal *31*
2.5 Debye's Model of a Crystal *35*
2.6 Spins on a Lattice *35*
2.7 Ideal Monatomic Gas. Quantal *37*
2.8 The Classical Limit for a System of Independent Indistinguishable Particles *43*
2.9 Statistical Thermodynamics of Gases *45*
2.10 Translation *46*
2.11 Rotation of Linear Molecules *49*
2.12 Linear and Nonlinear Rotators *52*
2.13 Vibration *53*
2.14 Electronic States *54*
2.15 Nuclear States *55*
2.16 Chemical Reactions *55*
2.17 The Equilibrium Constant *57*
2.18 The Limitations of Chemical Models and Other Independent Subsystem Models *58*
Problems *59*

3 *The Classical Limit*

3.1 Introduction *63*
3.2 An Alternative Formulation of the Partition Function *64*
3.3 The Classical Limit *67*
3.4 The Leading Quantum Correction *69*
Problems *71*

4 *Thermodynamics and Structure. Role of Intermolecular Forces*

4.1 Thermodynamic Functionals of the Potential U_N *75*
4.2 Structure: Spatial Correlation Functions *77*
4.3 Intermolecular Forces *83*
4.4 Some Properties of the Spatial Pair Correlation Function *86*
Problems *90*

5 *Simulation and Diffraction*

5.1 Introduction *93*
5.2 Periodic Boundary Conditions *93*
5.3 Molecular Dynamics *96*
5.4 Monte Carlo *101*
5.5 Diffraction *104*
Problems *108*

6 Cluster Expansions

6.1 The Virial Equation of State *111*
6.2 The Second Virial Coefficient, B_2 *112*
6.3 Functional Derivatives *114*
6.4 Functional Derivative Formulas for Pair Correlation Functions *116*
6.5 Graphs, Diagrams, Cluster Expansions *118*
6.6 Graph Theory *121*
6.7 The Virial Expansion of A^{ex} *129*
6.8 Virial Expansion of the Pair Correlation Function *132*
6.9 Virial Expansion for Hard-Sphere Fluids *132*
Problems *134*

7 Pair Correlation Function Theory

7.1 The Direct Correlation Function and the Ornstein-Zernicke Equation *137*
7.2 The Linearized Debye-Hückel Approximation *140*
7.3 The Stillinger-Lovett Moment Conditions *143*
7.4 The Percus-Yevick Approximation and the Hard-Sphere Fluids *144*
7.5 The Mean Spherical Approximation *148*
7.6 The Hypernetted Chain Approximation *150*
7.7 The Numerical Solution of Integral Equations *152*
Problems *154*

8 Molecular Correlation Functions

8.1 Atoms or Molecules *157*
8.2 Invariant Expansions *158*
8.3 MSA for Dipolar Hard Spheres *162*
8.4 The Dielectric Constant *167*
8.5 Site-Site and Atom-Atom Correlation Functions *171*
8.6 Interaction Site Models and the Site-Site Ornstein-Zernike Equation *174*
8.7 Interaction Site Approximations *178*
Problems *180*

9 Perturbation Theory

9.1 What and Why? *183*
9.2 The λ Expansion *184*
9.3 The Gibbs-Bogoliubov Inequality *185*
9.4 Blip Function Theory *186*
9.5 Topological Reduction *188*
9.6 Cluster Perturbation Theory *189*

9.7 Optimized Cluster Perturbation Theory *192*
9.8 WCA Theory *193*
9.9 Integral Equation Perturbation Theory *195*
Problems *197*

10 Equilibrium Theory of Solutions

10.1 Kirkwood-Buff Theory *201*
10.2 McMillan-Mayer Theory of Solutions *203*
10.3 Adelman's Theory of Solutions *207*
10.4 The Laws of Dilute Solutions *209*
10.5 Second Virial Coefficients in Solution *210*
Problems *212*

11 Time Dependence

11.1 Introduction *215*
11.2 Equations of Motion in Classical Mechanics *216*
11.3 Time Correlation Functions. Formal Theory *219*
11.4 Linear Response Theory *223*
11.5 Electrical Conduction *226*
11.6 Van Hove Correlation Functions *227*
11.7 Linear Response Theory for Local Fields *235*
Problems *237*

12 Langevin's Equation

12.1 Introduction *243*
12.2 Langevin's Equation *245*
12.3 Relaxation and Transport *246*
12.4 Evolution of Distribution Functions *249*
12.5 Rotational Brownian Motion *251*
12.6 The Noisy Oscillator *256*
12.7 Magnetic Resonance *257*
12.8 Chemical Rate Constant *258*
Problems *265*

13 Simulation and Inelastic Neutron Scattering

13.1 Introduction *269*
13.2 Molecular Dynamics Simulation of Rate Processes *269*
13.3 Brownian Dynamics *274*

13.4 Inelastic Neutron Scattering *278*
Problems *283*

14 Techniques in Linear Response Theory

14.1 Introduction *285*
14.2 The Generalized Langevin Equation *285*
14.3 Mori's Continued Fraction Theory *293*
14.4 Correlation Matrix Theory *296*
14.5 Generalized Hydrodynamics *299*
14.6 Perturbation Theory *300*
14.7 Kinetic Theory *306*
Problems *315*

Appendix 1 Fourier Transforms in Three Dimensions *319*

Appendix 2 Generalized Fourier Transforms for Orientation-Dependent Interactions *323*

References *327*

Index 335

Preface

The aim of this text is to help students prepare to follow the current research reports of advances in statistical mechanics ranging from the part of the field often called liquid-state physics to the part that is concerned with the molecular theories associated with a host of problems in solution physical chemistry. It is written for people who like mathematics. It will be easier for those who have learned the content of some other graduate courses in physics and chemistry with a strong mathematical content, but in terms of background, only thermodynamics seems quite necessary. It also helps to be able to consult one of the many excellent books on mathematical physics when more background in particular areas of mathematics is needed.

In relation to the one-semester graduate course that typically is available for statistical mechanics, I find that there is more material here than can be covered, so some selection of the topics by the students and instructor already is required. Hence, more complete coverage doesn't promise any obvious advantage. In selecting topics to be included, I was strongly influenced by the excellent monograph by Hansen and McDonald. Of course many of the topics that are omitted or scarcely covered here are also very important in statistical mechanics. I believe that the material offered here provides a strong background for further study of the missing topics, using the plentiful, less pedagogical materials.

In consideration of the superabundance of important topics in statistical mechanics per se, I also have omitted a lot of interesting and closely related results concerning the application of statistical mechanics to obtain molecular interpretations of the material properties of the various systems which have been studied. It is hoped that readers will make use of the references, especially references to the figures, to locate the reports where such aspects are discussed in detail, and, since the field is developing so rapidly, that they will expect to find more recent reports too. At any rate, my aim here is to offer a text on statistical mechanics rather than on the fruits of its application in various areas of liquid-state research.

Some of the problems at the end of each chapter serve to exercise the material in the text, while others—a significant fraction—are intended to guide the student along the lines begun in the text, but a few steps further. While this practice makes the book less useful as a reference, since some of the really important results are only found in the lists of problems, the advantages seem dominant. At any rate, the subject index covers the problem sets as well as the text in order to make reference easier.

Finally, I should explain the practice followed in references to the literature. With few exceptions, I have not tried to indicate priority but only to give entree to the literature at the point which, on balance, seemed likely to be most helpful to the reader. Where possible, I have referred to monographs or review articles, thus to places where priority is customarily given important attention, but where especially one can expect to find detailed references to the research literature.

The urge to write a book like this was something that I fought off for many years; looking back now on what happened after I committed myself to the effort, I am surprised at how much I enjoyed it. It has been said that a good theory is like a good joke: It is short and the last line is quite unexpected. There is so much good theory in the work of many minds that fills this book that in the writing I found myself crashing from one chapter to the next with sustained exuberance. I only hope that my readers catch some of this enthusiasm and are not too much put off by those passages which are not short, or the punch line not bright enough and, equally, are not put off by the errors inherent in this mode of writing. I have had assistance in finding many errors, notably from Professor J. E. Enderby, Professor W. G. Madden, Professor R. Thuraisingham, Dr. T. K. Lim, Mr. Xu Rienlang, and Dr. W. Sung, who also helped with writing several of the passages in Chapter 14. A special debt is owed to Professors Madden, D. K. Carpenter, and P. J. Rossky, who read the entire manuscript and provided many helpful suggestions. I am very grateful to all of them as well as to Mrs. S. Rudnick, who patiently and skillfully typed and retyped and retyped.

Harold L. Friedman

1

Fundamentals

1.1 INTRODUCTION

Statistical mechanics is the discipline concerned with interpreting the measurable properties of materials in terms of the properties and interactions of the constituent atoms and molecules. Like thermodynamics and quantum mechanics, statistical mechanics is a theoretical discipline in which one begins with a small number of axioms and then builds an enormous structure on them to do the characteristic work of the theory. It is interesting to contrast the main features of the three theories as shown in Table 1.1.

The fundamental results of statistical mechanics, some of which are derived in this chapter, relate to real materials. For such systems these results connect experimentally measurable properties to the microscopic description of the system, whether to the quantum states of the material system or to the Hamiltonian from which the quantum states may be derived. The exact relations, although they may not be difficult to derive, are so difficult to evaluate numerically that an essential role is played by approximation methods. By now many approximation methods are well developed and widely useful; their study will be a central part of this course.

Another, more subtle, kind of approximation is to replace the real material system of interest by a "model" system whose Hamiltonian is specified. Usually, the model has a simpler Hamiltonian than the real system. In some cases the properties of a Hamiltonian model system can be calculated as accurately as desired and, in a smaller number of cases, can even be calculated exactly. It turns out that a very deep understanding of the properties of real material systems, especially fluids, can be reached by studying Hamiltonian models. The study of some of these models also will be an important part of this course.

In the remainder of this chapter we summarize the ensemble theory of statistical

TABLE 1.1 THEORETICAL CHEMISTRY

	Thermodynamics	Quantum mechanics	Statistical mechanics
Axioms	About temperature, heat, work, energy, reversibility: all in macroscopic systems	About representing physical systems in terms of wave functions and operators	About averages and probabilities in physical systems
Objects treated	Macroscopic bodies of real materials at equilibrium	Assemblies of electrons and nuclei or other elementary particles	Real or model assemblies of particles which may be complex
Results	Relations among measurable macroscopic equilibrium properties (e.g., Clapeyron's equation)	Energy levels of stationary states, or probabilities of transitions between states	Macroscopic properties of the real or model systems; thermodynamic, transport, and relaxation coefficients

mechanics as a basis for the study of the modern developments referred to above. It should be noted that all of these results are exact and are applicable to real material systems, except for a few cases which are explicitly noted as being of limited validity.

1.2 BASIC CONCEPTS

One of the recurring problems in statistical mechanics is to find the exact theoretical formulation of quantities that may be measured experimentally. Consider that any experimentally determined material property of any system ultimately derives from the properties of the molecules (or other particles) from which the system is made. We ask what mathematical operations, applied to expressions for the molecular properties and intermolecular forces, result in a quantity that may be identified with the measured property.

The measurable coefficients of real systems need to be expressed as *averages* of processes or states that characterize the system at a molecular level. Consider a simple example. If we measure the pressure of a gas in one of the usual ways, we measure a quantity that appears to be quite steady except for noise introduced by the measuring device. On the other hand, in molecular terms, the pressure results from collisions of the molecules with the walls of the system and must be a fluctuating quantity, owing to the chaotic motions of the molecules. Undoubtedly, one could observe the fluctuations in a sufficiently delicate experiment. We need to formulate the steady pressure that one may measure in terms of the actual fluctuating pressure.

Now to analyze this situation more generally, let D be some dynamical variable of the system: that is, a variable that depends on the quantum state or, classically, on the locations and velocities of the particles. With a sufficiently delicate measuring system we might measure a fluctuating quantity $D(t)$, but the measured value of D is usually taken to be a time average, owing to the slow response of the measuring system when compared to the rapid changes in locations and velocities of the molecules that make up the system. Then the measured value is the time average $\langle D \rangle_t$ given by

$$\langle D \rangle_t \equiv \lim_{t\to\infty} \frac{1}{t} \int_0^t D(t')\, dt' \tag{1.1}$$

Equation (1.1) is just the mathematical expression for the time average of a fluctuating quantity.

Now we suppose that D is a property of a system S whose stationary quantum states are indexed 1, 2, . . . , j, ... Owing to the interaction of S with other systems, it passes from one quantum state to another as a function of time. The fluctuating quantity $D(t)$ is determined by the quantum state at the instant t, so that D_j or $D(j)$ may be written in place of $D(t)$. Thus the time dependence of $D(j)$ is implicit; D changes because as t changes to t', the system's quantum state may change from j to j'. Now it is easy to see that to evaluate the integral in Eq. (1.1), we must know the sequence of quantum states through which the system passes in time. The changes in quantum state are induced by the interaction of the system with its surroundings. If the time average is to be evaluated directly, this interaction has to be specified and the resulting dynamics have to be worked out.

Statistical mechanics really begins with the idea that almost all of the information about the interaction of a system with its surroundings can be overlooked in calculating measurable properties which characterize the interior of a system, such as bulk thermodynamic and transport coefficients. Although this idea is not easy to exploit in the direct evaluation of the time integral in Eq. (1.1), it fits in neatly with another way to average $D(t)$.

The other way is to construct a large number N_r of *replicas* of the system of interest, to measure D of each replica at the same instant, and to average the results. The collection of N_r replica systems is called an *ensemble.* The ensemble average of D is defined by the equation

$$\langle D \rangle_e \equiv \lim_{N_r\to\infty} \frac{1}{N_r} \sum_{a=1}^{N_r} D(j_a) \tag{1.2}$$

where j_a is the quantum number of replica system a. While each replica system a has reached its particular quantum state j_a at time t, the instant at which $\langle D \rangle_e$ is calculated, by passing through a sequence of quantum states as described above, there is no need to follow this sequence to calculate $\langle D \rangle_e$ if there is some other way to find *how many* of the replica systems of the ensemble are in each quantum state j. This can be done if we accept one of the axioms of statistical mechanics, as we shall see. Depending on your point of view, we need another axiom to tell us either that $\langle D \rangle_t$ is the same as $\langle D \rangle_e$ or that the measured value of D is just $\langle D \rangle_e$.

At first it may be helpful to consider $\langle D \rangle_t$ to be a thermodynamic variable, such as the pressure or energy, but in fact the considerations of this section are also applicable to the theoretical formulation of experimentally measurable transport and relaxation coefficients, to be treated later.

Now we consider the question: How may the replica systems of a given ensemble differ from one another and how must they resemble one another? This may be answered in many ways as long as we allow that each replica system may be in any of the quantum states that the system may pass through in the course of the time evolution implied in the measurement or in Eq. (1.1). Subject to this condition, it can be shown that all the different kinds of ensemble give the same $\langle D \rangle_e$ if the system is sufficiently large (i.e., if it is a macroscopic system). We describe here the three ensembles that are met most frequently. They have in common the specification that the replica systems have equal volumes V.

For simplicity, the ensembles are first described in terms that need amplification if the systems represented are made up of more than one species of molecules. Additional details for multicomponent systems are considered later.

In the *microcanonical ensemble* all the replica systems have the same energy E and number of particles N. This ensemble is directly applicable to calculating $\langle D \rangle_e$ for comparison with measurements on a system with specified N, V, and E, but if one can also carry through the calculation of the temperature T of the system, the microcanonical ensemble $\langle D \rangle_e$ can be compared with measurements on a system with specified N, V, and T. The microcanonical ensemble is not very simply applicable to the systems we usually study in the laboratory which are in thermal equilibrium with their surroundings, but it is centrally involved in one of the most important approximation methods of statistical mechanics: molecular dynamics simulation.

In the *canonical ensemble* (the petit canonical ensemble of Gibbs) all of the replica systems are in thermal equilibrium with a heat reservoir whose temperature is T. This ensemble is directly applicable to calculate $\langle D \rangle_e$ for comparison with measurements on systems with specified N, V, T. It often provides the simplest theoretical treatment of systems of noninteracting particles. It is central to another important approximation method, called Monte Carlo simulation.

In the *grand ensemble* (the grand canonical ensemble of Gibbs) all of the replica systems are in thermal equilibrium with a heat reservoir whose temperature is T. They are also in equilibrium with respect to exchange of particles with a "particle" reservoir in which the temperature is T and the chemical potential of the particles is μ. This ensemble is directly applicable to calculate $\langle D \rangle_e$ for comparison with measurements on systems with specified μ, V, T. Consideration of the grand ensemble often provides the simplest theoretical treatment of systems of interacting particles.

For any of these ensembles the sum in Eq. (1.2) can be written in another form, which is suggestive of methods of evaluation. The idea is to replace the sum over the replica systems by a sum over their quantum states.

For some definite number of replicas N_r we define n_j as the number of replica systems in quantum state j and note that the sum in Eq. (1.2), for this particular value of N_r, may be replaced by

$$\sum_j n_j D(j)$$

with no effect on the result. Notice that we have, of course,

$$\sum_j n_j = N_r$$

Now we define

$$P(j) \equiv \lim_{N_r \to \infty} \frac{n_j}{N_r} \tag{1.3}$$

which may be called the probability of state j. It is also called the *distribution function* for the specified ensemble. Finally, Eq. (1.2) may be written in the following form:

$$\langle D \rangle_e = \sum_j P(j)D(j) \tag{1.4}$$

The preceding paragraph applies very simply to the microcanonical and canonical ensembles. When it is applied to the grand ensemble we must replace the index j specifying the quantum state of the system by the compound index $J = N, j$, where N is the number of particles and j is the index for a particular quantum state of the N particles in the volume V.

We turn to calculating $P(j)$ in terms of microscopic characteristics of the system, but first make a formal statement of the axioms of statistical mechanics.

1.3 AXIOMS

Once again we remark that in statistical mechanics we build on the description of matter provided by quantum mechanics, whether in terms of the quantum states of a system or the Hamiltonian from which the states are derived. As indicated above, the theory of ensembles enables us to express measurable quantities of macroscopic systems in terms of the underlying quantum mechanical description, but with a minimum of detail. If we look carefully we can see that the ensemble-average technique depends on two postulates, statements that we do not know how to prove a priori, which are needed to supplement the quantum mechanical description. One is needed to equate the ensemble average to a more elementary description of what is being measured. A second postulate is needed to connect the attributes of a quantum state to its probability (i.e., to the fraction of the replica systems in an ensemble that will be found in that state). The two postulates together link the measurable properties of the macroscopic system to the microscopic (i.e., molecular) description provided by quantum mechanics. The two axioms are stated here.

First Postulate

$$\langle D \rangle_t = \langle D \rangle_e \tag{1.5}$$

Second Postulate

$$P(j) = P(j') \qquad \text{if } E(j) = E(j') \tag{1.6}$$

where j and j' are distinct quantum states of a given physical system. That is, their wave functions Ψ_j and Ψ'_j are two different stationary solutions of Schrödinger's equation for the same Hamiltonian operator $\hat{H}$.

$$\hat{H}\Psi_j = \Psi_j E(j) \qquad \text{and} \qquad \hat{H}\Psi_{j'} = \Psi_{j'} E(j')$$

The second postulate implies that in an isolated system (fixed energy and number of particles), all of the quantum states consistent with these constraints are equally probable.

There is a large literature going under the name "ergodic theory" which is related to the problem of justifying the first postulate.

The second postulate is employed in Sections 1.4 and 1.7 to obtain useful expressions for $P(j)$ in the canonical and grand ensembles, respectively. It is easy to rationalize it in simple cases in which distinct states of the same energy can be generated from one state by application of some rotation operator. Consider, for example, the two degenerate bending vibrational modes of a CO_2 molecule.

In applying the second postulate to the study of a mixture of H_2 and O_2 at room temperature, one should not consider those quantum states which correspond to some of the H_2 and O_2 having reacted to form H_2O unless the corresponding chemical process is allowed in the experimental system with which the theoretical results are to be compared. This aspect of the second postulate is closely related to the problem of metastability in thermodynamics as well as to considerations of ergodic theory, referred to above. Just as in thermodynamics, we must be careful in statistical mechanics in cases in which we wish to neglect a process whose rate is not quite zero.

1.4 CANONICAL ENSEMBLE AND DISTRIBUTION FUNCTION

We now consider a very large heat bath H in which two macroscopic systems A and B are immersed (Fig. 1.1). Our objective is to calculate the distribution function

$$P_A(j)$$

which is proportional to the chance to find system A in quantum state j when A is in thermal equilibrium with the heat bath whose temperature is T. Thus $P_A(j)$ depends on T, although this is not explicit in the notation. System B is included in the picture to help in the derivation, as we shall see.

We know (Section 1.3) that $P_A(j)$ can depend on the quantum state j only through $E_A(j)$, the energy of system A when it is in quantum state j. Thus

$$P_A(j') = P_A(j) \qquad \text{if } E_A(j') = E_A(j) \tag{1.7}$$

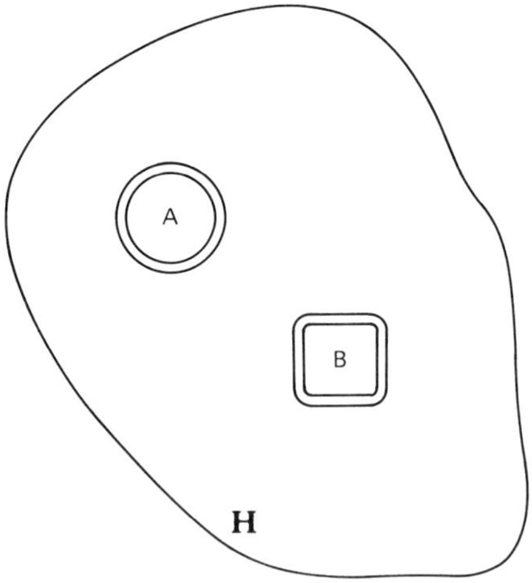

Figure 1.1 Systems A and B in thermal contact with a heat bath H.

For these concepts to be applicable to system A, its quantum states must be well defined. Thus system A must not interact too strongly with H. On the other hand, at least a weak interaction must be provided to enable heat transfer. For example, we might have an evacuated space between A and H, in which case the required interaction is provided by the radiation field.

System B can be chosen at will but, like A, interacts only weakly with the heat bath, so the quantum states $1, 2, \ldots, k, \ldots$ of system B may also be well defined. The distribution function

$$P_B(k)$$

is proportional to the chance to find system B in quantum state k when B is in thermal equilibrium with the heat bath.

The essential characteristic of the heat bath is that its temperature does not change appreciably when it exchanges energy with A or B. One way to assure this is to have H be very large compared to A and B.

So that we may define the distribution functions more precisely, we consider an ensemble (i.e., a collection) of N_r replicas of the composite system $H + A + B$ we have just described. In each replica it is assumed that thermal equilibrium has been reached. Systems A and B have been immersed in H long enough so that the heat flow to A or B from H is fluctuating back and forth with no net flow over long periods. Although the replica systems all have the same temperature, they will not have the same values of j and k at any given instant. The precise definition of $P_A(j)$ is that it is the fraction of the replicas in which, at any given instant, the A systems are in quantum state j, determined in the limit as the number of replicas increases to infinity [see Eq. (1.3)]. The definition of $P_B(k)$ is analogous.

Useful results may now be obtained by considering that A is physically far from B in H, so that the interaction between them is negligible, and looking at the behavior of AB, the composite system whose two noninteracting parts are A and B. From the preceding discussion it follows that the quantum state of AB is defined by an ordered pair of quantum numbers j, k, and that the energy of AB is

$$E_{AB}(j, k) = E_A(j) + E_B(k) \tag{1.8}$$

Furthermore, the fraction of the replicas in the ensemble at a given instant in which A is in quantum state j while B is in quantum state k is

$$P_{AB}(j, k) = P_A(j)P_B(k) \tag{1.9}$$

All of this follows from the assumed lack of interaction between A and B. We also have, in complete analogy to Eq. (1.7),

$$P_{AB}(j', k') = P_{AB}(j, k) \qquad \text{if } E_{AB}(j', k') = E_{AB}(j, k) \tag{1.10}$$

That is, all the states of AB with a given energy are equally probable, according to the second postulate.

Equations (1.7) to (1.10), which follow from various *physical* considerations, completely determine the way $P_A(j)$ depends on $E_A(j)$. To extract the result is a *mathematical* exercise that can be done in various ways. For example, we may begin by writing Taylor's series as follows.

$$\begin{aligned} \ln P_A(j) &= \alpha_A + \beta_A E_A(j) + \gamma_A [E_A(j)]^2 + \cdots \\ \ln P_B(k) &= \alpha_B + \beta_B E_B(k) + \gamma_B [E_B(k)]^2 + \cdots \\ \ln P_{AB}(j, k) &= \alpha_{AB} + \beta_{AB} E_{AB}(j, k) + \gamma_{AB} [E_{AB}(j, k)]^2 + \cdots \end{aligned} \tag{1.11}$$

[We might worry that the existence of these Taylor's series expansions requires some additional assumptions. Also note the discrete variables; for example, we are concerned with the dependence of P_A on E_A at $E_A(1)$, $E_A(2)$, . . . , $E_A(j)$, . . . and not at intermediate energies. Readers who are concerned by these questions should study derivations of the distribution law by more sophisticated mathematical methods.] The reason the expansions in Eq. (1.11) are defined for the logarithms of the distribution functions rather than for the distribution functions is that taking the logarithm of Eq. (1.9) yields

$$\ln P_{AB}(j, k) = \ln P_A(j) + \ln P_B(k) \tag{1.12}$$

which is like the corresponding relation for the energies, Eq. (1.8).

Now we may substitute Eqs. (1.11) and (1.8) in (1.12) to get

$$\alpha_{AB} + \beta_{AB}(E_A + E_B) + \cdots = \alpha_A + \beta_A E_A + \cdots + \alpha_B + \beta_B E_B + \cdots \tag{1.13}$$

where, to simplify the notation, we have written E_A for $E_A(j)$, and so on. If Eq. (1.13) is to hold for all values of E_A and E_B, we must have

$$\alpha_{AB} = \alpha_A + \alpha_B \tag{1.14.1}$$

$$\beta_{AB} = \beta_A = \beta_B \equiv -\beta \tag{1.14.2}$$

$$0 = \gamma_{AB} = \gamma_A = \gamma_B \tag{1.14.3}$$

and all the coefficients of higher terms must vanish as well.

The quantity β defined in Eq. (1.14) clearly does not depend on the nature of system A or of system B. The minus sign is included in the definition because it turns out that β is a positive quantity.

Finally we have, for system A,

$$\ln P_A(j) = \alpha_A - \beta E_A(j) \qquad \text{or} \qquad P_A(j) = e^{\alpha_A} e^{-\beta E_A(j)} \tag{1.15}$$

We recall that we must have

$$\sum_j P_A(j) = 1$$

so we find

$$e^{-\alpha_A} = \sum_j e^{-\beta E_A(j)} \equiv Q_A \tag{1.16}$$

The function Q_A defined in Eq. (1.16) is called the *partition function* of system A. The energy $E_A(j)$ depends on the volume V of the system and on the number N of particles in it; Q_A may also depend on the temperature T of the heat bath, presumably through β, since the $E_A(j)$ are determined without reference to T. Thus in terms of thermodynamical variables we have $Q_A(N, V, T)$. To summarize, for any system of specified N, V, T, the following distribution law of the canonical ensemble applies:

$$P(j) = Q^{-1}e^{-\beta E(j)}$$
$$Q \equiv \sum_j e^{-\beta E(j)} \tag{1.17}$$

Notice that an isolated sample of an ideal gas is an especially simple example of a canonical ensemble of the kind considered here. Then "system" is just one molecule, while the rest comprise the heat bath, and the quantum states j are states of a single molecule. Another very simple example of a canonical ensemble is Einstein's model of a crystal as a collection of independent harmonic oscillators. Then one oscillator may be taken as the "system" with quantum states j while the rest of the oscillators comprise the heat bath. These examples are among the systems discussed in Chapter 2.

DO PROBLEMS 1.1 TO 1.5.

To make full use of the canonical ensemble distribution law, we need to establish the relations of Q and β to thermodynamic functions. This is taken up in Section 1.6, following a summary of the necessary thermodynamics.

1.5 THERMODYNAMICS

This section is a brief summary of the thermodynamic equations for systems in which the independent variables are chosen in various ways. The thermodynamic notation to be used here is summarized in Table 1.2 for one-component systems. Multicomponent systems are discussed following Eq. (1.20). We summarize three parallel systematic formulations.

TABLE 1.2 NOTATION FOR THERMODYNAMIC VARIABLES

Intensive variables	Extensive variables
T is the temperature	V is the volume
p is the pressure	N is the number of particles
μ is the chemical potential	S is the entropy
	E is the energy
	H is the enthalpy
	A is the Helmholtz free energy
	G is the Gibbs free energy
$\rho = \dfrac{N}{V}$ is the concentration (number density)	$C_v = \left(\dfrac{\partial E}{\partial T}\right)_{V,N}$ is the heat capacity at constant volume
$\kappa_T = -\left(\dfrac{\partial \ln V}{\partial p}\right)_{N,T}$ is the isothermal compressibility	
$\beta = \dfrac{1}{k_B T}$	
$\lambda = e^{\beta\mu}$ is the absolute activity	

Conventional Thermodynamics

The combined first and second laws may be written in the form

$$dE = T\,dS - p\,dV + \mu\,dN \tag{1.18}$$

which is directly applicable to the changes of state of a system for which the independent variables are S, N, and V. For direct application to the familiar experimental situation in which the independent variables are N, p, and T it is convenient to replace this fundamental equation for E by the equation for the change in

$$G \equiv E + pV - TS$$

so we get the fundamental equation for G:

$$dG = -S\,dT + V\,dp + \mu\,dN \tag{1.19.1}$$

from which the various expressions for other thermodynamic coefficients in terms of G are derived:

$$S = -\left(\frac{\partial G}{\partial T}\right)_{N,p} \tag{1.19.2}$$

$$H = \left(\frac{\partial(G/T)}{\partial(1/T)}\right)_{N,p} \tag{1.19.3}$$

$$V = \left(\frac{\partial G}{\partial p}\right)_{N,T} \tag{1.19.4}$$

$$\mu = \left(\frac{\partial G}{\partial N}\right)_{p,T} \tag{1.19.5}$$

In the theory of thermodynamics it is also shown that

$$G = \mu N \tag{1.19.6}$$

and that this result together with the fundamental equation for G leads to the *Gibbs equation:*

$$0 = S\,dT - V\,dp + N\,d\mu \tag{1.20}$$

All of these thermodynamic equations (as well as those which follow) may be applied to a system of several component species (1, 2, . . . , s, . . .) provided only that we reinterpret N and μ and the products in which they appear in the following way:

$$\begin{aligned} N &\rightarrow \mathbf{N} \equiv N_1, N_2, \ldots, N_s, \ldots \\ \mu &\rightarrow \boldsymbol{\mu} \equiv \mu_1, \mu_2, \ldots, \mu_s, \ldots \\ \mu N &\rightarrow \boldsymbol{\mu} \cdot \mathbf{N} \equiv \Sigma_s \mu_s N_s \\ \mu\,dN &\rightarrow \boldsymbol{\mu} \cdot d\mathbf{N} \equiv \Sigma_s \mu_s\,dN_s \\ N\,d\mu &\rightarrow \mathbf{N} \cdot d\boldsymbol{\mu} \equiv \Sigma_s N_s\,d\mu_s \end{aligned} \tag{1.21}$$

For example, Eq. (1.20) becomes

$$0 = S\,dT - V\,dp + \mathbf{N} \cdot d\boldsymbol{\mu}$$

and Eq. (1.19.5) becomes

$$\mu_s = \left(\frac{\partial G}{\partial N_s}\right)_{p,T,N_1,\ldots,N_s-1,N_s+1,\ldots}$$

Thermodynamics for the Canonical Ensemble

In working with the canonical ensemble it is naturally convenient to use the thermodynamic fundamental equation for the independent variables N, V, T. This is the equation for the *Helmholtz free energy:*

$$A = E - TS$$

The fundamental equation and the derived expressions for several other thermodynamic coefficients follow.

$$dA = -S\,dT - p\,dV + \mu\,dN \tag{1.22.1}$$

$$S = -\left(\frac{\partial A}{\partial T}\right)_{N,V} \tag{1.22.2}$$

$$E = \left(\frac{\partial(A/T)}{\partial(1/T)}\right)_{N,V} \tag{1.22.3}$$

$$p = -\left(\frac{\partial A}{\partial V}\right)_{N,T} \tag{1.22.4}$$

$$\mu = \left(\frac{\partial A}{\partial N}\right)_{V,T} \tag{1.22.5}$$

Thermodynamics for the Grand Ensemble

In working with the grand ensemble it is most convenient to use the thermodynamic fundamental equation in which μ/T, V, and T are the independent variables. The ratio μ/T is proportional to the logarithm of the thermodynamic activity, so this choice is the same as choosing the activity, the volume, and the temperature as independent variables. The appropriate fundamental equation is not usually discussed in thermodynamics texts, so it is derived here.

We begin with the Gibbs equation (1.20), divided through by T,

$$0 = \frac{S}{T}\,dT - \frac{V}{T}\,dp + \frac{N}{T}\,d\mu$$

add the identity

$$d\left(\frac{pV}{T}\right) = \left(\frac{p}{T}\right) dV + \left(\frac{V}{T}\right) dp + (pV)\, d\left(\frac{1}{T}\right) \tag{1.23}$$

and obtain

$$d\left(\frac{pV}{T}\right)=\left(\frac{S}{T}\right)dT+\left(\frac{p}{T}\right)dV+(pV)\,d\left(\frac{1}{T}\right)+\left(\frac{N}{T}\right)d\mu$$
$$=(pV-TS-N\mu)\,d\left(\frac{1}{T}\right)+\left(\frac{p}{T}\right)dV+N\,d\left(\frac{\mu}{T}\right)$$

which is the desired form. The coefficient of $d(1/T)$ is apparently $-E$, so the fundamental equation for pV/T is

$$d\left(\frac{pV}{T}\right)=-E\,d\left(\frac{1}{T}\right)+\left(\frac{p}{T}\right)dV+N\,d\left(\frac{\mu}{T}\right) \tag{1.24.1}$$

with

$$E=-\left(\partial\frac{(pV/T)}{\partial(1/T)}\right)_{V,\mu/T} \tag{1.24.2}$$

$$N=\left(\partial\frac{(pV/T)}{\partial(\mu/T)}\right)_{T,V} \tag{1.24.3}$$

$$p=T\left(\partial\left(\frac{pV}{T}\right)\partial V\right)_{T,\mu/T} \tag{1.24.4}$$

The last of these equations is trivial. The appearance of μ/T as an independent variable in these equations reflects the fact that, in thermodynamics, $\exp(\mu/RT)$ is the fugacity or absolute activity.

1.6 THE CANONICAL ENSEMBLE AND THERMODYNAMICS

The objective of this section is to establish the thermodynamic interpretation of the parameters β and Q appearing in the distribution function for the canonical ensemble, Eq. (1.17). We begin with the following results obtained in Problems 1.1 and 1.2.

$$\langle E\rangle=-\frac{\partial\ln Q}{\partial\beta} \tag{1.25}$$

$$\langle p\rangle=\beta^{-1}\frac{\partial\ln Q}{\partial V} \tag{1.26}$$

We may use the notations E, p for the thermodynamical variables. According to the first postulate, we have $E=\langle E\rangle$ and $p=\langle p\rangle$. So these equations may be compared with the thermodynamic equations (1.22.3) and (1.22.4), respectively. It is easily verified that the comparisons reduce to identities provided that the following equations are both valid:

$$\beta=\frac{1}{k_B T} \tag{1.27a}$$

$$\ln Q=-\frac{A}{k_B T} \tag{1.27b}$$

where k_B, called Boltzmann's constant, is a factor with the units of energy/temperature, which does not depend on the system, the temperature, or any of the other variables [see Eq. (1.14.2]. It can be determined later by calculating $\langle p \rangle$ for a very dilute gas.

Beginning with Eq. (1.14.1),

$$\alpha_{AB} = \alpha_A + \alpha_B$$

which was derived for systems A and B in thermal equilibrium with the same heat bath, and rewriting it in terms of Eq. (1.16), we find that

$$\ln Q_{AB} = \ln Q_A + \ln Q_B \tag{1.28}$$

for systems A and B in thermal equilibrium. It is apparent that this is consistent with Eqs. (1.27) and the fact that the Helmholtz free energy is an extensive variable, which tends to confirm the assignment made in Eqs. (1.27).

Another check of this kind is provided by the result obtained in Problem 1.3: namely, that $d\beta/dT$ is negative. This is also consistent with Eqs. (1.27).

Of course, we have not proved that Eqs. (1.27) are correct. We only demonstrate that the validity of these equations is *sufficient* to make the canonical ensemble equations consistent with thermodynamics. A proof is difficult (see the discussion of the literature at the end of the chapter). This is an example of a common difficulty in statistical mechanics: It is difficult to establish the ensemble-average expression for any macroscopic measurable quantity which (like T) is not simply related to an ensemble average over a dynamical variable (like E).

Before proceeding, we write all the canonical ensemble equations corresponding to the thermodynamic equations summarized above for a system for which N, V, T are the independent variables. The subscripts in the partial derivative notation may be omitted; because if it is stated that N, V, T are the independent variables, then only one of them may change in any partial differentiation. Of course, the ensemble-average notation $\langle \cdot \cdot \cdot \rangle$ can be omitted here now that the derivations are over. For example, the ensemble average $\langle E \rangle$ is the same as the thermodynamic E in Table 1.3. The results are given in Table 1.3.

TABLE 1.3 EQUATIONS OF THE CANONICAL ENSEMBLE

$$Q = \sum_j e^{-\beta E(j)}$$

$$A = -\beta^{-1} \ln Q$$

$$S = k \ln Q - k_B \beta \frac{\partial \ln Q}{\partial \beta}$$

$$E = -\frac{\partial \ln Q}{\partial \beta}$$

$$p = \beta^{-1} \frac{\partial \ln Q}{\partial V}$$

$$\mu = -\beta^{-1} \frac{\partial \ln Q}{\partial N} \qquad \text{(one component)}$$

$$\mu_s = -\beta^{-1} \frac{\partial \ln Q}{\partial N_s} \qquad \text{(multicomponent)}$$

As these equations show, if we can calculate Q as a function of N, V, T for some model, we can calculate *all* of the thermodynamic properties of the model system. Some other examples are given in the problems.

DO PROBLEMS 1.6 TO 1.8.

1.7 GRAND ENSEMBLE AND DISTRIBUTION FUNCTION

Although the canonical-ensemble equations provide a complete basis for calculating the thermodynamic properties of model systems, the actual calculations are often simpler when they are carried out on the basis of a more complicated ensemble, the grand ensemble, in which the systems all have specified μ, V, T, where μ is the chemical potential.

To derive the equations for a grand ensemble, we consider a large heat bath in which two systems, A and B, are immersed. The situation is assumed to be similar to that described in Section 1.4 with one exception: Now systems A and B do not have a fixed number of molecules. Instead, the molecules are free to pass back and forth between the systems and the heat bath, which is now assumed to be a large reservoir for these molecules as well as for heat. We specify that the reservoir has a definite chemical potential μ for these molecules as well as a definite temperature T. At first we assume that there is only one species of molecules or particles involved.

When a molecule enters system A, the energy *and* the quantum state of the system must change. We introduce $J \equiv N, j$ as an index that specifies both N, the number of molecules in the system, and j, an index specifying the quantum state of the N-molecule system. We take K to be an index of the population and the quantum state of B. Then we have, provided that A and B are widely separated (see Section 1.4),

$$E_{AB}(J,K) = E_A(J) + E_B(K) \tag{1.29}$$

$$\ln P_{A,B}(J, K) = \ln P_A(J) + \ln P_B(K) \tag{1.30}$$

as before. We also have, for the number of molecules in the combined system,

$$N_{AB} = N_A + N_B \tag{1.31}$$

These three equations determine the way $\ln P_A(J)$ depends on N_A and $E_A(J)$. The result can be obtained by comparing the following Taylor's series, each in two independent variables E and N, which are written with $L^{n,m}$ as the coefficient in the term in E^nN^m.

$$\begin{aligned} \ln P_A(J) &= L_A^{0,0} + L_A^{1,0}E_A + L_A^{0,1}N_A + L_A^{2,0}E_A^2 + L_A^{1,1}E_AN_A + \cdots \\ \ln P_B(K) &= L_B^{0,0} + L_B^{1,0}E_B + L_B^{0,1}N_B + L_B^{2,0}E_B^2 + L_B^{1,1}E_BN_B + \cdots \\ \ln P_{AB}(J, K) &= L_{AB}^{0,0} + L_{AB}^{1,0}E_{AB} + L_{AB}^{0,1}N_{AB} + L_{AB}^{2,0}E_{AB}^2 + \cdots \end{aligned} \tag{1.32}$$

where

$$E_A \equiv E_A(J), \qquad E_B \equiv E_B(K), \qquad E_{AB} \equiv E_{AB}(J,K)$$

Now combining Eqs. (1.29) to (1.32) and comparing coefficients as in the corresponding problem in Section 1.4, we find that

$$L_{AB}^{0,0} = L_A^{0,0} + L_B^{0,0} \tag{1.33.1}$$

$$L_{AB}^{1,0} = L_A^{1,0} = L_B^{1,0} \equiv -\beta \tag{1.33.2}$$

$$L_{AB}^{0,1} = L_A^{0,1} = L_B^{0,1} \equiv \gamma \tag{1.33.3}$$

$$0 = L_{AB}^{2,0} = L_A^{2,0} = L_B^{2,0} \tag{1.33.4}$$

$$0 = L_{AB}^{1,1} = L_A^{1,1} = L_B^{1,1} \tag{1.33.5}$$

and so on.

The quantity β defined in Eq. (1.33.2) clearly does not depend on the nature of system A or of system B. The minus sign is included in the definition because it turns out that β is a positive quantity. Similarly, γ does not depend on the nature of system A or B. However, both β and γ may depend on the characteristics of the reservoir.

According to these results we have, for system A,

$$\ln P_A(J) = L_A^{0,0} - \beta E_A(J) + \gamma N_A \tag{1.34}$$

Now, in view of the fact that each term in this equation applies to the same system, we may drop the subscripts specifying the system. We also replace J by N, j and rearrange the equation slightly to get

$$P(N, j) = e^{L^{0,0}} e^{\gamma N} e^{-\beta E(N,j)} \tag{1.35}$$

Noting that

$$\sum_N \sum_j P(N, j) = 1$$

since $P(N, j)$ is the fraction of systems in the ensemble that have N particles and are in quantum state j, we find[1]

$$e^{-L^{0,0}} = \sum_N \sum_j \exp\left[-\beta E(N, j) + \gamma N\right] \equiv \Xi \tag{1.36}$$

which defines Ξ, called the *grand partition function* of the system. It depends on the volume of the system and on the temperature and the chemical potential of the particles in the reservoir. It is easy to guess from the form of Eq. (1.36) that Ξ depends on the temperature through β and on the chemical potential of the particles through γ. At any rate, we have

$$\Xi = \Xi(\mu, V, T) \tag{1.37}$$

which apparently is a thermodynamic function of state.

[1] Σ_N means $\Sigma_{N=0}^{\infty}$. Similarly, Σ_j means sum over all possible j. Ξ is the Greek capital letter xi, pronounced ksi.

To summarize, for any system of specified μ, V, T the following distribution law of the grand ensemble applies:

$$P(N, j) = \frac{e^{\gamma N} e^{-\beta E(N,j)}}{\Xi}$$
$$\Xi \equiv \sum_N \sum_j e^{\gamma N - \beta E(N,j)} \tag{1.38}$$

It is quite straightforward to generalize this law to apply to multicomponent systems. The result is obtained by changing the notation, just as in Eq. (1.21), but with the following changes in addition:

$$\gamma \rightarrow \boldsymbol{\gamma} \equiv \gamma_1, \gamma_2, \ldots, \gamma_s, \ldots$$
$$N \rightarrow \mathbf{N} \equiv N_1, N_2, N_3, \ldots \tag{1.39}$$
$$\gamma N \rightarrow \boldsymbol{\gamma} \cdot \mathbf{N} \equiv \gamma_1 N_1 + \gamma_2 N_2 + \cdots$$

In the next section we find how Ξ depends on the thermodynamic state of the system in order to express β, γ, and Ξ in terms of the usual thermodynamic coefficients.

DO PROBLEMS 1.9 AND 1.10.

1.8 THE GRAND ENSEMBLE AND THERMODYNAMICS

The thermodynamic significance of the parameters Ξ, β, and γ in Eqs. (1.38) may be found by a procedure like that used in Section 1.6. We begin with the following results from Problem 1.10:

$$\langle E \rangle = -\left(\frac{\partial \ln \Xi}{\partial \beta}\right)_{V,\gamma} \tag{1.40a}$$

$$\langle N \rangle = \left(\frac{\partial \ln \Xi}{\partial \gamma}\right)_{\beta,V} \tag{1.40b}$$

$$\langle p \rangle = \frac{1}{\beta}\left(\frac{\partial \ln \Xi}{\partial V}\right)_{\beta,\gamma} \tag{1.40c}$$

which we compare with the corresponding thermodynamic equations for a system in which the independent variables are μ/T, V, and T. First we compare Eq. (1.40c) with Eq. (1.24.4):

$$p = T\left(\frac{\partial(pV/T)}{\partial V}\right)_{T,\mu/T}$$

Because fixed T and μ imply the fixed state of the reservoir in Section 1.7, and hence fixed β and γ, this grand ensemble equation is the same as the thermodynamic equation if we have

$$\beta^{-1} \ln \Xi = pV \tag{1.40d}$$

Assuming that this equation is valid, we go on to compare Eq. (1.40a) with Eq. (1.24.2):

$$E = -\left(\frac{\partial(pV/T)}{\partial(1/T}\right)_{V,\mu/T}$$

The two equations are the same if we have

$$\beta = \frac{1}{k_B T} \tag{1.40e}$$

where k_B is independent of the system as well as of the temperature. Finally, we assume that Eqs. (1.40a) and (1.40c) are valid and compare Eq. (1.40b) with (1.24.3):

$$N = \left(\frac{\partial(pV/T)}{\partial(\mu/T}\right)_{T,V}$$

and conclude that the two equations are the same if we have

$$\gamma = \frac{\mu}{k_B T} \tag{1.40f}$$

It may also be noted that Eqs. (1.33.1) and (1.36) relate the grand partition function of a composite system AB to the grand partition functions of its parts, which have the same temperature and chemical potential, according to the equation

$$\ln \Xi_{AB} = \ln \Xi_A + \ln \Xi_B \tag{1.41}$$

Making substitutions according to Eq. (1.40d), we obtain

$$\left(\frac{pV}{T}\right)_{AB} = \left(\frac{pV}{T}\right)_A + \left(\frac{pV}{T}\right)_B \tag{1.42}$$

In view of the equilibrium conditions we know that p and T are the same throughout, so this reduces to

$$V_{AB} = V_A + V_B$$

Thus Eq. (1.41), interpreted according to Eq. (1.40d) is consistent with thermodynamics.

As in Section 1.6, we have found sufficient conditions [Eqs. (1.40d) to (1.40f) in the present case] for the ensemble equations to be consistent with thermodynamics, but we have not proved that these conditions are *necessarily* satisfied. The complete theory, which is more difficult, may be found in many texts.

Having identified $\gamma = \beta\mu$ and noticing that $\Sigma_j \exp[-\beta E(N, j)]$ in Eq. (1.38) is the canonical partition function $Q(N, V, T)$, we can write Eq. (1.38) in the form

$$\Xi(\mu, V, T) = \sum_N e^{\beta\mu N} Q(N, V, T) \tag{1.43}$$

The equations for the γ, V, T grand ensemble are collected in Table 1.4.

When classical mechanics is an adequate approximation, the partition function can be written in the form [see Eqs. (3.25) and (3.28)]

$$Q(N, V, T) = \frac{V^N}{\Lambda^{3N} N!} \cdot \frac{Z(N, V, T)}{V^N} \tag{1.44a}$$

TABLE 1.4 EQUATIONS OF THE GRAND ENSEMBLE WITH VARIABLES μ, β, V

$$Q(N, V, T) = \sum_j e^{-\beta E(j,N)}$$

$$\Xi = \sum_N e^{\beta\mu N} Q(N, V, T)$$

$$P(N, j) = e^{\beta\mu N} e^{-\beta E(j,N)}/\Xi$$

$$E = N\mu - \left(\frac{\partial \ln \Xi}{\partial \beta}\right)$$

$$p = \frac{\ln \Xi}{V\beta}$$

$$N = \beta^{-1}\left(\frac{\partial \ln \Xi}{\partial \mu}\right) \quad \text{(one component)}$$

$$N_s = \beta^{-1}\left(\frac{\partial \ln \Xi}{\partial \mu_s}\right) \quad \text{(multicomponent)}$$

where

$$\Lambda = h\left(\frac{\beta}{2\pi m}\right)^{1/2} \tag{1.44b}$$

is a factor called the thermal de Broglie wavelength.[1] The factor $V^N/\Lambda^{3N}N!$ is just the partition function of a classical ideal gas made up of N particles of mass m in volume V at temperature $T = 1/\beta k_B$. Therefore, the factor Z, called the *configuration integral,* carries the effects of the interactions among the N particles of the system. Because of the importance of Z in the theory of classical fluids, it is convenient to introduce the activity function

$$z = \frac{e^{\beta\mu}}{\Lambda^3} \tag{1.44c}$$

so we can rewrite the equation for the grand partition function in the form [see Eq. (1.38)]

$$\Xi = \sum_N \frac{z^N}{N!} Z(N, V, T) \tag{1.44d}$$

and then deduce the equations for the grand ensemble in which the independent variables are z, V, and T, as given in Table 1.5. In order to remove the classical limit consideration which motivates our interest in the z, V, T ensemble but which otherwise is out of place in this chapter, we may regard Eq. (1.44a) as the general definition of the configuration integral Z regardless of the applicability of the classical limit.

DO PROBLEMS 1.11 TO 1.14.

[1] h is Planck's constant.

TABLE 1.5 EQUATIONS OF THE GRAND ENSEMBLE WITH VARIABLES z, β, V

$$z = \frac{e^{\beta\mu}}{\Lambda^3}$$

$$Z(N, V, T) = Q(N, V, T)\Lambda^{3N}N!$$

$$\Xi = \sum_N \frac{z^N}{N!} Z(N, V, T)$$

$$P(N) = \frac{z^N Z(N, V, T)}{N!\Xi}$$

$$E - \tfrac{3}{2} Nk_B T = -\left(\frac{\partial \ln \Xi}{\partial \beta}\right)$$

$$p = \frac{\ln \Xi}{V\beta}$$

$$N = z\left(\frac{\partial \ln \Xi}{\partial z}\right) \qquad \text{(one component)}$$

$$N_s = z_s\left(\frac{\partial \ln \Xi}{\partial z_s}\right) \qquad \text{(multicomponent)}$$

1.9 FLUCTUATIONS. GENERAL ASPECTS

We begin by returning to the theory of the canonical ensemble and the aspect of it developed in Problem 1.1, where $\langle E\rangle$ and $\langle E^2\rangle$ are expressed in terms of the canonical ensemble distribution function $P(j)$. In that context we introduce a new distribution function $P(E)$ such that the fraction of systems in the ensemble with energies between E and $E + dE$ is $P(E)\,dE$. Now the replica systems of the ensemble are assumed to be big enough and the temperature high enough so that dE may be very small compared to $\langle E\rangle$ while still very large compared to the spacing of the energies of the discrete quantum states of the system; this condition enables us to discuss $P(E)$ as though E is a continuous variable. For example, we have

$$\langle E\rangle = \int_{-\infty}^{\infty} EP(E)\,dE$$

$$\langle E^2\rangle = \int_{-\infty}^{\infty} E^2P(E)\,dE$$

for the moments given in terms of the distribution of quantum states in Eqs. (1.72) and (1.74), respectively.

More generally, for any dynamical variable D we have ensemble average moments $\langle D^m\rangle$ which may be expressed either in terms of the distribution function of the states of the system, such as

$$\langle D^m\rangle = \sum_N \sum_j [D(N, j)]^m P(N, j) \tag{1.45}$$

to give a grand ensemble example, or in terms of $P(D)$, which is defined so $P(D)\,dD$ is the fraction of systems of the ensemble with D in the range from D to $D + dD$:

$$\langle D^m \rangle = \int_{-\infty}^{\infty} dD D^m P(D)$$

provided that the replica systems of the ensemble are large enough.

In a large system the dynamical variable D is determined by very many practically independent variables which determine the quantum state of the system. For example, we may think of the large system as partitioned into many small regions, each in its own quantum state and each making its own contribution to the system's D. Under these circumstances $P(D)$ is governed by the central limit theorem of probabilistic mathematics; that is, $P(D)$ must have the Gaussian form

$$P(D) = \frac{1}{\sqrt{2\pi}\,\sigma_D} \exp\left[-\frac{(D - \langle D \rangle)^2}{2\sigma_D^2}\right]$$

Here

$$\sigma_D \equiv (\langle D^2 \rangle - \langle D \rangle^2)^{1/2} \tag{1.46}$$

is called the *standard deviation.* Thus $P(D)$ is a bell-shaped curve (indeed, *the* bell-shaped curve) with a maximum at $D = \langle D \rangle$ and with width σ_D in the sense that we have

$$\frac{P(\langle D \rangle \pm \sigma_D)}{P(\langle D \rangle)} = \frac{1}{\sqrt{e}} \simeq 0.6$$

so that σ_D is very nearly the half-width at half-height of the $P(D)$ curve. Thus the fraction $\sigma_D/\langle D \rangle$ is a measure of the spread of the fluctuating values of D about the mean $\langle D \rangle$. The term "fluctuate" here refers to the fact that different replica systems have, as a rule, different values of D; it does not imply anything about rate processes in this context.

1.10 FLUCTUATIONS IN ENERGY

The distribution of values of the energy of a large system is, according to Section 1.9, given by

$$P(E) = (\sqrt{2\pi}\,\sigma_E)^{-1} \exp\left[-\frac{(E - \langle E \rangle)^2}{2\sigma_E^2}\right] \tag{1.47}$$

$$\sigma_E \equiv (\langle E^2 \rangle - \langle E \rangle^2)^{1/2} \tag{1.48}$$

Now in Problem 1.3 we found that for the canonical ensemble we have

$$C_v = \frac{\langle E^2 \rangle - \langle E \rangle^2}{k_B T^2} \tag{1.49}$$

So we find that the half-width of the $P(E)$ distribution relative to $\langle E \rangle$ is

$$\frac{\sigma_E}{\langle E \rangle} = \left\{\frac{k_B T^2 C_v}{\langle E \rangle}\right\}^{1/2} \frac{1}{\langle E \rangle^{1/2}} \tag{1.50}$$

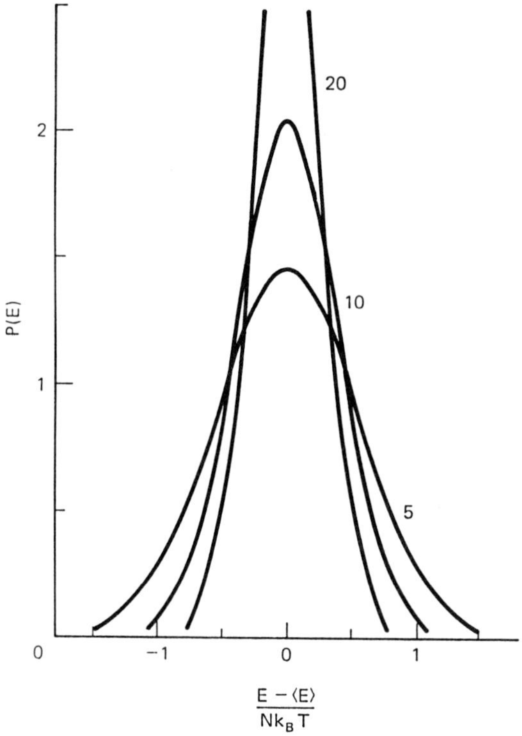

Figure 1.2 Energy distribution function $P(E)$ calculated from Eqs. (1.47) and (1.50) for the case $C_v = \frac{3}{2} Nk_B T$ for a system of 5, 10, or 20 particles.

The factor $\{\cdot \cdot \cdot\}$ is independent of the size of the system if we consider variations in V while holding N/V and T fixed, at least for macroscopic systems, so it is clear that the distribution gets sharper as V gets larger. Moreover, in the limit of this variation as $V \longrightarrow \infty$, called the *thermodynamic limit,* the distribution $P(E)$ becomes a Dirac delta function.

$$P(E) \longrightarrow \delta(E - \langle E \rangle) \cdot \cdot \cdot \text{ thermodynamic limit} \tag{1.51}$$

The tendency of $P(E)$ to sharpen up with increasing system size is evident even for very small systems, as shown in Fig. 1.2.

1.11 FLUCTUATIONS IN DENSITY

The distribution of the number of molecules in a large open system is, according to Section 1.9, given by

$$P(N) = (\sqrt{2\pi}\,\sigma_N)^{-1} \exp\left[-\frac{(N - \langle N \rangle)^2}{2\sigma_N^2}\right] \tag{1.52}$$

$$\sigma_N \equiv (\langle N^2 \rangle - \langle N \rangle^2)^{1/2} \tag{1.53}$$

In Problem 1.14 we find the following result for the grand ensemble:

$$\kappa_T = \frac{V\beta}{N^2}(\langle N^2 \rangle - \langle N \rangle^2) \tag{1.54}$$

so we find that the relative half-width of the peak in $P(N)$ is given by

$$\frac{\sigma_N}{\langle N\rangle} = \sqrt{\frac{k_B T \kappa_T}{V}} \tag{1.55}$$

If we consider variations in V, holding μ and T fixed we find that the $P(N)$ distribution gets relatively sharper as V gets larger. In the limit of this variation as $V \longrightarrow \infty$, the thermodynamic limit, we find that the distribution becomes a Dirac delta function:

$$P(N) \longrightarrow \delta(N - \langle N\rangle) \quad \cdots \quad \text{thermodynamic limit} \tag{1.56}$$

1.12 LARGE FLUCTUATIONS

There are some important cases in which even in macroscopic systems the fluctuations are enormous. For example, consider a liquid in equilibrium with its vapor. The system's pressure is a function of temperature alone, so we have $(\partial P/\partial V)_T = 0$ and $\kappa_T = \infty$. Therefore, the fluctuations in density are infinite. This reflects the fact that in a small but macroscopic test volume placed somewhere in the system we might find the density to be that of the liquid, or that of the gas, or that of a mixture of liquid and gas. Without gravity or effects due to the walls of the system there would be no correlation of the density in the test volume with its location within the system. We also find $\kappa_T = \infty$ in a one-phase system at the liquid–vapor critical point. Notice that in these examples C_p is also infinite but C_v is not, so the fluctuations in energy, although possibly quite large, are not infinite.

When two phases of different chemical composition are in equilibrium, for example in a two-phase mixture of phenol and water, there are infinite fluctuations in composition. This also happens at a critical mixing point. These phenomena may be discussed in terms of the theory in Section 1.13.

1.13 FLUCTUATIONS IN COMPOSITION

Consider N in Eqs. (1.52) and (1.53) to be N_s, the number of molecules of species s. Then we find that

$$\begin{aligned}\sigma^2_{s,s} &\equiv \langle N_s^2\rangle - \langle N_s\rangle^2 \\ &= \frac{(\partial N_s/\partial \mu_s)}{\beta} = \frac{V(\partial \rho_s/\partial \mu_s)}{\beta}\end{aligned} \tag{1.57}$$

Since $\partial\rho_s/\partial\mu_s$ is a thermodynamic coefficient that is independent of the size of a macroscopic system we find, for the relative width of $P(N_s)$,

$$\frac{\sigma_{s,s}}{\langle N_s\rangle} \simeq \frac{1}{\sqrt{\langle N_s\rangle}} \tag{1.58}$$

Proceeding in the same way we investigate the *cross* terms in the composition fluctuations (see Eq. 1.87):

$$\langle N_s N_{s'} \rangle - \langle N_s \rangle \langle N_{s'} \rangle = \frac{V(\partial \rho_s / \partial \mu_{s'})}{\beta} = \frac{V(\partial \rho_{s'} / \partial \mu_s)}{\beta} \tag{1.59a}$$

The new thermodynamic coefficients in Eqs. (1.57) and (1.59a), namely

$$M_{ss'} \equiv \left(\frac{\partial \rho_s}{\partial \mu_{s'}} \right)_{V,\, T,\, \boldsymbol{\mu}/\mu_{s'}} \tag{1.59b}$$

where $\boldsymbol{\mu}/\mu_{s'}$ means the chemical potential of every component except s', need to be treated carefully. We might be tempted to identify them with the reciprocals of the more familiar coefficients

$$L_{ss'} \equiv \left(\frac{\partial \mu_s}{\partial \rho_{s'}} \right)_{V,\, T,\, \boldsymbol{\rho}/\rho_{s'}} \tag{1.60}$$

but the actual relation is more complicated. Thus consider an arbitrary small variation in composition at fixed V and T. In view of Eq. (1.60) the change in chemical potential of component s may be written

$$d\mu_s = \sum_{s'} L_{ss'}\, d\rho_{s'}, \qquad s = 1, 2, \ldots \tag{1.61}$$

where the sum is over all component species 1, 2, ... The same set of equations may also be written in matrix algebra notation[1]:

$$|d\mu) = |L|d\rho) \tag{1.62}$$

where $|d\mu)$ and $|d\rho)$ are column matrices with one element for each component species 1, 2, . . . and $|L|$ is a square matrix:

$$\begin{matrix} L_{11} & L_{12} & \cdots \\ L_{21} & L_{22} & \cdots \\ \cdot & \cdot & \\ \cdot & \cdot & \\ \cdot & \cdot & \cdots \end{matrix}$$

In the same variation the chemical potentials may be thought of as the independent variables and then the changes in the particle number densities ρ_s are given by

$$|d\rho) = |M|d\mu) \tag{1.63}$$

where $|M|$ is the matrix whose elements are the $M_{ss'}$ coefficients defined in Eq. (1.59). If we substitute Eq. (1.63) into (1.62), we obtain

$$|d\mu) = |L|M|d\mu) \tag{1.64}$$

[1] In a more standard linear algebra notation, Eq. (1.61) would be $\mathbf{x} = \mathbf{L} \cdot \mathbf{y}$, where $\mathbf{x}$ is the vector $d\mu_1, d\mu_2, \ldots$ and y is the vector $d\rho_1, d\rho_2, \ldots$, but it seems marginally less suited for explaining the relation of the $M_{ss'}$ to the $L_{ss'}$ than the notation in Eq. (1.62).

so the matrix product $|L|M|$ equals the unit matrix and we have

$$|M| = |L|^{-1} \tag{1.65}$$

For example, for a system of two component species we have

$$M_{11} = \frac{L_{22}}{\det |L|}, \qquad M_{12} = \frac{-L_{21}}{\det |L|} = M_{21}, \qquad M_{22} = \frac{L_{11}}{\det |L|}$$

where det means determinant,

$$\det |L| = L_{11}L_{22} - L_{12}^2 \tag{1.66}$$

(Both the $|M|$ and the $|L|$ matrices are always symmetrical.)

In Chapter 4 it will be shown that the derivatives $M_{ss'}$ are simply related to certain distribution functions that characterize the structure of the system at the molecular level. Thus Eq. (1.66) provides a relation between the thermodynamic coefficients $\partial\mu_s/\partial\rho_{s'}$ and the molecular structure. This connection is the basis of the Kirkwood–Buff solution theory (Chapter 10).

DO PROBLEMS 1.16 AND 1.17.

1.14 EQUIVALENCE OF ENSEMBLES

The theory of fluctuations may be applied to see whether in a given system the ensemble average of a dynamical variable depends on which ensemble is used. Here we show that the canonical and grand ensembles give the same average in the thermodynamic limit (i.e., the limit of infinite system size for given N/V). The method illustrated here can be applied to other ensembles with similar results. In this way it can be established that the choice of ensemble is a matter of convenience unless we are interested in the effects of finite system size.

Let D be any variable, such as the pressure or the energy per unit volume, which does not change when the size of a macroscopic system is changed (e.g., in approaching the thermodynamic limit) and which is also a dynamical variable (like p or E/V, but unlike S). The observable value of D in the grand ensemble is [see Eq. (1.38)]

$$\langle D\rangle_\mu = \sum_{\mathbf{N},j} D(\mathbf{N}, j)\, P(\mathbf{N}, j) \tag{1.67}$$

while in the canonical ensemble for specified $\mathbf{N}$ it is [see Eq. (1.17)]

$$\langle D\rangle_\mathbf{N} = \sum_{j} D(\mathbf{N}, j)\, P(j) \tag{1.68}$$

Now we may pose the question: If the two ensembles are chosen to represent a given system, do we have

$$\langle D\rangle_\mu = \langle D\rangle_\mathbf{N} \quad ? \tag{1.69}$$

We investigate the problem by writing out Eq. (1.67) and then manipulating it as follows:

$$\langle D\rangle_\mu = \frac{\sum_{\mathbf{N}} \lambda^{\mathbf{N}} \sum_j D(\mathbf{N}, j) \exp(-\beta E_j)}{\Xi} = \frac{\sum_{\mathbf{N}} \lambda^{\mathbf{N}} Q(\mathbf{N}, V, T)\langle D\rangle_{\mathbf{N}}}{\Xi}$$
$$= \sum_{\mathbf{N}} P(\mathbf{N})\langle D\rangle_{\mathbf{N}} \tag{1.70}$$

For the rest we consider only the one-component case, so $\mathbf{N} \longrightarrow N$. The basic idea is that $P(N)$, in the neighborhood of $N = \langle N\rangle$, is very close to the Gaussian form shown in Eq. (1.52). Moreover, for extremely large $\langle N\rangle$, as appropriate to a macroscopic system, the Gaussian function is well approximated by a Dirac delta function. This approximation becomes exact in the thermodynamic limit ($N \longrightarrow \infty$ at fixed $\rho = N/V$) and in that limit the following sequence of analytical relations is justified.

$$\langle D\rangle_\mu = \sum_N P(N)\langle D\rangle_N \simeq \sum_N \bar{P}(N)\langle D\rangle_N \simeq \int_0^\infty \bar{P}(N)\langle D\rangle_N \, dN$$
$$\simeq \int_0^\infty \delta(N - \langle N\rangle)\langle D\rangle_N \, dN = \langle D\rangle_N \tag{1.71}$$

where $\bar{P}(N)$ means a suitable Gaussian distribution. It is not difficult to generalize this argument to apply to the multicomponent case.

THE LITERATURE

T. Hill (1956) gives a very clear presentation of the topics of this chapter, although mostly by different methods. Jackson (1968) is useful for elementary and detailed discussion of ensembles, averages, and probabilities, while Schrödinger (1960) discusses many of the trickiest points in a detailed and very clear way, especially material of Section 1.6. Landau and Lifschitz (1958) is especially useful for detailed treatment of the quantum-mechanical aspects of the various averages as well as fluctuation theory, including multivariate fluctuations. Fowler and Guggenheim (1952) is especially complete on the thermodynamics of the grand potential. Kirkwood and Buff (1951) and Buff and Brout (1955) show how one can use fluctuation theory to calculate solution thermodynamic functions in terms of molecular interactions. The book by Mathews and Walker (1965) has simple explanations of Dirac delta functions, the central limit theorem, similarity transformations, and other mathematical aspects encountered in this chapter and later in the course. It is typical of the excellent books of this type that may be helpful.

PROBLEMS

1.1. By applying Eq. (1.4) and the canonical ensemble equations (1.17), obtain the following important results ($\langle \cdots \rangle \equiv \langle \cdots \rangle_e$):

(a) $$\langle E\rangle = Q^{-1} \sum_j E(j) e^{-\beta E(j)} \tag{1.72}$$

(b) $\langle E \rangle = -\dfrac{\partial \ln Q}{\partial \beta}$ (1.73)

(c) $\langle E^2 \rangle = Q^{-1} \sum_j E(j)^2 e^{-\beta E(j)}$ (1.74)

(d) $\langle E^2 \rangle = \dfrac{\partial^2 \ln Q}{\partial \beta^2} + \langle E \rangle^2$ (1.75)

1.2. For a system in a given quantum state j in a volume V, the pressure $p(j)$ is given by the equation

$$p(j) = -\frac{\partial E(j)}{\partial V} \tag{1.76}$$

Extend the results in Problem 1.1 to include the derivation of

$$\langle p \rangle = \beta^{-1} \frac{\partial \ln Q}{\partial V} \tag{1.77}$$

1.3. Show that C_v, the extensive heat capacity at constant volume, is given by the canonical ensemble equation

$$\langle C_v \rangle = -(\langle E^2 \rangle - \langle E \rangle^2) \frac{\partial \beta}{\partial T} \tag{1.78}$$

Now show that $\langle E^2 \rangle - \langle E \rangle^2 = \langle (E - \langle E \rangle)^2 \rangle$, so that $\langle E^2 \rangle - \langle E \rangle^2$ is a positive quantity and therefore $\partial\beta/\partial T$ is negative, since it is known from thermodynamics that C_v is positive.

1.4. In Problem 1.2 we calculate the pressure as the ensemble average of $p(j)$, while in Problem 1.3 we do not calculate the heat capacity as the ensemble average of the heat capacity of a quantum state. Explain.

1.5. The quantities $E(j)$ (or E_j) in Eq. (1.17) derive from the underlying Hamiltonian operator H via the time-independent Schrödinger equation

$$\hat{H}\Psi_j(\mathbf{q}) = E_j \Psi_j(\mathbf{q}) \tag{1.79}$$

given the wave functions Ψ_j for the quantum states of the system. Here $\mathbf{q} = q_1, q_2, \ldots$ is the set of coordinates on which $\hat{H}$ is defined. Assuming that the set of Ψ_j for all j comprise a complete orthonormal set of wave functions, we have

$$\int \Psi_j^* \, \Psi_k \, d\mathbf{q} = \delta_{jk} \tag{1.80}$$

and

$$\int \Psi_j^* \hat{H} \Psi_j \, d\mathbf{q} = E_j \tag{1.81}$$

where the left side is the expectation value for the energy in quantum state j; it is the energy to use in Eq. (1.17). We also note the definition of the exponential

$$e^Z = 1 + Z + \frac{1}{2!} ZZ + \frac{1}{3!} ZZZ + \cdots \tag{1.82}$$

whether Z is a simple number, a matrix, or an operator. Show that we have

(a) $\int \Psi_j^* \hat{H}\hat{H} \Psi_j \, d\mathbf{q} = E_j^2$

(b) $\int \Psi_j^* \hat{H}\hat{H}\hat{H}\Psi_j \, d\mathbf{q} = E_j^3$

(c) $Q = \sum_j \int \Psi_j^* e^{-\beta \hat{H}} \Psi_j \, d\mathbf{q}$ (1.83)

1.6. Write out all of the equations in Section 1.5 for a system of two species of molecules, a and b, in order to see the implications of the shorthand notation for multicomponent systems introduced in Eqs. (1.21).

1.7. Fill in the graphs in Fig. 1.3 for a system having just two states, $j = 1$ and $j = 2$, with $[E(2) - E(1)]/k_B = 500$ K.

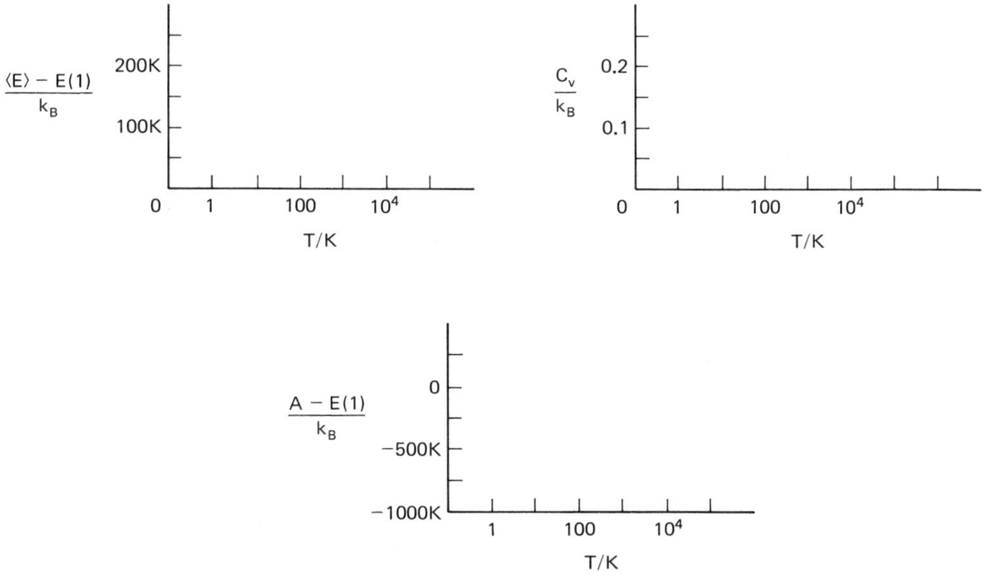

Figure 1.3 Figures for Problem 1.7.

1.8. A given system has just two quantum states, $j = +$ and $j' = -$. Also, $E(+) = -E(-) = 1$ erg. When the system is placed in thermal contact with a particular heat reservoir it is found that $P(+) = 0.25$ and $P(-) = 0.75$. Calculate Q, β, and $\langle E \rangle$ for the system in this case.

1.9. Generalize the derivation of the grand ensemble equations, Section 1.7, to a system of two components, species a and b. Assume that the reservoir has definite values of T, μ_a, and μ_b.

1.10. By proceeding as in Problem 1.1 but using the grand canonical equations (1.38), derive Eqs. (1.40a) to (1.40c).

1.11. Derive the equations in Table 1.5 from those in Table 1.4 and the definitions in Eq. (1.44).

1.12. In Chapter 3 we find that for a classical ideal gas we have $Z(N, V, T) = V^N$. From Table 1.5 show that in this case for a one-component gas, $z = N/V$ and $E = \frac{3}{2}Nk_BT$.

1.13. See Problem 1.12. Find the activity z_s for a component in a mixture of ideal gases.

1.14. The isothermal compressibility κ_T is given in Table 1.2.

(a) Using Gibbs equation (1.20), show that an equivalent expression for κ_T is

$$\kappa_T = \left(\frac{V}{N^2}\right)\left(\frac{\partial N}{\partial \mu}\right)_{V,\,T} \tag{1.84}$$

(b) Now use the grand ensemble equations to derive the important results

$$\kappa_T = \left(\frac{V}{\beta N^2}\right)\left(\frac{\partial^2 \ln \Xi}{\partial \mu^2}\right) \tag{1.85}$$

$$\kappa_T = \left(\frac{V\beta}{N^2}\right)(\langle N^2\rangle - \langle N\rangle^2) \tag{1.86}$$

(c) Using the grand ensemble equations for a mixture, show that the multicomponent generalization of Eq. (1.86) is

$$\frac{1}{\beta}\left(\frac{\partial\langle N_s\rangle}{\partial \mu_{s'}}\right)_{V,T} = \langle N_s N_{s'}\rangle - \langle N_s\rangle\langle N_{s'}\rangle\,. \tag{1.87}$$

(The mathematics is similar to the derivation of Eq. (1.75).)

1.15. Using the equations in Sections 1.10 and 1.11, estimate $\sigma_E/\langle E\rangle$ and $\sigma_N/\langle N\rangle$ for 1 cm^3 of argon gas at standard temperature and pressure (STP).

1.16. Write out Eqs. (1.61) to (1.65) in detail for the two-component system specified in Problem 1.11. Finally, express each of the four coefficients $\partial\mu_s/\partial\rho_{s'}$ in terms of fluctuations such as $\langle N_s N_{s'}\rangle - \langle N_s\rangle\langle N_{s'}\rangle$.

1.17. **(a)** Beginning with Eq. (1.20) for a mixture, derive

$$V\left(\frac{\partial p}{\partial V}\right)_{\mathbf{N}} = \sum_s N_s\left(\frac{\partial \mu_s}{\partial V}\right)_{\mathbf{N}} \tag{1.88}$$

(b) Then, using the identity $(\partial\rho_s/\partial V)_{\mathbf{N}} = -N_s/V^2$, derive

$$\frac{1}{\kappa_T} = \sum_{ss'} \rho_s L_{ss'} \rho_{s'} \tag{1.89}$$

a thermodynamic relation that is used with the Kirkwood–Buff theory.

2

Independent Subsystems

2.1 INTRODUCTION

In this chapter we treat the special case in which the system of interest is made up of a number N of noninteracting subsystems, that is, subsystems with no forces acting between them. In such cases the energy $E(\mathbf{j})$ of a state $\mathbf{j}$ of the system is made up of a sum of terms

$$E(\mathbf{j}) = E(j_1) + E(j_2) + \cdots + E(j_n) + \cdots + E(j_N) \tag{2.1}$$

where j_n is a quantum number of the nth subsystem and $\mathbf{j}$ is the set (list) j_1, j_2, . . . of all of the subsystem quantum numbers. Typically, each j_n can be any nonnegative integer. Equation (2.1) can be deduced from Schrödinger's equation,

$$\hat{H}\Psi(\mathbf{j}) = E(\mathbf{j})\Psi(\mathbf{j}) \tag{2.2}$$

if the Hamiltonian operator of the full system is a sum of subsystem contributions

$$\hat{H} = \hat{H}_1 + \hat{H}_2 + \cdots + \hat{H}_n + \cdots + \hat{H}_N \tag{2.3}$$

More generally there are cross-terms $\hat{H}_{nn'}$, connecting the mechanics of subsystem n to the mechanics of subsystem n'; the general case is assumed in Chapter 3. But here we assume that Eq. (2.3) is valid as written, so we have a system whose parts are mechanically independent of each other, the *independent subsystems.* Such a system is an idealization, a Hamiltonian model, of a gas (with the subsystems taken to be the molecules) or a crystal (with the subsystems taken to be the modes of vibration). It leads easily to molecular interpretations of some of the dominant equilibrium properties of gases and crystals, as described in this chapter.

2.2 DISTINGUISHABLE SUBSYSTEMS

If the N subsystems are distinguishable from each other, as is the case for particles of N different species or particles at N distinct locations, two lists of quantum states of the subsystems

$$\begin{aligned} \mathbf{j} &= j_1, j_2, \ldots, j_N \\ \mathbf{j}' &= j_1', j_2', \ldots, j_N' \end{aligned} \tag{2.4}$$

specify distinct quantum states of the systems as a whole unless $j_n = j_n'$ for every subsystem n. If the subsystems are independent *and* distinguishable, the canonical partition function $Q(N, V, T)$ can be simplified as follows:

$$\begin{aligned} Q &= \sum_{\mathbf{j}} \exp[-\beta E(\mathbf{j})] \\ &= \sum_{\mathbf{j}} \exp\{-\beta[E_1(j_1) + \cdots + E_N(j_N)]\} \\ &= \sum_{j_1} \cdots \sum_{j_N} \exp\{-\beta[E_1(j_1) + \cdots + E_N(j_N)]\} \\ &= \sum_{j} \exp[-\beta E_1(j)] \times \sum_{j} \exp[-\beta E_2(j)] \times \cdots \times \sum_{j} \exp[-\beta E_N(j)] \\ &= q_1 \cdot q_2 \cdots q_n \cdots q_N \end{aligned} \tag{2.5}$$

The first equality is the definition. The second equality follows from Eq. (2.1). The third equality follows from Eq. (2.4) together with the fact that every distinct list $j_1, j_2, \ldots, j_n, \ldots, j_N$ corresponds to a distinct quantum state of the system. (It is this step that is not valid if the particles are not distinguishable, for then lists $j_1, \ldots, j_N$ which differ only in the *order* in which the integers appear do not correspond to distinct quantum states.) The fourth equality is merely a rearrangement of the terms, as in the example

$$\sum_i \sum_j \sum_k \cdots A(i)B(j)C(k) \cdots = \sum_i A(i) \sum_j B(j) \sum_k C(k) \cdots$$

The fifth equality involves the definition of the "one-particle" partition function

$$q_n = \sum_{j_n} \exp[-\beta E_n(j_n)] \tag{2.6}$$

where we sum over all quantum states accessible to subsystem n.

2.3 A COLLECTION OF SIMPLE HARMONIC OSCILLATORS

A simple example of a system of distinguishable, independent subsystems is a collection of N simple harmonic oscillators of given frequency ν but with each one at a distinct location. In this case $q_1 = q_2 = \cdots = q_N$, so we have

$$Q = q_1^N \tag{2.7}$$

For a simple harmonic oscillator each quantum state j has the energy

$$E_j = (j + \tfrac{1}{2})h\nu, \qquad j = 0, 1, \ldots, \infty \tag{2.8}$$

We are free to pick the zero of the energy scale to suit our convenience, as we shall verify in Section 2.17. Then we take $E_0 = 0$ so that we have

$$\begin{aligned} q &= \sum_j = \exp(-\beta E_j) \\ &= \sum_j x^j \\ &= \frac{1}{1-x} \end{aligned} \tag{2.9}$$

where we define

$$x = e^{-\beta h\nu} \tag{2.10}$$

Then from Eq. (2.7) we find the partition function for the system of N independent simple harmonic oscillators

$$Q = \frac{1}{(1-x)^N} \tag{2.11}$$

It may be verified that if we retain an arbitrary value for E_0, we would have, instead of Eqs. (2.9) and (2.11),

$$q = \frac{e^{-\beta E_0}}{1-x} \tag{2.12}$$

$$Q = \frac{e^{-N\beta E_0}}{(1-x)^N} \tag{2.13}$$

DO PROBLEMS 2.1 TO 2.3.

2.4 EINSTEIN'S MODEL OF A CRYSTAL

Dulong and Petit reported in 1819 that the heat capacities are the same for the solid elements, the constant value being near 6 cal/(mol deg). This "rule" was important in the early determination of atomic weights. Later it was found that specific heats are always smaller at low temperatures and that the lighter elements had heat capacities much smaller than the Dulong and Petit value, even at room temperature.

In 1871, Boltzmann showed how a classical model for a crystal leads to the heat capacity at constant volume $C_v = 3R \simeq 6$ cal/(mol deg), where R is the gas constant, even while experimental evidence was accumulating which showed that C_v was less than this value for any element at low enough temperature.

In 1907, Einstein showed how the quantization concept that explained Planck's distribution law for blackbody radiation also explained the low-temperature behavior of the heat capacities of the solid elements. We give a version of this simple theory which illustrates a number of the points that we have discussed.

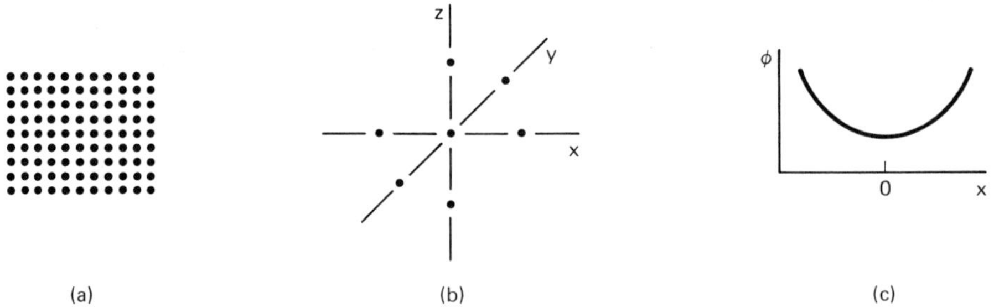

Figure 2.1 (a) Projection of three-dimensional simple cubic lattice. (b) Coordinate system in the three-dimensional lattice. The six neighbors of a given lattice point are shown. (c) Potential for movement of the central lattice point along the x axis.

Consider a three-dimensional simple cubic lattice whose projection on the xy plane is shown in Fig. 2.1a. At each lattice point there is a point of mass m, representing an atom. The atoms are assumed to exert forces on each other such that at classical mechanical equilibrium each atom is at rest on its lattice point.

Now choose some particular lattice point and a coordinate system such that this lattice point is at the origin and its neighbors are on the x, y, or z axes (Fig. 2.1b). The forces acting on the atom at this lattice point tend to push it back to the lattice point if it is displaced. We may think of this restoring force as having x, y, and z components. The potential ϕ of the x component of the force must have a shape like the graph shown in Fig. 2.1c. It has a Taylor's series expansion

$$\phi(x) = \phi(0) + x\phi_1(0) + \tfrac{1}{2}x^2\phi_2(0) + \cdots$$

where

$$\phi_n(0) \equiv \left(\frac{\partial^n \phi}{\partial x^n}\right)_{x=0}$$

Since $-\phi_1$ is the force in the x direction, we have $\phi_1(0) = 0$, as one can also see from the figure. In a model in which we neglect higher-order terms in the potential, we have

$$\phi(x) - \phi(0) = \tfrac{1}{2}x^2\phi_2(0) \tag{2.14}$$

which is the potential-energy function for a simple harmonic oscillator. The quantum-mechanical energy levels for the simple harmonic oscillator are given in Eq. (2.8) if

$$h\nu = \hbar\sqrt{\frac{\phi_2(0)}{m}}$$

How many oscillators are there in a model for a crystal having N atoms? In this model the x, y, and z directions are equivalent, so there are three oscillators at each lattice point, each distinct and acting independently of the others. Then we have for the system partition function

$$Q = q^{3N}, \qquad q = \frac{1}{1 - e^{-\beta h\nu}} \tag{2.15}$$

where we assume an energy scale with $E_0 = 0$.

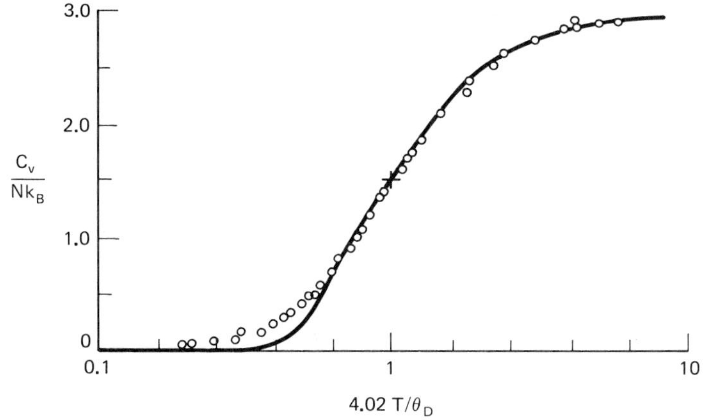

Figure 2.2 Circles, heat-capacity data for various metals; line, equation from Einstein model.

Notice that this result has been obtained by a trick. To begin with, one tends to think of the atoms as the particles that make up the crystal. However, the crystal thought of in this way is nothing like a collection of independent particles to which we may apply the elementary theory. But when we think of the crystal as a collection of oscillators whose dynamics are similar to that of the real crystal, these particles, the oscillators, are independent of each other in a model in which the restoring force is independent of the displacements of the neighboring oscillators. The oscillators, rather than the atoms, are the independent subsystems of Section 2.1. Neglecting their interaction is one idealization in Einstein's model. Neglecting the higher $\phi_n(0)$ is the other.

Experimental heat capacity data are compared with Einstein's model in Fig. 2.2. The parameter Θ_D is chosen for each substance to make its curve pass through the point +, while for the Einstein model we define

$$\Theta_D = 1.34 T\beta h\nu = 1.34 \frac{\hbar}{k_B}\sqrt{\frac{\phi_2(0)}{m}} \tag{2.16}$$

The experimental value of Θ_D, the *Debye temperature,* is an important material property which depends on the intermolecular forces through ϕ_2 as shown in this equation. It tends to be highest for crystals of light atoms held together by strong bonds. Other important features of this figure are:

1. The data approach the Dulong and Petit limit at high T.
2. The universal shape of the temperature dependence of the experimental $C_v(T)$ curve is well approximated, except at low temperatures, by Einstein's model.

The Debye temperatures of a few elements are given in Table 2.1.

DO PROBLEMS 2.4 TO 2.6.

TABLE 2.1 DEBYE CHARACTERISTIC TEMPERATURES FOR SOME CRYSTALLINE ELEMENTS

Element	C[a]	Cu	Ag	Au	Be	Mg	Ca	Li	Na	K	Zn	Cd	Hg
Θ_D (K)	1860	315	215	185	1030	330	230	430	160	99	240	221	90

[a] Diamond.

2.5 DEBYE'S MODEL OF A CRYSTAL

Einstein's model is perhaps most unrealistic in its neglect of the force on an atom due to the displacements of its neighbors from their equilibrium positions at their respective lattice sites. A much better model developed by Debye will be described briefly here because it is rather similar, although it will not be carried very far because this would require additional preparation in mechanics and mathematics.

We first consider the crystal as an elastic solid object and ask for the *normal modes* of its vibrational motion. For a violin string the normal modes of vibration are standing sinusoidal waves with nodes at the ends of the string. They have wavelength $\lambda = 2L/m$, where L is the length of the string and m is an integer. An important property of the normal modes is that they are mutually independent as long as the vibrations are harmonic; that is, the restoring force at each point is proportional to the displacement from equilibrium. Therefore, it is even more appropriate to choose the independent subsystems to be the normal modes rather than the vibrations of each atom about its lattice point.

The largest-wavelength vibration of a crystal has a wavelength $\lambda \simeq 2L$, where L is the length of an edge of the crystal. There is also a shortest-wavelength vibration with $\lambda = 2a$, where a is the interatomic spacing; this results from the discrete nature of the crystal. There are $3N$ normal modes altogether in a crystal of N atoms. It is quite difficult to calculate the number of normal modes at each wavelength λ in the range $2a \leq \lambda \leq 2L$ in order to calculate the partition function. Most often one employs an approximation, due to Debye, for the frequency distribution of the normal modes. This gives close agreement with experimental data such as those in Fig. 2.2. In particular, it gives for the limiting T dependence of C_v at low T

$$C_v \longrightarrow \frac{12Nk_B\pi^4}{5}\left(\frac{T}{\Theta_D}\right)^3 \tag{2.17}$$

which is in good agreement with experiment.

In this theory $k_B\Theta_D/h$ is the frequency of the shortest-wavelength normal mode. It can be determined experimentally from the low-temperature behavior [Eq. (2.17)], from the medium-temperature behavior as described for Fig. 2.2, and also, with the help of the theory of the vibrations of an elastic solid, from the elastic constants of the material. The best experiments indicate that Θ_D determined in different ways is not quite the same; even Debye's model does not *exactly* represent the real crystals. However, it is interesting to note that for graphite, which may be thought of as a stack of two-dimensional crystals which are only weakly coupled to each other, the low-temperature heat capacity is nearly proportional to T^2 (rather than T^3), just as one would expect from Debye's theory adapted to two-dimensional crystals.

2.6 SPINS ON A LATTICE

A model for discussing some magnetic phenomena consists of a lattice with independent spins at the lattice points. A *spin*, for this purpose, is a particle that can be in one of a small number of quantum states, each of which has a characteristic energy.

We consider only the model where the spin has just two states, one of energy E_+ and the other of energy $E_- = -E_+$.

Suppose that the spin is associated with a magnetic moment m_j which can be either parallel or opposed to an external magnetic field H. Then the source of the energy is the interaction with the field and

$$E_j = m_j H, \qquad j = + \text{ or } -$$

and

$$m_+ = -m_- \equiv m$$

The one-particle partition function in this model is

$$q = e^{\beta m H} + e^{-\beta m H} = 2 \cosh \beta m H \tag{2.18a}$$

and for a system of N spins

$$\ln Q = N \ln 2 + \ln \cosh \beta m H \tag{2.18b}$$

In this case, in which the system is subject to a magnetic field, we write the fundamental equation for the Helmholtz free energy as

$$dA = -S\,dT - p\,dV + \mu\,dn - M\,dH$$

where M is the average total magnetic moment of the sample

$$M \equiv \left\langle \sum_{\text{spins}} m_{\text{spin}} \right\rangle$$

and $M\,dH$ is a term for non-pV work associated with changing the magnetic field. We have

$$M = \frac{\partial A}{\partial H} = mN \tanh \beta m H \tag{2.19}$$

we also have

$$M = m\langle N_+ - N_- \rangle \tag{2.20}$$

where N_+ is the number of spins "up" and then

$$\tanh \beta m H = \frac{\langle N_+ - N_- \rangle}{N_+ + N_-} \tag{2.21}$$

The energy is also easy to calculate:

$$\langle E \rangle = -\frac{\partial \ln Q}{\partial \beta} = NmH \tanh \beta m H = MH \tag{2.22}$$

Now if we attempt to get the entropy from $S = (E - A)/T$ using the canonical ensemble equations, we get quite a complicated combination of logarithms and hyperbolic functions. On the other hand, a very simple expression comes from the microcanonical ensemble equation $S = k_B \ln \Omega$, where Ω is the number of distinct states accessible to the system of specified energy E. For a state of specified E we have

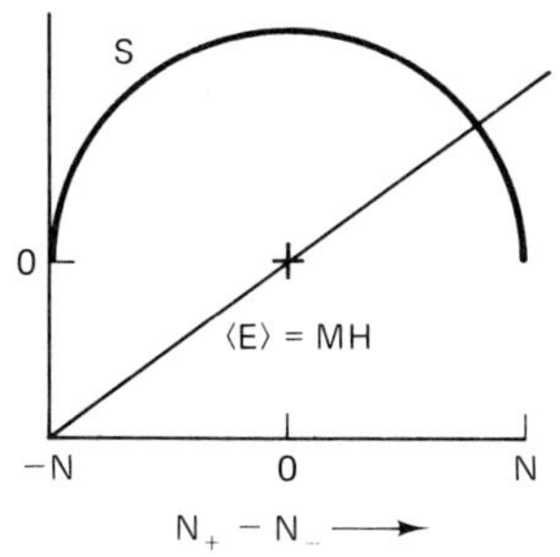

Figure 2.3 Energy and entropy of independent spins on a lattice.

(see above) specified N_+ and N_-, so we can at once calculate the number of quantum states that have this energy and find

$$S = k_B \ln \binom{N}{N_+} \tag{2.23}$$

Where $\binom{a}{b}$ is the binomial coefficient $a!/b!(a-b)!$. Figure 2.3 represents the equation of state of the spin system. As the energy increases (left to right), the polarization changes from negative values (in equilibrium with the field) to positive values (opposed to the field). However, the entropy has a maximum value where $\langle E \rangle = 0$. Now, from the fundamental equation for E in thermodynamics we have

$$T = \left(\frac{\partial E}{\partial S}\right)_{N,V} \tag{2.24}$$

Thus it follows that states on the left half of the figure are states of positive temperature, whereas those on the right are states of negative temperature.

The nuclear spins in a material sample are often very weakly coupled to the rest of the world. For nuclei of spin $\frac{1}{2}$ ($I = \frac{1}{2}$) the only coupling is through magnetic fields, and for such nuclei, particularly in crystals at low temperature, the model described here is realistic enough to be interesting. For example, spin system states of negative temperature actually have been obtained by the following steps:

1. Let the spin system come to equilibrium with magnetic field H. This takes some time, which usually is expressed in terms of a spin-lattice relaxation time T_1.
2. In a time short compared to T_1, change H to $-H$. This has the effect of changing N_+ into N_-, and vice versa, since the system cannot relax so quickly.

Now the state of the system is represented by a point on the right side of the diagram, hence a state of negative temperature.

2.7 IDEAL MONATOMIC GAS. QUANTAL

Equations (2.1) to (2.3) certainly apply to an ideal gas, but now a new problem appears. Since a particle of a gas or other fluid is free to move about the volume V occupied by the sample, we cannot use its location to distinguish one particle from another as we could in a crystal. But then we cannot distinguish particles of a given species from each other at all and the third equality in Eq. (2.5) is not applicable to the ideal gas. It permits far too many terms in the sum because it counts a new system quantum state whenever indistinguishable particles interchange particle quantum numbers. As we shall see, *if* the ideal gas is not too cold or too dense, the correct result for indistinguishable independent particles is

$$Q = \frac{q^N}{N!} \tag{2.25}$$

in place of (2.7).

To handle this problem we first calculate Q with explicit allowance for the indistinguishability of the particles. To proceed, we define:

The quantum state of any one particle: $1, 2, 3, \ldots, j, \ldots$.

The energy of this state: $E_1, E_2, E_3, \ldots, E_j, \ldots$.

The number of particles in this state: $s_1, s_2, s_3, \ldots, s_j, \ldots$.

The maximum number of particles allowed in each state: m(m is the same for every state; thus $s_j \leq m$ for every j).

The distribution for a particular state of the N-particle system: $\mathbf{s} = s_1, s_2, \ldots, s_j, \ldots$. That is, $\mathbf{s}$ is the list of the number of particles in each state.

From quantum mechanics we learn that there are only two possibilities for m: (1) $m = 1$, for particles called *fermions* (e.g., electrons, protons, He^3), or (2) $m = \infty$, for particles called *bosons* (e.g., photons, He^4, closed-shell molecules with closed-shell nuclei). It is clear that for a given distribution $\mathbf{s}$ we have the energy and the number of particles given by

$$E(\mathbf{s}) = \sum_j E_j s_j \tag{2.26}$$

$$N(\mathbf{s}) = \sum_j s_j \tag{2.27}$$

Then the canonical partition function is

$$Q(N, V, T) = \sum_{\mathbf{s}} e^{-\beta E(\mathbf{s})} \delta\left(N - \sum_j s_j\right) = \sum_{\mathbf{s}} \delta\left(N - \sum_j s_j\right) \prod_j e^{-\beta s_j E_j} \tag{2.28}$$

The sum follows because each different $\mathbf{s}$ labels a distinct quantum state of the entire system of indistinguishable particles, while the *Kronecker delta function,*

$$\delta(N - N') = \begin{cases} 1 & \text{if } N = N' \\ 0 & \text{Otherwise} \end{cases}$$

assures that the only distributions that contribute are those for which the total number of particles is N.

At this stage the whole problem of calculating the distribution function of a system of indistinguishable particles shows up in the delta function in Eq. (2.28). It is only this which makes the problem more complicated than the one dealt with in

Eq. (2.5). However, we can get around this difficulty by passing to the grand canonical ensemble, for then we have

$$
\begin{aligned}
\Xi(\mu, V, T) &= \sum_N e^{\beta\mu N} Q(N, V, T) \\
&= \sum_N e^{\beta\mu N} \sum_{\mathbf{s}} \delta\left(N - \sum_j s_j\right) \prod_j e^{-\beta s_j E_j} \\
&= \sum_{\mathbf{s}} \prod_j (e^{-\beta s_j E_j}) \sum_N e^{\beta\mu N} \delta\left(N - \sum_j s_j\right) \\
&= \sum_{\mathbf{s}} \left(\prod_j e^{-\beta s_j E_j}\right) \exp\left(\beta\mu \sum_j s_j\right) \\
&= \sum_{\mathbf{s}} \prod_j \exp\,[\beta s_j(\mu - E_j)] \\
&= \prod_j \xi_j
\end{aligned}
\tag{2.29}
$$

where we define

$$\xi_j \equiv \sum_{s_j=0}^{m} \exp\,[\beta s_j(\mu - E_j)] \tag{2.30}$$

The ξ_j has the form of the grand partition function of a very simple sort of system: It has just one energy level E_j but may be occupied by up to m particles. When it is occupied by s_j particles, it has energy $s_j E_j$. So it is clear that the particles in this simple system do not interact with each other. For example, if state j is occupied by 0, 1, or 2 particles, it has energy 0, E_j, or $2E_j$, respectively.

The simple system for which ξ_j is the grand partition function is in equilibrium with other simple systems which are similar but have different j numbers. Since the system is in a grand ensemble, the average number of particles in the system is dependent on μ, the chemical potential of the particles in the ensemble as a whole. By the usual operation for the grand partition function, we find the mean occupation numbers of the simple systems.

$$
\begin{aligned}
\langle s_j \rangle &= \beta^{-1}\left(\frac{\partial \ln \xi_j}{\partial \mu}\right)_{V,T} \\
&= \sum_{s_j} \exp\,[\beta s_j(\mu - E_j)]\,\frac{s_j}{\xi_j}
\end{aligned}
\tag{2.31}
$$

It should be noted that the most detailed description of the macroscopic equilibrium system of indistinguishable particles we can hope to construct is the distribution of the average occupation numbers of the quantum states:

$$\langle s_1 \rangle, \langle s_2 \rangle, \langle s_3 \rangle, \ldots, \langle s_j \rangle, \ldots$$

As the reader can verify by employing Eq. (2.27), this distribution satisfies

$$\sum_j \langle s_j \rangle = \langle N \rangle \tag{2.32}$$

so it is clear that $\langle s_j \rangle / \langle N \rangle$ is the average fraction of the *particles* in quantum state j. It may be compared with P_j, the fraction of *systems* in quantum state j in a canonical ensemble [Eq. (1.17)].

For fermions, Eqs. (2.30) and (2.31) become

$$\xi_j = 1 + e^{\beta(\mu - E_j)} \tag{2.33}$$

$$\langle s_j \rangle = \frac{e^{\beta(\mu - E_j)}}{1 + e^{\beta(\mu - E_j)}} \tag{2.34}$$

while for bosons,

$$\xi_j = \frac{1}{1 - e^{\beta(\mu - E_j)}} \tag{2.35}$$

$$\langle s_j \rangle = \frac{e^{\beta(\mu - E_j)}}{1 - e^{\beta(\mu - E_j)}} \tag{2.36}$$

These results may be combined to give

$$\langle s_j \rangle = \frac{1}{e^{\beta(E_j - \mu)} \pm 1} \qquad \begin{matrix} + \text{ for fermions} \\ - \text{ for bosons} \end{matrix} \tag{2.37}$$

Isn't it remarkable that whether one chooses $m = 1$ or ∞ in Eq. (2.30) only changes a term from $+1$ to -1 in Eq. (2.37)?

Now we examine the limiting case in which all $\langle s_j \rangle \ll 1$. Then Eq. (2.37) reduces to the same form, whether for fermions or for bosons:

$$\langle s_j \rangle = e^{\beta(\mu - E_j)} \qquad \text{boltzons} \tag{2.38a}$$

No real particles are *boltzons* (this name is not widely used), but the properties of a system of boltzons will be a good approximation to the behavior of a system of fermions or a system of bosons provided that $\langle s_j \rangle \ll 1$ for all j. This condition, which we may call the *classical limit,* is easily seen to be equivalent to the condition

$$\beta(E_j - \mu) \gg 1 \tag{2.38b}$$

Thus boltzons are a good approximation to real particles if the chemical potential μ is much smaller than the lowest energy level, E_0 or E_1, depending on the system. Recall that we can always make μ less positive or more negative by making the system more dilute.

The dependence of the occupation number $\langle s_j \rangle$ on the chemical potential is shown in Fig. 2.4a for fermions, bosons, and boltzons. The figure clearly shows that in the classical limit fermions, bosons, and boltzons become indistinguishable from each other. The figure also shows the strikingly different behavior of the three kinds of particles at high chemical potentials. One sees for bosons that for any state j, we cannot have μ as big as E_j or $\langle s_j \rangle$ would be infinite. Therefore, μ must be less than E_0. At low temperatures all the particles tend to crowd into the $j = 0$ state. This is related to the unique properties of liquid He(II).

For fermions at low temperature a graph of $\langle s_j \rangle$ as a function of j has the form shown in Fig. 2.4b. At $T = 0$ K the distribution is represented by the solid

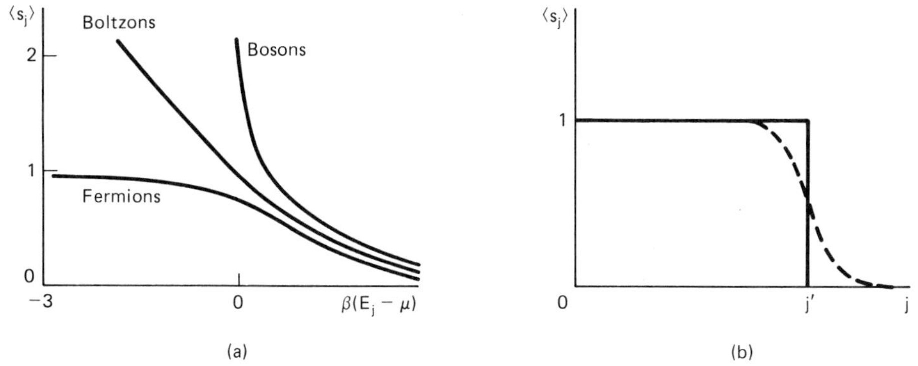

Figure 2.4 (a) Average occupation number of quantum state j as a function of the reduced β. (b) Average occupation number in a fermion fluid as a function of quantum number j (in order of increasing E_j as j increases). Solid line, $T = 0$; dashed line $T > 0$.

line. The energy of the j' level is equal to the chemical potential, the *Fermi level* as it is sometimes called. The dashed line shows the distribution at somewhat higher temperatures. This is a fair representation of the distribution of the conduction electrons in a metal such as copper at room temperature.

To apply these equations to a system of particles in a box of volume $V = L^3$, we use the quantum-mechanical result for the energy levels of a particle of mass M in a cubic box of edge L:

$$E_{klm} = \frac{h^2(k^2 + l^2 + m^2)}{8ML^2} \tag{2.39a}$$

For this system we interpret the index j of the quantum state in the equations above as a set of three positive integers $\mathbf{j} = k, l, m$ and then we can write an equation showing explicitly how $\langle s_{\mathbf{j}} \rangle$ depends on β, μ, L, and M. Given this function we can use it to calculate ensemble averages such as the one in Eq. (2.32) as well as [see Eq. (2.26)]

$$\langle E \rangle = \sum_{\mathbf{j}} E_{\mathbf{j}} \langle s_{\mathbf{j}} \rangle \tag{2.39b}$$

The evaluation of these formulas is not easy although the results are well known. We shall consider only two cases which involve relatively simple calculations. One is the classical limit, discussed in the following section, and the other is the other extreme, the $T = 0$ limit for a system of electrons, but with the Coulombic interactions neglected, to which we now turn.

We shall exploit the fact that for a system of electrons at $T = 0$ the solid curve in Fig. 2.4b gives the distribution function:

$$\langle s_{\mathbf{j}} \rangle = \begin{cases} 1 & \text{if } E_{\mathbf{j}} < \mu \\ 0 & \text{if } E_{\mathbf{j}} > \mu \end{cases} \tag{2.40a}$$

The first step is to calculate the number of electrons $N(\mu)$ in the system. This is twice the number of the states $\mathbf{j}$ for which $\langle s_{\mathbf{j}} \rangle = 1$ because electrons have spin $\frac{1}{2}$, so in each state $\mathbf{j}$ we may have two electrons with different spin quantum numbers.

For any given energy E the largest value of k (or of l or m) is evidently

$$k_E = \left(\frac{8ML^2E}{h^2}\right)^{1/2}$$

The condition for other states to have this same energy is

$$k_E^2 = k^2 + l^2 + m^2 \tag{2.40b}$$

Therefore, provided only that E is very large, so that we may represent Eq. (2.40b) as a spherical surface in the k, l, m coordinate system, the number $N(E)$ of k, l, m states with energy equal to or less than E is just $(4\pi/3)k_E^3/8$; the $\frac{1}{8}$ factor appears because we allow only positive values of k, l, and m. So we find that

$$N(E) = \frac{2^{9/2}\pi(ME)^{3/2}V}{3h^3} \tag{2.40c}$$

which implies that

$$E = \frac{[3N(E)/\pi V]^{2/3}h^2}{8M} \tag{2.40d}$$

showing how the energy E is related to the concentration $N(E)/V$ of electrons in states of energy less than or equal to E. Finally, we can calculate the thermodynamic total energy of the system for which the chemical potential or Fermi level μ is given.

$$\langle E \rangle = \int_0^\mu E[N(E + dE) - N(E)] \tag{2.40e}$$

$$= \int_0^\mu E\left[\frac{\partial N(E)}{\partial E}\right] dE$$

$$= \int_0^\mu \frac{3}{2}E\left[\frac{N(E)}{E^{3/2}}\right] E^{1/2}\, dE \tag{2.40f}$$

where $[\cdot\ \cdot\ \cdot]$ does not depend on E, as one can see from Eq. (2.40c). So we have

$$\begin{aligned}\langle E \rangle &= \frac{N(\mu)}{\mu^{3/2}}\left(\frac{3}{2}\right)\left(\frac{2}{5}\right)\mu^{5/2}\\ &= \frac{3\mu N(\mu)}{5}\end{aligned} \tag{2.40g}$$

At $T = 0$ the Helmholtz free energy A is equal to $\langle E \rangle - TS = \langle E \rangle$. Then the pressure is given by

$$\begin{aligned}p &= -\left(\frac{\partial A}{\partial V}\right)_{N,T}\\ &= -\frac{3}{5}N(\mu)\left(\frac{\partial \mu}{\partial V}\right)_N \qquad (2.40h)\\ &= \left(\frac{2}{5}\right)N(\mu)\frac{\mu}{V}\end{aligned}$$

It is interesting that p is proportional to N/V, just as for an ideal boltzon gas, but the proportionality constant is rather different.

We now consider the application of these formulas to the theory of metals. In a typical metal the metal atoms are disposed as positive ions at lattice sites, while the remaining electrons, the conduction electrons, fill the space between the ions. Because of the long range of the Coulombic forces it is not a bad approximation to represent the effect of the positive ions as merely providing a box that holds the electrons. The walls of this box (at the surface of the metal) represent the attractive force of all of the ions acting on any electron. In this free electron model for a metal, Coulombic interactions of the electrons with each other are also neglected. Then the equations derived above can be applied to the conduction electrons.

A typical metal has a density of about 0.1 mol of positive ions per cubic centimeter. We assume that there is just one conduction electron per positive ion. Then the equations above are readily evaluated, giving

$$\mu = 130 \text{ kcal/mol}$$

$$\frac{\langle E(\mu)\rangle}{\langle N(\mu)\rangle} = 80 \text{ kcal/mol}$$

$$p \simeq 2 \times 10^5 \text{ atm}$$

These numbers correspond to very large forces which derive, not from the Coulombic interactions, which have been neglected, but from the fact that to have so many electrons in such a small place, many of them must have very high kinetic energies, so that the requirements of the Pauli exclusion principle can be satisfied. In the metal these forces are in some sense matched by the Coulombic attraction of the ions for the electrons. One finds similar high densities of electrons in chemical bonds in molecules. It is impressive that the comparison of predictions of the electron gas theory with the experimental coefficients for a metal is generally quite satisfactory at the semiquantitative level. Typically, if the concentration $N(\mu)/V$ is adjusted to bring some measurable coefficient to agreement with some data, the result is within about a factor of 2 of what may be obtained from other data.

The present theory depends on the assumption that the number of electrons in the system is very large, so it applies even less precisely to a molecule than to a macroscopic metallic body. However, the qualitative picture of balance between the electron gas pressure and the Coulombic interaction with the nuclear centers is still valid. Indeed, a quantitative electron gas theory along these lines gives a very satisfactory account of the strongly repulsive part of the force between molecules (Gordon and Kim, 1972).

2.8 THE CLASSICAL LIMIT FOR A SYSTEM OF INDEPENDENT INDISTINGUISHABLE PARTICLES

In this section we obtain additional useful relations for systems of boltzons. To begin, we sum both sides of Eq. (2.38a) to get

$$\langle N\rangle = qe^{\beta\mu} \tag{2.41}$$

where q is the one-particle partition function as in Eq. (2.6). We also note that for fermions or bosons we have, when Eq. (2.38b) is satisfied,

$$\ln \xi_j = e^{\beta(\mu - E_j)}, \qquad \beta(E_j - \mu) \gg 1 \tag{2.42}$$

From this limiting form together with the general equation

$$\beta pV = \ln \Xi = \sum_j \ln \xi_j \tag{2.43}$$

we find that for either fermions or bosons in the classical limit, we have

$$\beta pV = qe^{\beta\mu} \tag{2.44}$$

In the same limit (2.41) applies, so we have

$$\beta pV = \langle N \rangle \qquad \text{or} \qquad pV = \langle N \rangle k_B T \tag{2.45}$$

Thus a boltzon fluid is a perfect gas. Now to relate q to the system partition function Q, we employ the completely general relation for the latter,

$$\beta A = -\ln Q$$

and the thermodynamic equation for the chemical potential of a one-component system with specified N, V, T,

$$\beta A = \beta(N\mu - pV) = \beta N\mu - N \tag{2.46}$$

where we have used the fact that for an ideal gas

$$\beta pV = N$$

Taken together these equations give

$$\beta\mu = -\frac{1}{N} \ln Q + 1 \tag{2.47}$$

To employ this canonical-ensemble result together with Eq. (2.41), we may use the fact that in a macroscopic system, the fluctuations do not lead to different results for the thermodynamic quantities calculated from different ensembles. Therefore, the input μ in the grand ensemble is the same as the calculated μ in the canonical ensemble; the input N in the latter is the same as the output $\langle N \rangle$ in the former. So we find, by eliminating μ between Eqs. (2.41) and (2.47),

$$\ln Q = N \ln q - N \ln N + N$$

or, using Stirling's approximation for $\ln N! \simeq N \ln N - N$,

$$Q = \frac{q^N}{N!} \tag{2.48}$$

Therefore, our system of independent, indistinguishable particles in the classical limit has a partition function Q related to the particle partition function q by Eq. (2.48) and it is a perfect gas.

These deductions were all made for particles with no internal structure. We now indicate how it can be shown that this is not a necessary condition.

We assume that Eq. (2.3) is valid, as before, and that the one-particle Hamiltonian H_n for particle n is separable into a translational part $H_{n,x}$ and an internal part $H_{n,i}$. Then in Eq. (2.1) the one-particle energy will comprise a translational part and an internal part, while the one-particle quantum number, say j_n, will represent a set of quantum numbers, some corresponding to the translational state and some to the internal state. Then the one-particle partition function q appearing in Eqs. (2.41) to (2.48) may be expanded as follows:

$$q = \sum_{j} e^{-\beta E_j} = \sum_{jx} \sum_{ji} e^{-\beta(E_{jx} + E_{ji})} = q_x q_i \tag{2.49}$$

where E_{jx} is the energy of the particle in a translational quantum state with index jx, where E_{ji} is the energy of the particle in an internal quantum state with index ji, and

$$q_x \equiv \sum_{jx} e^{-\beta E_{jx}}, \qquad q_i \equiv \sum_{ji} e^{-\beta E_{ji}} \tag{2.50}$$

When this classification is made, we also often use the definitions

$$Q_{\text{trans}} \equiv \frac{q_x^N}{N!}, \qquad Q_{\text{int}} \equiv q_i^N \tag{2.51}$$

so we have

$$Q = Q_{\text{trans}} \cdot Q_{\text{int}} \tag{2.52}$$

Notice that the "price" of indistinguishability of the particles is paid only once, conventionally in the translational factor of Q, as we have done here.

DO PROBLEMS 2.7, 2.8.

2.9 STATISTICAL THERMODYNAMICS OF GASES

In this section we describe the elementary methods that may be used to calculate the thermodynamic properties of ideal gases with molecules of varying complexity. Entropies calculated in this way have been very important in connection with third-law studies in thermodynamics. They continue to be useful in combination with measurements of free energies (i.e., equilibrium constants) to get heats, and in combination with measured heats to get free energies. Heat capacities calculated in this way are valuable for deducing thermodynamic quantities at one temperature from measurements at another and, especially, for extrapolating gas properties to absolute zero where the molecular interpretation is often easiest. Finally, calculations of ideal gas thermodynamic properties, when compared with experimental values, sometimes elucidate problems of molecular structure and dynamics (e.g., the internal rotation problem).

We study systems of boltzons in which the energy of a one-particle state j can be divided

$$E_j = E_{jx} + E_{ji} \tag{2.53}$$

as discussed in connection with Eq. (2.49). We further assume that the internal part can be divided among rotational, vibrational, electronic, and nuclear contributions:

$$E_{ji} = E_{jr} + E_{jv} + E_{je} + E_{\text{nuc}} \tag{2.54}$$

Thus the quantum number j in Eq. (2.53) corresponds to a set of quantum numbers specifying the translational, rotational, vibrational, and electronic states of the molecule. Of course, in a given system not all of these states may be relevant; for example, a monatomic gas can be treated without attention to rotation or vibration of the molecules. Also, the separability of the Hamiltonian which gives rise to Eq. (2.54) is at best an approximation, although a useful one. Its limitations are discussed a little more below, but generally may be neglected in an introductory account such as this.

A consequence of Eq. (2.54) is that the one-particle partition function [Eqs. (2.6) and (2.49)] is given by the product

$$q = q_x q_r q_v q_e q_{\text{nuc}} \tag{2.55}$$

so the system partition function has the product structure

$$Q = Q_{\text{trans}} \cdot Q_{\text{rot}} \cdot Q_{\text{vib}} \cdot Q_{\text{el}} \cdot Q_{\text{nuc}} \tag{2.56}$$

$$Q_{\text{trans}} = \frac{q_x^N}{N!}, \quad Q_{\text{rot}} = q_r^N, \quad Q_{\text{vib}} = q_v^N, \quad Q_{\text{el}} = q_e^N, \quad Q_{\text{nuc}} = q_{\text{nuc}}^N \tag{2.57}$$

and the system free energy is a sum of contributions from the various modes:

$$A = A_{\text{trans}} + A_{\text{rot}} + A_{\text{vib}} + A_{\text{el}} + A_{\text{nuc}} \tag{2.58}$$

The other thermodynamic coefficients are derivatives of the free energy, so it is clear that the royal road to the statistical thermodynamics of ideal gases is through the study of q_x, q_r, q_v, q_{el}, and q_{nuc}. We now turn to these matters.

2.10 TRANSLATION

For a single particle of mass M in a cubic box of volume L^3, we find from quantum mechanics that the state is fixed by three integers, call them k, l, m, and has energy

$$E_{klm} = \frac{h^2(k^2 + l^2 + m^2)}{8ML^2} \tag{2.59}$$

It is clear that for thermodynamic properties that are independent of the shape of the sample, this is all we need. The corresponding partition function for the particle is

$$\begin{aligned} q_x &= \sum_{k=1}^{\infty} \sum_{l=1}^{\infty} \sum_{m=1}^{\infty} e^{-\beta E_{klm}} \\ &= \left(\sum_k e^{-\alpha^2 k^2} \right)^3 \end{aligned} \tag{2.60}$$

$$\alpha^2 \equiv \frac{\beta h^2}{8ML^2} \tag{2.61}$$

The sum in (2.60) is not so easy to do, but when α is small it may be approximated by an integral over a Gaussian function

$$q_x^{1/3} \simeq \int_{k=0}^{\infty} e^{-\alpha^2 k^2}\, dk = \frac{\sqrt{\pi}}{2\alpha} \tag{2.62}$$

So we find, when α is small enough,

$$\begin{aligned} q_x &= \tfrac{1}{8}\left(\frac{8\pi ML^2}{\beta h^2}\right)^{3/2} \\ &= \frac{V}{\Lambda^3} \end{aligned} \tag{2.63}$$

where

$$V \equiv L^3$$

$$\Lambda \equiv \frac{h}{\sqrt{2\pi M k_B T}} \tag{2.64}$$

Λ is called the *de Broglie thermal wavelength* because the de Broglie wavelength of a particle of mass M and velocity v is

$$\frac{h}{Mv}$$

while the root-mean-square (rms) velocity of a molecule at thermal equilibrium at temperature T is fixed by the relation

$$\tfrac{3}{2} k_B T = \tfrac{1}{2} M\langle v^2\rangle$$

so the wavelength of a particle with the rms velocity is

$$\frac{h}{\sqrt{3Mk_BT}}$$

which is rather close to Λ as defined in Eq. (2.64).

Now let us see under what conditions the approximations leading to Eq. (2.63) are valid. We find that

$$\alpha = \frac{1}{2}\sqrt{\frac{\pi\Lambda}{L}} \tag{2.65}$$

so α is much less than unity, as required, when

$$\Lambda \ll L \tag{2.66}$$

Now we inquire under what conditions it is acceptable to treat the particles as boltzons rather than as fermions or bosons. From Eq. (2.40) we see that when boltzons are a good representation of the physical system, we have

$$\beta(E_j - \mu) \gg 1, \qquad j = 0, 1, 2, \ldots \tag{2.67}$$

for which a sufficient condition is

$$\beta\mu \ll 0 \tag{2.68}$$

For boltzons we also have [Eq. (2.41)]

$$e^{\beta\mu} = \frac{\langle N \rangle}{q} \tag{2.69}$$

Combining Eqs. (2.63), (2.68), and (2.69), we have

$$e^{\beta\mu} = \frac{\Lambda^3 \langle N \rangle}{V} \ll 1 \tag{2.70}$$

or

$$\Lambda \ll \left(\frac{V}{N}\right)^{1/3} = \rho^{-1/3} \tag{2.71}$$

where ρ is the particle number density

$$\rho \equiv \frac{N}{V} \tag{2.72}$$

Equation (2.70) shows that boltzons are a good approximation to real particles when Λ is much less than the mean distance between particles. Clearly, this condition is much more restrictive than the one justifying Eq. (2.62), so that for boltzons, replacing the sum by the integral in Eq. (2.62) is always permitted.

Table 2.2 shows how well the boltzon approximation is satisfied for various systems. For the rest, we shall *only* discuss systems for which it is satisfied. We also have the following thermodynamic functions for the translational motions of boltzons. Readers should verify the derivation of these equations from Eqs. (2.57) and (2.63).

$$A_{\text{trans}} = Nk_B T \left(\ln \frac{\Lambda^3}{e} + \ln \rho \right) \tag{2.73}$$

$$p = k_B T \rho \tag{2.74a}$$

$$\langle E_{\text{trans}} \rangle = \tfrac{3}{2} Nk_B T = \tfrac{3}{2} pV \tag{2.74b}$$

$$\langle C_{v,\,\text{trans}} \rangle = \tfrac{3}{2} Nk_B \tag{2.74c}$$

$$S_{\text{trans}} = -Nk_B \left(\ln \left(\frac{\Lambda^3}{e^{5/2}} \right) - \ln \rho \right) \tag{2.74d}$$

These are all extensive quantities. The contributions per molecule A_{trans}/N, $\langle E_{\text{trans}} \rangle/N$, are required for many applications. With species labels these per-molecule contributions $A_{\text{trans},s}/N_s$, $\langle E_{\text{trans},s} \rangle/N_s$, and so on, also apply in mixtures of independent particles. It is very convenient now to introduce the notation

$$a_{\text{trans},s} = \frac{A_{\text{trans},s}}{N_s k_B T} \tag{2.75}$$

$$a_s = \frac{A_s}{N_s k_B T} \tag{2.76}$$

TABLE 2.2 DE BROGLIE THERMAL WAVELENGTH FOR SEVERAL SYSTEMS

			$\rho^{-1/3}$ (Å) in:	
Substance	T (K)	Λ (Å)	1 atm gas	Liquid
e^-	300	43	74	—
He	4.2	4.25	18	3.71
H_2	20	2.72	30	3.6
Ne	27	0.75	33	3.05
A	87	0.29	49	3.95

and generally, for any mode m of motion of species s,

$$\begin{aligned} a_{m,s} &= \frac{A_{m,s}}{N_s k_B T} \\ &= -\ln q_{m,s} + [\ln N_s]_{\text{if } m=\text{trans}} \end{aligned} \tag{2.77}$$

DO PROBLEMS 2.9 TO 2.12.

2.11 ROTATION OF LINEAR MOLECULES

The quantum state of a linear rotator is fixed by two quantum numbers, say j and m, and has the energy

$$E_j = \frac{j(j+1)h^2}{8\pi^2 I} \tag{2.78}$$

where j is an integer, m is an integer with a value in the range from $-j$ to $+j$, and I is the moment of inertia

$$I = \sum_i r_i^2 M_i \tag{2.79}$$

where r_i is the distance of particle i in the molecule from the center of mass. For a diatomic molecule

$$I = a^2 M_r \tag{2.80}$$

where a is the bond length and M_r is the *reduced mass*

$$M_r \equiv \frac{M_1 M_2}{M_1 + M_2} \tag{2.81}$$

If the rotator is unsymmetrical (e.g., CO, CCO, HCC—Cl), then j may have any value from zero to infinity. If the molecule is symmetrical (e.g., N_2, CO_2, H_2C_2), then j may have only even values. In either case the energy depends only on the j number of the quantum state, not on the m number. If we define the *symmetry number* σ to be 2 for symmetrical linear molecules and 1 for unsymmetrical linear molecules, we have in either case

$$\begin{aligned} q_r &= \sum_j (2j+1)e^{-\beta E_j} \qquad \begin{cases} \text{unsymm.: } j = 1, 2, 3, \ldots \\ \text{symm.: } \quad j = 0, 2, 4, \ldots \end{cases} \\ &\simeq \sigma^{-1} \sum_j (2j+1)e^{-\mathrm{B} E_j} \qquad j = 0, 1, 2, 3, \ldots \\ &= \sigma^{-1} \sum_j (2j+1)e^{-j(j+1)\Theta_r/T} \end{aligned} \tag{2.82}$$

where

$$\Theta_r \equiv \frac{h^2}{8\pi^2 I k_B} = \frac{24.30 \text{ K}}{I \text{ amu}^{-1}\text{Å}^{-2}} \tag{2.83}$$

is the characteristic temperature for rotation. The approximation in Eq. (2.82) is good when Θ_r/T is small, as the reader may verify. In the same limit the final sum in (2.82) may be replaced by an integral that is easy to evaluate:

$$\begin{aligned} q_r &= \int_0^\infty (2j+1)e^{-j(j+1)\Theta_r/T} \frac{dj}{\sigma} \\ &\simeq \int_0^\infty e^{-x\Theta_r/T} \frac{dx}{\sigma} \\ &= \frac{T}{\Theta_r \sigma} \end{aligned} \tag{2.84}$$

To elucidate the role of the symmetry number it is interesting to express this result in a different notation. We define the thermal de Broglie "wavelength" for rotation by the equation

$$\Lambda_r \equiv \left(\frac{4\pi\Theta_r}{T}\right)^{1/2} = \frac{h}{\sqrt{2\pi I k T}} \tag{2.85}$$

(note its units). This is similar to Eq. (2.64) and in place of Eq. (2.63),

$$q_x = \frac{V}{\Lambda^3}$$

it gives

$$q_r = \frac{4\pi}{\Lambda_r^2 \sigma} \tag{2.86}$$

Now the analogy between these expressions may be developed in the following way. Just as V is the volume accessible to the particle-in-a-box, so 4π, the surface area of a unit sphere, is the "area" accessible to the rotator, for we may think of the rotator as centered at the center of the sphere with a line extending from the rotator intersecting the sphere at a point. As the rotator turns, the point moves on the surface of the sphere. If the rotator is unsymmetrical, every point on the surface of the unit sphere represents a distinct orientation of the rotator, but if the rotator is symmetrical, this is no longer the case, for then every orientation corresponds to a point in one hemisphere, of area $4\pi/2$. Thus the symmetry number reflects the fact that there are fewer distinct orientations for the symmetrical rotator, hence fewer quantum states.

Rotational constants are often expressed in terms of the rotational wave number (units: cm^{-1})

$$B = \frac{h}{8\pi^2 Ic} = \frac{k_B\Theta_r}{hc} = \frac{\Theta_r}{1.43879 \text{ cm K}} = \frac{16.858 \text{ cm}^{-1}}{I \text{ amu}^{-1}\text{Å}^{-2}} \tag{2.87}$$

or the rotational frequency (units: sec^{-1})

$$B = \frac{h}{8\pi^2 I} = \frac{k_B\Theta_r}{h} \tag{2.88}$$

where c is the velocity of light. The same notation is used in either case, as shown here, so you have to watch for the units. Some rotational constants, obtained either from spectroscopic experiments or from calculations from the known geometries of the molecules, are given in Table 2.3.

An interesting source of unsymmetrical rotational behavior is found in homonuclear diatomic molecules in which the nucleus can be in more than one spin state. The best known example is H_2, in which the proton has spin quantum number $\frac{1}{2}$ and can be in either of two spin states. If both nuclei in an H_2 molecule are in the same spin state (*para*-hydrogen), it can have only even rotational quantum numbers j, whereas if they are in different states (one spin "up" and the other "down," *ortho*-hydrogen), it can have both even and odd rotational quantum numbers j. Because changes in nuclear spin states are slow in many circumstances and because Θ_r is large for H_2, the *ortho–para* difference has important thermodynamic consequences.

TABLE 2.3 SOME MOLECULAR PARAMETERS

Species	Θ_v (deg)	Θ_r (deg)	r_e (Å)	D_0 (eV)
H_2	6210	85.4	0.740	4.477
HD	5350	64	—	—
CO	3070	2.77	1.13	9.14
NO	2690	2.42	1.15	5.29
N_2	3340	2.86	1.10	9.76
O_2	2230	2.07	1.204	5.08
HF	5840	30.3	0.917	5.86
HCl	4140	15.2	1.28	4.43
HBr	3700	12.1	1.41	3.60
HI	3200	9.0	1.60	3.06
F_2	1328	1.29	1.41	1.59
Cl_2	810	0.346	1.99	2.48
Br_2	470	0.116	2.28	1.97
I_2	310	0.054	2.67	1.54
CO_2	1932, 3380, 960, 960	0.5616		
H_2O	5262, 2294, 5404	39.3, 20.9, 13.7		

2.12 LINEAR AND NONLINEAR ROTATORS

In general, rigid molecules have three moments of inertia, call them I_x, I_y, and I_z, which are defined, for example, by the expression

$$I_x = \sum_i M_i(y_i^2 + z_i^2) \tag{2.89}$$

where the Cartesian axes x, y, and z intersect at the center of mass of the molecule and are oriented in such a way that off-diagonal moments of inertia such as

$$I_{xy} = \sum_i M_i x_i y_i$$

vanish. In these equations M_i is the mass of the atom located at I_i, y_i, z_i. If they are applied to a linear molecule whose axis runs in the z direction, we have

$$I_x = I_y \equiv I, \qquad I_z = 0$$

A nonlinear molecule in which two of the moments are equal is called a *symmetric top* (e.g., BF_3, PCl_5). If all the moments are equal, it is called a *spherical top* (e.g., CH_4, UF_6), while if all the moments are different, it is called an *asymmetric top* (e.g., H_2O, N_2O, *trans*-dichloroethylene).

The quantum state of a nonlinear rotator is fixed by three quantum numbers; the energy depends on two of them. The result for the partition function is

$$q_r = \frac{\sqrt{\pi}}{\sigma}\sqrt{\frac{T}{\Theta_x}}\sqrt{\frac{T}{\Theta_y}}\sqrt{\frac{T}{\Theta_z}} \tag{2.90}$$

where Θ_x is related to I_x as Θ_r is related to I in Eq. (2.83). The symmetry number σ may be defined in the following way. Label the n atoms in the molecule from 1 to n. Now σ is the number of different orientations of the labeled molecule that would look the same if the labels were erased. Some examples are shown in Table 2.4. Again, there is an effect of nuclear spins which is neglected in these examples, as is customary in all the cases in which one cannot separate nuclear spin isomers. Another interesting complication appears with NH_3, in which there is a low-frequency vibration related to the motion in which the N "tunnels" through the plane of the H atoms. If this vibration were higher in frequency than the rotations, NH_3 would be planar with symmetry number 6. Actually, we can get correct results in calculating the rotational partition function in either way as long as we handle the vibrational partition function in a matching way. The thermodynamic functions for linear and nonlinear rotators are summarized in Table 2.5.

To use these equations when given rotational constants in terms of moments of inertia I or rotational frequencies B requires, of course, that we first calculate the characteristic temperatures from Eq. (2.83), (2.87), or (2.88). Although there

TABLE 2.4 SYMMETRY NUMBERS OF SEVERAL MOLECULES

Compound	H_2O	NH_3	BF_3	CH_4	CH_3Cl	UF_6	C_6H_6	C_6H_5Cl
σ	2	3	6	12	3	24	12	2

TABLE 2.5 THERMODYNAMIC FUNCTIONS FOR ROTATORS WHEN $\Theta_r \ll T$

Function	Linear	Nonlinear
$\beta A_{\text{rot}}/N$	$-\ln(T/\Theta_r\sigma)$	$-\frac{1}{2}\ln(\pi T^3/\sigma^2\Theta_x\Theta_y\Theta_z)$
$\beta E_{\text{rot}}/N$	2/2	3/2
C_{rot}/Nk	2/2	3/2
S_{rot}/Nk	$1+\ln(T/\sigma\Theta_r)$	$\frac{3}{2}+\frac{1}{2}\ln(\pi T^3/\sigma^2\Theta_x\Theta_y\Theta_z)$

are many opportunities for error in the unit conversions involved, we can check that the characteristic temperatures are reasonable by comparison with those in Table 2.3.

2.13 VIBRATION

The vibrational motions of a mechanical system, if they are small enough vibrations to be harmonic [Eq. (2.14)], can be described as combinations of noninteracting motions called *normal modes of vibration.* For a linear molecule of n atoms the number of such vibrations is $3n - 5$, for nonlinear molecules it is $3n - 6$. Thus a diatomic molecule has but a single mode. A linear triatomic molecule has four, of which two are said to be degenerate (i.e., they have the same force constant and the same frequency). For example, in CO_2 the modes are as shown in Fig. 2.5. Such diagrams for a variety of molecular geometries may be found in books on molecular spectroscopy.

←O · · · · C · · · · O→ O→ ←C · · · · O→ O↑ · · · · C↓ · · · · O↑ , perpendicular

Figure 2.5 Normal modes of CO_2. One phase of each mode is shown except that the fourth mode is not shown. It is the same as the third except that the displacements are perpendicular to the plane of the page.

Now if we take m to be an index denoting a normal mode of a molecule, then [see Eq. (2.15)] the vibrational part of the molecular partition function is

$$q_{\text{vib}} = \frac{1}{\prod_m (1 - e^{-\Theta_m/T})} \tag{2.91}$$

where Θ_m is the characteristic temperature of the mode:

$$\Theta_m = \frac{h\nu_m}{k} = \frac{h\omega_m c}{k} = \left(\frac{\hbar}{k}\right)\sqrt{\frac{\phi''_m(0)}{M_m}} \tag{2.92}$$

where ν_m = frequency of the mode, cycles/sec
ω_m = frequency of the mode, cm^{-1}
$\phi''_m(0)$ = force constant of the mode, $\phi_2(0)$ in Eq. (2.14)
M_m = reduced mass of the mode

In the case of diatomic molecules, the reduced mass is given by Eq. (2.81).

TABLE 2.6 CONTRIBUTIONS OF A SIMPLE HARMONIC OSCILLATOR TO THE THERMODYNAMIC FUNCTIONS

Θ_m/T:	0.2	0.4	0.6	0.8	1.0	1.5	2.0	2.5	3.0	4.0	5.0
$-\beta A/N$	1.71	1.11	0.80	0.60	0.46	0.25	0.14	0.09	0.05	0.02	0.01
$\beta E/N$	0.90	0.81	0.73	0.65	0.58	0.43	0.31	0.22	0.16	0.08	0.03
C/Nk_B	1.00	0.99	0.97	0.95	0.92	0.83	0.72	0.61	0.50	0.30	0.17
S/Nk_B	2.61	1.92	1.53	1.25	1.04	0.68	0.46	0.31	0.21	0.09	0.04

The zero-point energy factors do not appear in Eq. (2.91) because the zero of the vibrational energy is chosen as the ground state. The consequences of this are worked out in Sections 2.16 and 2.17.

It is clear from Eq. (2.91) that the various thermodynamic functions for vibration will be sums of contributions from each mode, for example [see Eq. (2.77)],

$$a_{\text{vib}} = \frac{\beta A_{\text{vib}}}{N} = \sum_m \ln\,(1 - e^{-\Theta_m/T}) \tag{2.93}$$

In practical calculations it is seldom possible to simplify this equation, so one has to do tedious computations to evaluate the contributions to A_{vib} or else use a published table of the thermodynamic functions of a harmonic oscillator. A short form of such a table is given in Table 2.6.

2.14 ELECTRONIC STATES

In many cases in which excited electronic states are involved, they are not at equilibrium with respect to electronic excitation and deexcitation processes. Then it may be convenient to treat each electronic energy level as a distinct chemical species. In the other limit in which there is thermal equilibrium among the electronic states, we may employ the electronic partition function

$$q_{\text{el}} = \sum_k g_k e^{-\beta E_k} \tag{2.94}$$

where g_k is the number of electronic quantum states in the kth level of energy E_k. Information about these energy levels can be obtained from spectroscopic experiments. The spacing between electronic energy levels is usually so large that the corresponding characteristic temperature is measured in thousands of kelvin, so the upper levels need be taken into account only if they lie exceptionally low (e.g., Cl·, NO) or if the temperature of the system is very high. In the latter case one usually finds only monatomic and perhaps a few diatomic ions and molecules present at appreciable concentrations at equilibrium. The ground state in paramagnetic molecules has multiplicity $g_0 > 1$. For example, for Cl· or O_2^- it is 2; for O_2 it is 3.

It is often convenient to pick the zero of the electronic energy scale so that $E_0 = 0$ in writing q_{el} and to account in a different place for the effect this has in computations of the thermodynamic changes in chemical reactions. For example, then, for an electron gas we have $q_{\text{el}} = 2$.

2.15 NUCLEAR STATES

The situation is like that for electronic states except that the characteristic temperatures corresponding to the excitation energies of nuclear states are in the range of billions of kelvin. Therefore, we simply have that q_{nuc}, the nuclear contribution to the one-particle partition function, is the multiplicity of the nuclear ground state of the molecule.

Because the states of the various nuclei in a molecule are independent of each other under almost all circumstances, we usually have

$$q_{\text{nuc}} = \prod_i g_{\text{nuc},i} \tag{2.95}$$

where $g_{\text{nuc},i}$ is the multiplicity of the ground state of the ith nucleus in the molecule. Now all the thermodynamic properties are obtained from

$$\ln q_{\text{nuc}} = \sum_i \ln g_{\text{nuc},i} \tag{2.96}$$

(i.e., from the sum of nuclear contributions) and since this sum does not change in a chemical reaction, one can neglect q_{nuc} in most cases. An exception would be the *ortho-para* hydrogen system, in which, by definition, the spins in a given molecule are not independent of each other. That is, they are parallel in *ortho*-hydrogen and antiparallel in *para*-hydrogen.

DO PROBLEM 2.13.

2.16 CHEMICAL REACTIONS

We represent a general chemical reaction by the equation

$$\sum_s \nu_s Y_s = 0 \tag{2.97}$$

where Y_s is the chemical formula of species s and ν_s is the stoichiometric coefficient, a positive integer for the products of the reaction and negative for the reactants. According to the standard thermodynamic analysis as the reaction slowly advances by an amount $d\xi$ in a closed system at fixed V and T the Helmholtz free energy evolves according to

$$\mathrm{dA} = \Sigma_s \nu_s \mu_s \, d\xi$$

where μ_s is the chemical potential of species s. The process reaches equilibrium when $\mathrm{dA}/d\xi = 0$ and then we have

$$\Sigma_s \nu_s \mu_s = 0 \tag{2.98a}$$

With the help of Table 1.3 and Eq. (2.5) we find

$$-\beta\mu_s = \partial \ln \mathrm{Q}/\partial \mathrm{N}_s = \ln \mathrm{q}_s$$

for our system of molecules that do not interact with each other except through the chemical transformation. Therefore the condition of chemical equilibrium for the reaction (Eq. 2.97) can also be written

$$\Sigma_{\mathcal{S}} \nu_{\mathcal{S}} \ln q_{\mathcal{S}} = 0 \tag{2.98b}$$

It is convenient to reduce this equation by separating the effects of the choice of the zero of energy for each species $\mathcal{S}$. A convenient choice is that the zero of energy is identified with the sum of the energies of the atomic constituents, each in its ground state and widely separated from the others. With this energy scale in mind we write [cf. Eq. (2.12)]

$$q_{\mathcal{S}} = e^{-\beta E_{0\mathcal{S}}} e^{-a_{\mathcal{S}}}$$

where $\exp(-a_{\mathcal{S}})$ is the partition function of molecules of species $\mathcal{S}$ on the energy scale that makes the ground state energy of the species equal to zero. The various one-mode partition functions q_x, $q_{m,\mathcal{S}}$, q_r, and so on, which were analyzed in Sections 2.10 to 2.15, are all defined for this zero of energy; the corresponding reduced Helmholtz free energy function $a_{m\mathcal{S}}$ is introduced in Eq. (2.77). Finally, Eq. (2.99) becomes

$$\Sigma_{\mathcal{S}} \nu_{\mathcal{S}} a_{\mathcal{S}} + \beta(\Delta E)_{T=0} = 0 \tag{2.99}$$

where

$$(\Delta E)_{T=0} \equiv \Sigma_{\mathcal{S}} \nu_{\mathcal{S}} E_{0,\mathcal{S}}$$

is the energy change at $T = 0$ for the amount of reaction in Eq. (2.97).

An example is offered to elucidate this point. In the dissociation of H_2,

$$H_2 \longrightarrow 2H$$

we have

$Y_{\mathcal{S}}$	H_2	H
$\nu_{\mathcal{S}}$	-1	2
$E_{0\mathcal{S}}$	$-D_0$	0

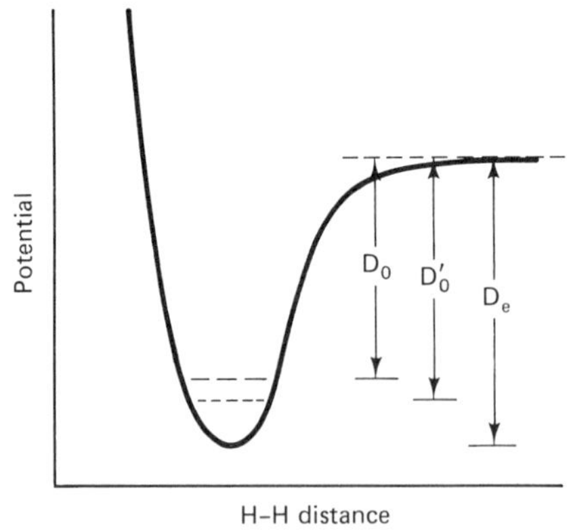

Figure 2.6 Schematic H—H bond potential as a function of internuclear distance. The dissociation energies D_0 (for H_2), D_0' (for D_2), and the equilibrium dissociation energy D_e are shown.

where D_0 is the dissociation energy of H_2 from its ground state (Fig. 2.6). In this case Eq. (2.99) becomes

$$2a_{trans,H} + 2a_{el,H} - a_{trans,H_2} - a_{rot,H_2} - a_{vib,H_2} - a_{el,H_2} + \beta D_0 = 0 \tag{2.100}$$

The following remarks are important.

1. The translational terms for H and H_2 are given by Eq. (2.73).
2. Ordinarily, one has only the ground electronic state of H, but this is twofold degenerate owing to the electron spin, so

$$A_{el,H} = -k_B T \ln 2$$

3. The nuclear terms for H and H_2 are assumed to mutually cancel, which implies equilibrium with respect to the *ortho* $\rightleftharpoons$ *para* transformation in H_2.
4. The vibrational term for H_2 is given by Eq. (2.91) in the harmonic oscillator approximation.
5. Given spectroscopic data for D_0, for the ground-state vibrational frequency ω_m for H_2, and for the rotational constant B for H_2, everything in Eq. (2.100) is known except for the concentrations ρ_H and ρ_{H_2} carried in the translational contributions to A. Therefore, from the spectroscopic data we can calculate the position of equilibrium in the chemical reaction.

Before going on to complete the theory of the equilibrium constant, we also remark that Eq. (2.100) enables us to predict the isotope effects on the equilibrium. Thus for the corresponding reaction

$$D_2 \longrightarrow 2D$$

the dissociation energy would be greater by the difference in zero-point vibrational energies:

$$\tfrac{1}{2}k_B(\Theta_{v,\,H_2} - \Theta_{v,\,D_2})$$

Since the potential curve is the same for both species, at least in the Born–Oppenheimer approximation (Fig. 2.6), one can easily estimate ω_m and B for D_2 from the corresponding data for H_2. Of course, one can easily generalize the calculation of the terms in Eq. (2.100) to include an harmonic terms in the potential for the bond in order to make use of the complete spectroscopic data that are available for a well-studied molecule like H_2. In tables such as Table 2.3, we often find that $D_e = D_0 + \frac{1}{2}k_B\Theta_v$ rather than D_0 is given.

2.17 THE EQUILIBRIUM CONSTANT

In Eq. (2.99) the concentration ρ_s of species s appears only in the translational terms [see Eq. (2.73)]. Now we write

$$a_s = a_s^0 + \ln \frac{\rho_s}{\rho_s^0} \tag{2.101}$$

where ρ_s^0 is a reference concentration, often called a *hypothetical standard-state concentration,* usually chosen to be 1 in some units. Then Eq. (2.99) becomes

$$0 = \sum_s \nu_s \ln \frac{\rho_s}{\rho_s^0} + \sum_s \nu_s a_s^0 + \frac{(\Delta E)_{T=0}}{k_B T} \tag{2.102}$$

One of the familiar definitions of the equilibrium constant for the reaction in Eq. (2.97) is

$$K = \prod_s \left(\frac{\rho_s}{\rho_s^0}\right)^{\nu_s} \tag{2.103}$$

Comparison with Eq. (2.102) shows that the statistical-mechanical theory of this equilibrium constant is expressed by the equation

$$-\ln K = \frac{(\Delta E)_{T=0}}{k_B T} + \sum_s \nu_s a_s^0 \tag{2.104}$$

It enables the equilibrium constant to be calculated from spectroscopic data when these are sufficiently complete. Alternatively, sometimes K can be measured directly and then used to test or refine the spectroscopic data needed to calculate the right-hand side of Eq. (2.104).

Variants of Eq. (2.103) in which the equilibrium constant is defined in terms of partial pressures, fugacities, and other partial free-energy functions are often encountered. The transformations equations relating these alternative equilibrium constants can readily be found by the methods of chemical thermodynamics.

DO PROBLEMS 2.14 TO 2.21.

2.18 THE LIMITATIONS OF CHEMICAL MODELS AND OTHER INDEPENDENT SUBSYSTEM MODELS

In this chapter we describe only a tiny fraction of the important and extensive results that have been obtained by the methods of statistical thermodynamics. By this term we mean methods in which partition function theory is exploited for systems in which Eq. (2.3) is a good starting point for model calculations in the context of the available spectroscopic data. However, it is easy to see that there are many problems that are beyond the scope of statistical thermodynamics.

An example is the chemical reaction $H_2 = 2H$ under conditions of high enough pressure and temperature so that the classification of the atoms as either H or part of H_2 becomes arbitrary. This happens when unpaired H atoms are found close to each other or close to an H_2 so often that these configurations make a significant contribution to the state of the system. An experimental indication of this complication might be a broad distribution of bond lengths in the H_2 molecules as measured by electron diffraction.

In some cases in which independent particle models are inadequate, progress can be made by a different, often less obvious choice of the subsystems, and in others it can be made by elaborating a chemical model to include activity coefficients and other refinements. But in many cases the theoretical method of choice is to exploit some of the more general techniques of statistical mechanics which are described in the remainder of this book.

Although the limitations of statistical thermodynamics may become important for the dissociation of H_2 only under extreme conditions, there is no shortage of examples of other processes for which similar complications occur under very mild conditions. A gas-phase example is the dissociation of hydrogen-bonded species such as water dimers. Also, many phenomena in solids can be understood only in terms of interactions among the phonons, the "independent" modes of the Debye theory mentioned earlier. But it is especially for phenomena in liquids and other dense fluids, where relatively little has been elucidated on the basis of the independent subsystem picture, that one has the most to gain by taking up more powerful methods, even if they are also more abstract.

THE LITERATURE

The few quantum-mechanical results on which this chapter depends can be found in many simple and clear introductory texts such as that by Hanna (1981). Also, they may be found in Flygare (1978) together with much information about the spectroscopic aspects. More complete accounts of statistical thermodynamics are given by McQuarrie (1974), Davidson (1962), Hill (1960), and Landau and Lifschitz (1958), as well as many older works.

PROBLEMS

2.1. Verify the third equality in Eq. (2.5) in the case $N = 2$ and $j_n = 0$, 1, or 2; that is, prove that

$$\sum_{j_1=0}^{2} \sum_{j_2=0}^{2} f(j_1)g(j_2) = [f(0) + f(1) + f(2)][g(0) + g(1) + g(2)]$$

2.2. For a system of N independent, distinguishable, but similar particles, as in Problem 2.3, we may write

$$A = Na, \quad S = Ns, \quad E = N\langle e \rangle, \quad \ldots$$

where a, s, and so on, are one-particle contributions to the respective thermodynamic functions. Prove that the relation of a to q is the same as the relation of A to Q, the relation of s to q is the same as the relation of S to Q, and so on.

2.3. For a system of N independent, distinguishable harmonic oscillators with a given frequency, find which properties in this list are affected by the choice of the zero of the energy.

$$A(T), \quad A(T'') - A(T'), \quad E(T), \quad C_v(T), \quad S(T)$$

Can you think of a *measurement* on the system whose outcome is affected by the choice of the zero of the energy? Explain after working Problem 2.6.

2.4. Show that for Einstein's model of a monatomic crystal,

$$C_v = 3Nk_B q^2 y^2 e^{-y} \tag{2.105}$$

where

$$y \equiv \beta h\nu, \qquad q = (1 - e^{-y})^{-1}$$

(*Hint:* Make the work easier by first deriving the *thermodynamic* relation $C_v = -k_B\beta^2 \, \partial E/\partial\beta$.)

2.5. Derive the equations for p and μ from Einstein's model for a crystal assuming that $\phi(0)$ and $\phi''(0)$ depend only on N/V.

2.6. With the same assumptions and assuming that the vapor is an ideal gas, derive the expression for the vapor pressure of an Einstein crystal. Note that to do this you must figure out the relation between $\phi(0)$ and the energy of vaporization of the crystal.

2.7. Extend the result in Eq. (2.48) to show that for a mixture of boltzons of various species,

$$Q(N, V, T) = \prod_{\text{species}} \frac{q_s^{N_s}}{N_s!} \tag{2.106}$$

2.8. From the result in Problem 2.7, derive the expression for the ideal entropy of mixing,

$$\Delta S = -Nk_B(x_1 \ln x_1 + x_2 \ln x_2) \tag{2.107}$$

where x_1 and x_2 are the mole fractions of the two species in the mixture.

2.9. From the distribution functions for fermions, bosons, and boltzons, show that in every case we have, in view of Eq. (2.59)

$$3\langle p\rangle V = 2\langle E\rangle \tag{2.108}$$

Derive the same equation from the ideal gas law and the heat capacity of a monatomic gas. Does the same result in all cases mean that systems of fermions and bosons are ideal gases? Explain.

2.10. Convert Eq. (2.74d) to an expression for S_{trans} in which the pressure (atm) rather than the particle number density is an independent variable.

2.11. Calculate the temperature at which $\Lambda^3\rho = 0.001$ for Ne gas at 5 atm pressure, assuming an ideal gas.

2.12. **(a)** Beginning with the thermodynamic relation

$$\mu = \left(\frac{\partial A}{\partial N}\right)_{V,T} \tag{2.109a}$$

extend the list of boltzon translation thermodynamic functions [Eq. (2.74)] to get

$$\mu_{\text{trans}} = k_B T(\ln \Lambda^3 + \ln \rho) \tag{2.109b}$$

(b) Then show that

$$N\mu_{\text{trans}} = G_{\text{trans}} = A_{\text{trans}} + pV \tag{2.109c}$$

(c) Finally, you can now see that

$$G_{\text{rot}} = A_{\text{rot}} \tag{2.109d}$$

and similarly for other internal contributions, since $G = A + pV$ for the total thermodynamic functions. Verify Eq. (d) directly from the equations for the rotational contributions. Can you find the underlying cause of the difference between Eqs. (c) and (d)?

2.13. Calculate the equilibrium constant of the reaction

$$^{17}O_2 + {}^{16}O_2 = 2\ {}^{17}O^{16}O$$

within 5% at 300 K. Verify that the nuclear spins ($I = 0$ for ^{16}O, $I = \frac{5}{2}$ for ^{17}O) make no net contribution to K.

2.14. Compute C_v for HCl at 500 K using the equations and data in this chapter.

2.15. From the data in this chapter, compute D_0 for HD.

2.16. Add a row for D_2 in Table 2.3, assuming that the potential is harmonic and that the mass of D is twice the mass of H.

2.17. The halogen atoms have low-lying electronic states as follows:

	F	Cl	Br	I	
$^2P_{1/2}$ state, degeneracy = 2:	0.050	0.11	0.46	0.94	eV
$^2P_{3/2}$ state, degeneracy = 4:	0	0	0	0	eV

(a) At what temperature will the electronic term make a 5% contribution to the heat capacity of Cl·?

(b) At about 1000°C the dissociation of I_2 into atoms is readily measurable. If the I_2 concentration is 0.05 M (M = mol/liter), what is the concentration of I atoms? What fraction of the I atoms are in the $^2P_{1/2}$ state?

2.18. CH_4 is a tetrahedral molecule with $r_{CH} = 1.094$ Å. Calculate Θ_{rot} for it.

2.19. Calculate the entropy of a mole of CO_2 at 1 atm and 298 K. The calorimetric third-law value is 51.1 cal/(mol deg).

2.20. Use the data in Table 2.3 to calculate the concentration of HI in an equilibrium equimolar mixture of I_2 and H_2 at a total concentration of 1 M at 700 K.

2.21. What is the equilibrium constant at 298 K for the gas reaction

$$H_2 + CO_2 \rightarrow H_2O + CO$$

given that $\Delta E°$ for the reaction is 9640 cal/mol at 0 K?

3

The Classical Limit

3.1 INTRODUCTION

For many interesting problems in statistical mechanics it is quite satisfactory to use classical mechanics to describe the states and motions of the system of interest. In particular, classical mechanics seems sufficient for the study of the ways in which the relatively weak forces between molecules are manifested in the measurable properties of a system. To develop this point with sufficient precision, we turn to the theory of the classical limit for systems of interacting particles.

Unfortunately, the theory of the classical limit for interacting particles is relatively difficult. In a first encounter with this material it is feasible to follow Sections 3.2 and 3.4 at a merely superficial level before going on to the study of classical systems on the basis of Section 3.3. This approach seems possible because the classical limit equations can be justified heuristically as indicated in the discussions of Eq. (3.24). Also, they can be verified in the limit of noninteracting particles by comparison with the theory of boltzons in Chapter 2. However, Problems 3.2 and 3.3 should not be skipped because they introduce some mathematics that is used in the following chapters.

To adapt the general statistical mechanical theory of systems at equilibrium to pertain to systems governed by classical rather than quantal mechanics, we may modify the general theory in such a way as to suppress the quantum effects. The discreteness that gives quantum mechanics its name is illustrated for several systems in Eqs. (2.8), (2.59), and (2.78). As we can see the discreteness is gauged by Planck's constant h. The de Broglie wavelength of a particle and the uncertainty principle are also gauged by h in the same sense. So, even without going more deeply into the subject, it seems logical to seek the classical limit for the statistical-mechanical equations by letting h vanish.

Letting h vanish is not quite the whole story, as one may see by trying to derive Eq. (2.63) by first setting $h = 0$ in Eq. (2.60). But there is a systematic and straightforward way to proceed along these lines by developing a power series in h to represent the operator that generates the partition function in Eq. (1.83). This development is our aim for this chapter.

3.2 AN ALTERNATIVE FORMULATION OF THE PARTITION FUNCTION

To find an expansion of Q in powers of h, we begin with the result in Eq. (1.83):

$$Q = \sum_j \int \psi_j^* \, e^{-\beta \hat{H}} \, \psi_j \, d\mathbf{q}$$

where the wave functions are eigenfunctions of $\hat{H}$. Next, we derive a relation of Q to $\hat{H}$ which is valid for *any* complete set of wave functions for the system

$$\Phi_j = \sum_k S_{jk} \Psi_k, \qquad j = 0, 1, 2, \ldots \tag{3.1}$$

We use the fact that any complete set of wave functions (here the Φ_j) can be expressed as linear combinations of a basis set (here the Ψ_j) which is itself a complete set of wave functions. We also require that the Φ_j set be orthonormal,

$$\int \Phi_j^*(q) \, \Phi_k(q) \, dq = \delta_{jk} \tag{3.2}$$

By substituting Eqs. (3.1) and (1.80) in this equation, we find an important condition on the transformation matrix $\mathbf{S}$, whose elements are the S_{jk}.

$$\begin{aligned} \delta_{jk} &= \sum_m S_{jm}^* S_{km} \\ &= \sum_m S_{km} (\mathbf{S}\dagger)_{mj} \\ &= (\mathbf{S} \cdot \mathbf{S}\dagger)_{kj} \end{aligned} \tag{3.3}$$

In the second equality we use the definition of the adjoint (indicated by the dagger), which is the complex conjugate of the transposed matrix. The third equality merely uses the definition of the scalar product of two matrices. It follows that $\mathbf{S}\dagger$ is the inverse of $\mathbf{S}$. Thus $\mathbf{S}$ is a unitary matrix.

Equation (1.83) is an expression for the partition function Q in terms of the matrix $\mathbf{A}$ whose elements are

$$A_{jk} = \int \Psi_j^* \, e^{-\beta \hat{H}} \, \Psi_k \, d\mathbf{q} \tag{3.4}$$

namely

$$\begin{aligned} Q &= \operatorname{tr} \mathbf{A} \\ &= A_{11} + A_{22} + \cdots \end{aligned} \tag{3.5}$$

where tr **A** is the conventional notation for the trace of **A**, the sum of the elements on its principal diagonal.

Now we examine the trace of the matrix $\mathbf{A}'$ whose elements are

$$\begin{aligned} A'_{mn} &= \int \Phi^*_m \, e^{-\beta \hat{H}} \, \Phi_n \, dq \\ &= \sum_{jk} S^*_{mj} S_{nk} A_{jk} \end{aligned} \tag{3.6}$$

where we have used Eq. (3.1) and the fact that the elements of **S** do not depend on the coordinates **q**. We find, in view of Eq. (3.3), that

$$\begin{aligned} \text{tr } \mathbf{A}' &= \sum_m A'_{mm} \\ &= \sum_{jk} A_{jk} \sum_m S^*_{mj} S_{mk} \\ &= \sum_{jk} A_{jk} (\mathbf{S}^\dagger \cdot \mathbf{S})_{jk} \\ &= \sum_{jk} A_{jk} \delta_{jk} \\ &= \text{tr } \mathbf{A} \end{aligned} \tag{3.7}$$

Thus $Q = \text{tr } \mathbf{A}'$ for $\mathbf{A}'$ defined by any basis Φ_0, Φ_1, Φ_2, . . . , and it follows that

$$Q = \text{tr } e^{-\beta \hat{H}} \tag{3.8}$$

where we need not and do not specify the basis in which the operator is represented as a matrix. Moreover, Q is here expressed directly in terms of the Hamiltonian operator $\hat{H}$ in a form that does not imply that one has solved Schrödinger's equation for the wave functions of the stationary states of the system. Whether one calculates Q from H by first calculating the quantum states or proceeds more directly is merely a matter of technique.

Next we evaluate the trace in Eq. (3.8) for a system of N structureless particles, not necessarily all of the same species, in volume V. The quantum-mechanical Hamiltonian operator needed to apply Eq. (3.8) to this case is

$$\hat{H} = \sum_{a=1}^{3N} \frac{\hat{p}_a^2}{2m_a} + U_N(q_1, \ldots, q_a, \ldots, q_{3N}) \tag{3.9}$$

where the sum is over the $3N$ Cartesian coordinates needed to fix the locations of all of the particles, where U_N is the potential of the interparticle forces, and where the momentum operator is

$$\hat{p}_a = \left(\frac{\hbar}{i}\right) \frac{\partial}{\partial q_a} \tag{3.10}$$

In a given configuration (i.e., set of locations $q_1, \ldots, q_{3n}$) the force on particle 1 is

$$\mathbf{F}_1 = -\hat{x}\frac{\partial U_N}{\partial q_1} - \hat{y}\frac{\partial U_N}{\partial q_2} - \hat{z}\frac{\partial U_N}{\partial q_3} \tag{3.11}$$

where $\hat{x}$, $\hat{y}$, and $\hat{z}$ are unit vectors on the Cartesian axes and q_1, q_2, and q_3 are the corresponding coordinates of particle 1.

To analyze the trace in Eq. (3.8) when h is small, we choose a basis set of wave functions

$$\Phi_{\mathbf{p}}(\mathbf{q}) = \Phi_{p_1, p_2, \ldots}(\mathbf{q}) = |\mathbf{p}) \tag{3.12}$$

which is the product of wave functions of N noninteracting particles,

$$|\mathbf{p}) = \prod_{a=1}^{3N} \frac{\exp(iq_a p_a/\hbar)}{h^{1/2}} \tag{3.13}$$

Here

$$\mathbf{p} \equiv p_1, p_2, \ldots, p_a, \ldots, p_{3N} \tag{3.14}$$

is a continuous $3N$-dimensional vector momentum variable with each component in the range

$$-\infty < p_a < \infty \tag{3.15}$$

We also can write Eq. (3.13) in the more compact form

$$|\mathbf{p}) = h^{-3N/2} \exp[(i/\hbar)\, \mathbf{p} \cdot \mathbf{q}] \tag{3.16}$$

Although the $3N$-dimensional continuous variable $\mathbf{p}$ in Eq. (3.14) has replaced the integer index j of Φ_j in Eq. (3.1), this change makes only a formal change in the theory of the trace; we replace sums over j and k by integrals over $\mathbf{p}$ and we replace the Kronecker delta function δ_{jk} by the Dirac delta function $\delta(\mathbf{p} - \mathbf{p}')$. We illustrate these operations by showing that the wave functions $|\mathbf{p})$, $|\mathbf{p}')$, . . . form a mutually orthonormal set [see Eq. (3.2)]

$$\begin{aligned}(\mathbf{p}|\mathbf{p}') &= \int dq_1 \cdots dq_a \cdots dq_{3N} \prod_{a=1}^{3N} \frac{\exp(iq_a(p_a' - p_a)/\hbar)}{h} \\ &= \prod_{a=1}^{3N} \delta(p_a' - p_a)\end{aligned} \tag{3.17}$$

where we exploit one of the representations

$$(1/2\pi)\int_{-\infty}^{\infty} dx\, e^{ixy} = \delta(y) \tag{3.18}$$

of the Dirac delta function.

In place of Eq. (3.4) or (3.6) we now have a sort of generalized matrix whose elements are

$$(\mathbf{p}|\hat{A}|\mathbf{p}') = h^{-3N}\int d\mathbf{q}\, e^{-(i/\hbar)\mathbf{p}\cdot\mathbf{q}} e^{-\beta\hat{H}} e^{(i/\hbar)\mathbf{p}'\cdot\mathbf{q}} \tag{3.19}$$

and whose trace is

$$\operatorname{tr} \mathbf{A} = \int d\mathbf{p}\,(\mathbf{p}|\hat{A}|\mathbf{p})$$
$$= \frac{1}{h^{3N}} \int' d\mathbf{q} \int' d\mathbf{p}\; e^{-(i/\hbar)\mathbf{p}\cdot\mathbf{q}} e^{-\beta \hat{H}} e^{(i/\hbar)\mathbf{p}\cdot\mathbf{q}} \tag{3.20}$$

The primes on the integrals mean that we restrict the range of integration so that each quantum state is counted only once. Thus we must notice that if the positions and momenta of two *identical* particles are interchanged, the quantum state of the system is not changed, although the compound variable **p, q** has changed. Thus a whole series of points in **p, q** space, in *phase space,* correspond to a given quantum state of the system, while in the formulation of the partition function Q each quantum state is counted only once [Eq. (1.17)].

In practice it is usually convenient to enlarge the region of integration by introducing an appropriate correction factor. In a system of N distinct particles no correction is required, while if the N particles are identical, the correction is given by dividing by the number of interchanges of the particle labels that do not change the quantum state:

$$\int' d\mathbf{p} \int' d\mathbf{q} \cdots = \frac{1}{N!} \int d\mathbf{p} \int d\mathbf{q} \cdots \tag{3.21}$$

For a mixture of N_α particles of species α, N_β particles of species β, and so on, the correction is given by

$$\int' d\mathbf{p} \int' d\mathbf{q} \cdots = \frac{1}{N_\alpha! N_\beta! \cdots} \int d\mathbf{p} \int d\mathbf{q} \cdots \tag{3.22}$$

As illustrated, for example, by Jackson (1968), a more sophisticated theory based on the occupation numbers of the quantum states in the grand ensemble leads to the $N!$ factors in a more natural way. In any case we find that

$$Q = \left(\prod_{\text{species}} N_\alpha! h^{3N_\alpha} \right)^{-1} \int d\mathbf{p} \int d\mathbf{q}\; e^{(i/\hbar)\mathbf{p}\cdot\mathbf{q}} e^{-\beta \hat{H}} e^{(-i/\hbar)\mathbf{p}\cdot\mathbf{q}} \tag{3.23}$$

which is a convenient form for studying the classical limit. Notice that because Q is real, it makes no difference in this equation if i is replaced by $-i$ in both places. Also, we remark that in other chapters we designate particle species by subscripts s, s', . . . rather than by α, β, ...

DO PROBLEMS 3.1 TO 3.4.

3.3 THE CLASSICAL LIMIT

From the results of Problem 3.1 we see that $e^{-\beta \hat{H}}$ in a sense becomes $e^{-\beta E}$ in the $h = 0$ limit. Here E is the classical Hamiltonian function, obtained from the Hamiltonian operator in Eq. (3.9) by replacing $\hat{p}_a$ by p_a for every a. (In other chapters the

Hamiltonian function is denoted H rather than E.) So we find, since $e^{-\beta E}$ and the wave function commute,

$$Q_{\text{classical}} = \left(\prod_{\text{species},\alpha} N_\alpha! h^{3N_\alpha} \right)^{-1} \int d\mathbf{p} \int d\mathbf{q} \exp\left[-\beta E(p, q)\right] \tag{3.24}$$

where the original expression for Q as a sum over discrete quantum states has been transformed into an integral over phase space. It is apparent that if we interpret this "phase integral" as a sum over small regions of phase space, each with a "volume"

$$\Delta\mathbf{p} \cdot \Delta\mathbf{q} = h^{3N}$$

we have one such region for each quantum state. This relation is quite consistent with Heisenberg's uncertainty principle. The $N_\alpha!$ factors were discussed above.

It is conventional to factor Eq. (3.24) as follows:

$$Q_{\text{classical}} = Q^{\text{id}} \frac{Z}{V^N} \tag{3.25}$$

where we define

$$Q^{\text{id}} \equiv \left(\frac{V}{h^3}\right)^N \prod_{\text{species}} \frac{\left[\int_{-\infty}^{\infty} dp \exp\left(-\beta p^2/2m_\alpha\right)\right]^{3N_\alpha}}{N_\alpha!} \tag{3.26}$$

$$Z = \int_V d^{3N}q \exp\left(-\beta U_N\right) \tag{3.27}$$

Q^{id} is the partition function for the corresponding "ideal" system, while Z, the *configuration integral,* carries the effect of the force interactions. In terms of the thermal de Broglie wavelength [Eq. (2.64)] we have

$$Q^{\text{id}} = \frac{1}{N!}\left(\frac{V}{\Lambda^3}\right)^N \tag{3.28}$$

for one species or

$$Q^{\text{id}} = \prod_{\text{species}} \frac{(V/\Lambda_\alpha^3)^{N_\alpha}}{N_\alpha!} \tag{3.29}$$

for a mixture of several species.

To investigate the classical limit for systems of particles with internal structure, notably molecules with more than one atom, we may use Eq. (3.9) while considering that the N particles are the atoms of the system and that U_N carries the forces within molecules as well as the forces between molecules. Then the classical limit is formally the same as derived above, with the consequence that the internal coordinates, the molecular orientations, and the vibrational coordinates are treated in the classical approximation. Alternatively, one may introduce, in place of Eq. (3.9), a Hamiltonian having $3 + n_\alpha$ coordinates (and conjugate momenta) for each molecule, where n_α is the number of internal coordinates for molecules of species α. We might take the classical limit with respect to the translational and orientational coordinates while neglecting the others, as would be appropriate if the others did not depend much on interactions between the molecules. Unfortunately, this is not always the case. Consider, for example, the electronic coordinates associated with molecular polarizability. Nevertheless, model systems with orientation-dependent interactions among rigid molecules have been found to be adequate for the molecular interpretation of many of the properties of real molecular fluids and have been extensively studied. Except for the detailed treatment of such systems in Chapter 8, we choose to keep the development of the statistical mechanical theory as simple as possible by limiting ourselves to systems of structureless particles except as otherwise noted in a few cases. Fortunately the resulting theory is easy to extend to models with orientation-dependent interactions, and with other types of internal degrees of freedom as well, so the main question in all such cases is whether classical mechanics is an acceptable approximation for the new modes. This question can often be answered by applying the theory of the leading quantum corrections.

DO PROBLEMS 3.5 TO 3.10.

3.4 THE LEADING QUANTUM CORRECTION

The leading term in $Q - Q_{\text{classical}}$ is a useful tool for the study of real fluids. It may be most simply derived with the help of the perturbation expansion in Eq. (3.56). Thus we have

$$\begin{aligned} e^{-\beta\hat{H}} = e^{-\beta E} &- \int_0^\beta d\beta'\ e^{-(\beta-\beta')E}[\hat{H} - E]e^{-\beta' E} \\ &+ \int_0^\beta d\beta' \int_0^{\beta'} d\beta''\ e^{-(\beta-\beta')E}[\hat{H} - E]e^{-(\beta'-\beta'')E}[\hat{H} - E]e^{-\beta'' E} \\ &+ \text{terms in } [\hat{H} - E]^3 \text{ and higher powers} \end{aligned} \tag{3.30}$$

Also, we have

$$\hat{H} - E = -\sum_a \left[\frac{\hbar^2}{2m_a}\left(\frac{\partial^2}{\partial q_a^2}\right) + \frac{p_a^2}{2m_a}\right] \tag{3.31}$$

$$[\hat{H} - E]e^{-\beta' E}|\mathbf{p}) = \sum_a \frac{\hbar^2\beta'}{2m_a}\left(U_{aa} - \beta' U_a^2 + 2U_a \frac{i}{\hbar} p_a\right)e^{-\beta' E}|\mathbf{p}) \tag{3.32}$$

$$\begin{aligned} &[\hat{H} - E]e^{-(\beta'-\beta'')E}[\hat{H} - E]e^{-\beta'' E}|\mathbf{p}) \\ &\quad = \sum_a \left[\left(\frac{\hbar p_a}{m_a}\right)^2 \beta''(U_{aa} - \beta' U_a^2) + \cdots\right]e^{-\beta' E}|\mathbf{p}) \end{aligned} \tag{3.33}$$

where the suppressed terms have only higher powers of $\hbar$ and where U_a and U_{aa} are defined in Problem 3.1.

Now we substitute Eqs. (3.30) to (3.33) in (3.23) to get

$$Q = Q_{\text{classical}}\left[1 - \int_0^{\beta} d\beta' \sum_a \frac{\hbar^2\beta'}{2m_a}(\langle U_{aa}\rangle - \beta'\langle U_a^2\rangle) \right. \tag{3.34}$$
$$\left. + \int_0^{\beta} d\beta' \int_0^{\beta'} d\beta'' \sum_a \left(\frac{\hbar}{2m_a}\right)^2 \beta'' \langle U_{aa}\rangle \frac{m_a}{\beta} + \cdots \right]$$

where we define the canonical-ensemble average of any phase-dependent function at "temperature" β as

$$\langle F\rangle \equiv \frac{\int d^{3N}p \int d^{3N}q\, e^{-\beta E} F(\mathbf{p}, \mathbf{q})}{Q_{\text{classical}}\, h^{3N}} \tag{3.35}$$

Moreover, integration by parts, using (3.35), shows that we have

$$\beta\langle U_a^2\rangle = \langle U_{aa}\rangle \tag{3.36}$$

and Eq. (3.34) reduces to[1]

$$Q = Q_{\text{classical}}\left[1 - \frac{\beta^2\hbar^2}{24}\sum_{a=1}^{3N}\frac{\langle U_{aa}\rangle}{m_a} + O(\hbar^3)\right] \tag{3.37}$$

or, taking the logarithm and using Table 1.3, we have

$$A = A_{\text{classical}} + \frac{\beta\hbar^2}{24}\sum_{a=1}^{3N}\frac{\langle U_{aa}\rangle}{m_a} + O(\hbar^3) \tag{3.38}$$

which is the basic equation for the first quantum correction to the thermodynamic properties of a classical system. It is also the basis for the theory of mass effects on thermodynamic properties.

To make a connection with the theory of mass effects in chemical equilibria it should be noticed that $\langle U_{aa}\rangle$ is an effective force constant, so $\omega_a = (\langle U_{aa}\rangle/m_a)^{1/2}$ is the corresponding angular frequency. In terms of the corresponding frequency ν_a expressed in wave numbers, Eq. (3.38) is

$$\beta A = \beta A_{\text{classical}} + \left(\frac{298.16\ \text{K}}{T}\right)^2 \sum_{a=1}^{3N}\left(\frac{\nu_a}{1015.1\ \text{cm}^{-1}}\right)^2 \tag{3.39}$$

Thus at room temperature or higher the quantum corrections are large only if the mean curvature of the potential corresponds to a frequency of 1000 cm^{-1} or more.

DO PROBLEM 3.11.

[1] $O(x^n)$ means a function of x, namely $f(x)$, such that $f(x)/x^n$ has a finite value, possibly zero, in the $x = 0$ limit.

THE LITERATURE

The theory of the classical limit and the first quantum corrections may also be found in Kirkwood (1933), Hirschfelder et al. (1954), Landau and Lifschitz (1958), and Jackson (1968). The theory of the Wigner distribution function, which is often useful in classical limit problems in dynamics as well as equilibrium, is given by Mori et al. (1962).

PROBLEMS

3.1. We define the classical Hamiltonian or total energy

$$E = E(\mathbf{p}, \mathbf{q}) \equiv \sum_{a=1}^{3N} \frac{p_a^2}{2m_a} + U_N(q_1, \ldots, q_{3N}) \tag{3.40}$$

and the derivatives of the potential energy.

$$U_a = \frac{\partial U_N}{\partial q_a}$$
$$U_{ab} = \frac{\partial^2 U_N}{\partial q_a\, \partial q_b} \tag{3.41}$$

(a) Prove that

$$\beta \hat{H}|\mathbf{p}) = \beta E|\mathbf{p}) \tag{3.42}$$

(b) Prove that

$$(\beta \hat{H})^2|\mathbf{p}) = \left[(\beta E)^2 + \frac{\beta^2 \hbar}{i} \sum_a \frac{U_a p_a}{m_a} - \frac{1}{2}(\beta\hbar)^2 \sum_a \frac{U_{aa}}{m_a}\right]|\mathbf{p}) \tag{3.43}$$

(c) Show that for any n greater than 1,

$$(\beta\hat{H})^n|\mathbf{p}) = (\beta E)^n|\mathbf{p}) + [\text{terms containing powers of } \hbar]|\mathbf{p}) \tag{3.44}$$

(d) Show that

$$e^{-\beta\hat{H}}|\mathbf{p}) = e^{-\beta E}|\mathbf{p}) + [\text{terms containing powers of } \hbar]|\mathbf{p}) \tag{3.45}$$

3.2. To derive Eq. (3.18) notice that

$$f(b, y) \longrightarrow \int dx\, e^{ixy} \qquad \text{as } b \longrightarrow 0 \tag{3.46}$$

where we define

$$f(b, y) \equiv \int_{-\infty}^{\infty} dx \exp\left[-\left(\frac{xb}{2} - \frac{iy}{b}\right)^2 - \frac{y^2}{b^2}\right] \tag{3.47}$$

Notice that to reduce $f(b, y)$ to a simple form you need only the integral over a Gaussian function, which we give here:

$$\int_{-\infty}^{\infty} dt\, e^{-t^2} = \pi^{1/2} \tag{3.48}$$

Evaluate $f(b, y)$ and graph it as a function of y for $b = 1, \frac{1}{3}$, and $\frac{1}{10}$. See the approach to $\delta(y)$.

3.3. By using Eqs. (3.18) and (3.46), show that

$$\int_{-\gamma}^{\gamma} \delta(y)\, dy = 1$$

for any positive number γ.

3.4. Verify in detail that Eq. (3.23) follows directly from Eqs. (3.9), (3.19), and (3.22).

3.5. Verify in detail that Eqs. (3.25) to (3.27) follow from Eq. (3.24) with E defined in Eq. (3.40).

3.6. The set of variables $\mathbf{p}, \mathbf{q} = \Gamma$ is called the *phase* of a classical system. On the basis of Eq. (1.17) and (3.24) show that you expect that the probability density to find a given system at Γ in the canonical ensemble is

$$P(\Gamma) = \frac{\exp\left[-\beta E(\Gamma)\right]}{Q_{\text{classical}} \prod_{\alpha} N_\alpha! h^{3N}} \tag{3.49}$$

Thus $P(\Gamma)\, d\Gamma$ is the fraction of systems in the ensemble that are in the phase-space "volume" $d\Gamma$ around the phase point Γ, and $P(\Gamma)$ is the phase-space distribution function.

3.7. Given Eq. (3.49), show that the average energy of a classical system is

$$\langle E \rangle = \int d\Gamma\, E(\Gamma)\, P(\Gamma) \tag{3.50}$$

Show that $\langle E \rangle$ also is given by

$$\langle E \rangle = -\frac{\partial \ln Q_{\text{classical}}}{\partial \beta} \tag{3.51}$$

3.8. From Eq. (3.29) derive the famous result for the "ideal entropy of mixing":

$$\Delta S = -Nk_B(x_1 \ln x_1 + x_2 \ln x_2) \tag{3.52}$$

where x_s is the mole fraction of species s. Also derive Dalton's law of partial pressures.

3.9. Show that in the classical limit the Helmholtz free energy A is given by

$$\beta A = -\ln Q^{\text{id}} - \ln \frac{Z}{V^N} \tag{3.53}$$

and $A^{\text{ex}} \equiv A - A^{\text{id}}$ is given by

$$e^{-\beta A^{\text{ex}}} = \frac{Z}{V^N} \tag{3.54}$$

This important relation can also be written in the form

$$e^{-\beta A^{\text{ex}}} = \frac{\int_V dq^{3N} e^{-\beta U_N}}{\int_V dq^{3N}} \tag{3.55}$$

which shows that A^{ex} is a particular average of U_N.

3.10. The equation for operators A and B,

$$e^{x(A+B)} = e^{xA} + \int_0^x dx'\ e^{(x-x')A} B e^{x'(A+B)} \tag{3.56}$$

is widely useful in cases in which

$$e^{x(A+B)} = e^{xA} e^{xB} \tag{3.57}$$

is not valid. We assume that every exponential is defined by its power series expansion as in Eq. (1.82) and that x is an ordinary (real or complex) number so that $xA = Ax$ and $xB = Bx$.

(a) Prove that Eq. (3.57) is valid only if $AB = BA$.

(b) Show that the operation d/dx applied to Eq. (3.56) gives an identity even if $AB \neq BA$.

(c) Since $de^{x(A+B)}/dx$ obeys Eq. (3.56) for all x, we need verify Eq. (3.56) at only one value of x to know that it is correct. Try $x = 0$.

(d) Show that Eq. (3.56) is the generalization of Eq. (3.57) in the sense that Eq. (3.56) reduces to Eq. (3.57) if $AB = BA$.

(e) Derive the following perturbation expansion from Eq. (3.56):

$$e^{x(A+B)} = e^{xA} + \int_0^x dx'\ e^{(x-x')A} B e^{x'A} + \int_0^x dx'\ e^{(x-x')A} B \int_0^{x'} dx''\ e^{(x'-x'')A} B e^{x''(A+B)} \tag{3.58}$$

3.11. For real fluids of particles with negligible internal structure we often assume that the potential is a sum of pair potentials

$$U_N(q_1, \ldots, q_{3N}) = \sum_{i=1}^{N} \sum_{j=i+1}^{N} u_{ij}(r_{ij}) \tag{3.59}$$

where r_{ij} is the distance between i and j in the given configuration, and also that the pair potential has the Lennard-Jones form

$$u(r) = 4\epsilon\left[\left(\frac{\sigma}{r}\right)^{12} - \left(\frac{\sigma}{r}\right)^{6}\right] \tag{3.60}$$

where ϵ is the energy parameter and σ is the size parameter. For some substances we have the experimental data in Table 3.1 (Hirschfelder et al., 1954, p. 1110).

For each substance estimate the quantum correction to $\beta A/N$ at 298 K and at $T = \epsilon/k_B$. Assume in each case that $\langle U_{aa} \rangle \simeq \partial^2 u/\partial r^2)$ at $r = 2^{1/6}\sigma$.

TABLE 3.1 LENNARD-JONES PARAMETERS FOR SEVERAL FLUIDS

Fluid	H_2	He	Ne	Ar	Kr	Xe	N_2	CH_4	CO_2
ϵ/k_B (K)	37	10	36	120	171	221	95	148	189
σ (Å)	2.9	2.6	2.8	3.4	3.6	4.1	3.7	3.8	4.5

4

Thermodynamics and Structure. Role of Intermolecular Forces

4.1 THERMODYNAMIC FUNCTIONALS OF THE POTENTIAL U_N

In this chapter we develop some of the consequences of Eqs. (3.27) and (3.54):

$$Z = \int_V d^{3N}q \exp\left[-\beta U_N(q_1, \ldots, q_{3N})\right] \tag{4.1}$$

$$\beta A^{\text{ex}} = \beta A - \beta A^{\text{id}}$$
$$= -\ln \frac{Z}{V^N} \tag{4.2}$$

which show how the Helmholtz free energy $A(N, V, T)$ is a "functional" of the potential $U_N(q_1, \ldots, q_{3N})$. That is, A depends on the whole function $U_N(q_1, \ldots, q_{3N})$. These are canonical ensemble equations. Thus A, A^{ex}, and Z depend on N, V, and T, the independent variables for the canonical ensemble.

The corresponding grand ensemble equations are [see Table 1.5 and Eqs. (3.25) and (3.28)]

$$\Xi = \Xi(z, V, T)$$
$$= \sum_{N=0}^{\infty} \frac{z^N}{N!} Z(N, V, T) \tag{4.3}$$

where z is the activity defined by

$$z = \frac{e^{\beta\mu}}{\Lambda^3} \tag{4.4}$$

In the canonical ensemble the thermodynamic energy $\bar{E}$ can be derived from the thermodynamic equation

$$\bar{E} = \left(\frac{\partial[\beta A]}{\partial \beta}\right)_{V,N} \tag{4.5}$$

In view of Eq. (4.2) we may write $\bar{E} = \bar{E}^{\text{id}} + \bar{E}^{\text{ex}}$ with

$$\begin{aligned}\bar{E}^{\text{ex}} &= -\frac{\partial \ln Z}{\partial \beta} \\ &= \frac{\int d^{3N}q\, U_N e^{-\beta U_N}}{Z}\end{aligned} \tag{4.6a}$$

$$\bar{E}^{\text{ex}} = \langle U_N \rangle \tag{4.6b}$$

The passage from Eq. (4.6a) to (4.6b) involves the idea that $\exp[-\beta U_N(q_1, \ldots, q_{3N})]/Z$ is the probability density for the configuration (set of locations) $q_1, \ldots, q_{3N}$. [Compare with Eq. (3.49)]. So for any function of configuration $B(q_1, \ldots, q_{3N})$ the canonical ensemble average value is

$$\langle B \rangle = \frac{\int d^{3N}q\, B(q_1 \cdots q_{3N}) e^{-\beta U_N}}{Z} \tag{4.7}$$

Corresponding to Eqs. (4.5) and (4.6) are the equations for the pressure:

$$p = -\left(\frac{\partial A}{\partial V}\right)_{T,N} \tag{4.8a}$$

$$= -\frac{\partial}{\partial V}\left(A^{\text{id}} - \beta^{-1} \ln \frac{Z}{V^N}\right) \tag{4.8b}$$

$$= \beta^{-1} \frac{\partial}{\partial V} \ln Z \tag{4.8c}$$

where we have noticed that $\beta A^{\text{id}} + N \ln V$ is independent of volume, as may be verified from Eqs. (3.28) and (1.27). So we have

$$\beta Z p = \frac{\partial Z}{\partial V} = \frac{\partial}{\partial V} \int_V d^{3N}q\, e^{-\beta U_N} \tag{4.9}$$

An easy way to evaluate this derivative is to suppose that V is the volume of a cube with edge L, in which case $dV = 3L^2\, dL$ and

$$3L^2 \beta Z p = \frac{\partial}{\partial L} \int_0^L dq_1 \cdots \int_0^L dq_{3N} e^{-\beta U_N(q_1, \ldots, q_{3N})} \tag{4.10a}$$

With the definition $s_a = q_a/L$ for $a = 1$ to $3N$, this becomes

$$3L^2\beta Zp = \frac{\partial}{\partial L}\left[\int_0^1 ds_1 \cdots \int_0^1 ds_{3N} e^{-\beta U_N(s_1 L, \ldots, s_{3N} L)} L^{3N}\right]$$

$$= \frac{3NZ}{L} - \beta \int_0^1 ds_1 \cdots \int_0^1 ds_{3N} \sum_{a=1}^{3N} \frac{\partial U_N}{\partial q_a}\left(\frac{\partial q_a}{\partial L}\right) e^{-\beta U_N} L^{3N} \tag{4.10b}$$

which reduces to

$$p = \frac{\rho}{\beta} - \frac{1}{3VZ}\int_V d^{3N}q \sum_a q_a \frac{\partial U_N}{\partial q_a} e^{-\beta U_N} \tag{4.10c}$$

so we have

$$p = \frac{\rho}{\beta} - \frac{\left\langle \sum_{a=1}^{3N} q_a(\partial U_N/\partial q_a)\right\rangle}{3V} \tag{4.11}$$

The function averaged is called the *virial* and Eq. (4.11) is one form of what is usually called the *virial equation.*

DO PROBLEMS 4.1 TO 4.4.

4.2 STRUCTURE: SPATIAL CORRELATION FUNCTIONS

Ensemble averages such as $\langle U_N \rangle$ and $\langle q_a\, \partial U_N/\partial q_a \rangle$ give thermodynamic functions but no direct structural information. One may say that the structural information that is implicit in the Hamiltonian has been lost by integrating over all of the coordinates. If we hold back on some of the integrations, we will get information about the average spatial relationships of the particles. In the limit we may hold back on all of the integrations over locations and study the weight function

$$\frac{1}{Z}\exp\left[-\beta U_N(q_1, \ldots, q_{3N})\right]$$

but in a sense, this gives more structural information than we can handle, while at the same time being rather far from experimentally accessible measures. It turns out that holding back on the integration over the locations of very few particles, even just one or two, gives structural functions which are more useful.

In place of the usual procedure, in which the one-component version of a theory is derived in the text and readers are asked to develop the multicomponent version as an exercise, we take the other tack here and present the theory for the multicomponent system together with the recommendation that readers should, as they work through the theory from Eq. (4.12) to (4.23), transcribe the equations to one-component form as an aid to understanding the development. To keep the equations as simple as possible within this context, we consider explicitly only systems of structureless particles (see Section 3.3).

Therefore, we now consider a mixture which, in the canonical ensemble, comprises N_1 structureless particles of species *1*, N_2 particles of species *2*, and so on.

We write the *local* density of species s (also called the density field of species s at location $\mathbf{R}$ as

$$\rho_s(\mathbf{R}) = \langle \hat{\rho}_s(\mathbf{R}) \rangle \tag{4.12}$$

It is the ensemble average over the density "operator" defined as

$$\hat{\rho}_s(\mathbf{R}) = \sum_{j=1s}^{Ns} \delta(\mathbf{r}_j - \mathbf{R}) \tag{4.13}$$

where $j = 1s$ means that j is particle 1 of species s. (As illustrated here, we use s for the species index.) In the canonical ensemble one has

$$\rho_s(\mathbf{R}) = Z^{-1} \int_V d^{3N}q\, e^{-\beta U_N} \hat{\rho}_s(\mathbf{R}) \tag{4.14a}$$

$$= Z^{-1} \int_V e^{-\beta U_N} \hat{\rho}_s(\mathbf{R})\, d^3 r_1 \cdots d^3 r_N \tag{4.14b}$$

where $N = N_1 + N_2 + \cdots$, and where we have introduced an alternative notation for $d^{3N}q = dq_1 \cdots dq_{3N}$. The N_s terms in Eq. (4.13) each give the same numerical result in Eq. (4.14), namely

$$Z^{-1} \int_V \exp\,[-\beta U_N(\mathbf{r}_1, \ldots, \mathbf{r}_N)]\delta(\mathbf{r}_{1s} - R) d^3 r_1 \cdots d^3 r_N = \frac{1}{V} \tag{4.14c}$$

where we noticed that

$$\int_V e^{-\beta U_N} d^3 r_2 \cdots d^3 r_N = \frac{Z}{V} \tag{4.14d}$$

if V is large enough and no external fields are acting, since the integral is then independent of $\mathbf{r}_1$. We finally have

$$\rho_s(\mathbf{R}) = \frac{N_s}{V} \qquad \text{in a homogeneous system} \tag{4.15}$$

Thus if we assume that U_N has contributions only from the interactions among the particles of the system, external forces being negligible, then the equilibrium state will be homogeneous (spatially invariant), and the local density will equal the overall (or stoichiometric) density $\rho_s = N_s/V$. An independent derivation of Eq. (4.14c) is given below in Eq. (4.18).

Next we consider a very particular and important example of an inhomogeneous system, namely a system in which one particle of species s is fixed at location $\mathbf{R}$. We may consider this particle as the source of an external field acting on $N - 1$ other particles labeled 2, 3, . . . , N. The system made up of the $N - 1$ particles has the potential function

$$U_{N-1}^{\star}(\mathbf{r}_2, \mathbf{r}_3, \ldots, \mathbf{r}_N) = U_N(\mathbf{R}, \mathbf{r}_2, \mathbf{r}_3, \ldots, \mathbf{r}_N) \tag{4.16}$$

where the superscript star in $U_{N-1}^{\star}$ is a device to remind us that an external field is acting. The corresponding configuration integral is

$$Z^\star(N-1,\ V,\ T) = \int e^{-\beta U^\star_{N-1}}\, d^3\mathbf{r}_2 \cdots d^3\mathbf{r}_N \tag{4.17}$$

It is interesting to notice that when V is large and $\mathbf{R}$ is well inside the volume V, $Z^\star(N-1, V, T)$ tends to be independent of $\mathbf{R}$. The reason is that the effect of a particle at $\mathbf{R}$ on the others has a limited range, at least on a macroscopic scale. It follows that the integral is independent of $\mathbf{R}$ as long as the distance from $\mathbf{R}$ to the wall of the vessel of volume V exceeds this range. Then we have

$$\begin{aligned} Z(N,\ V,\ T) &= \int_V d^3R\, Z^\star(N-1,\ V,\ T) \\ &= VZ^\star(N-1,\ V,\ T)\left[1 + O\left(\frac{1}{V}\right)\right] \end{aligned} \tag{4.18}$$

where $O(1/V)$ means a term that varies like constant$/V$ as $V \longrightarrow \infty$. This important result can be used to derive Eq. (4.14c), since the integral in the latter is, in fact, $Z^\star(N-1, V, T)$ in the field of particle 1_s at $\mathbf{R}$.

In a system that is inhomogeneous due to a fixed s particle at $\mathbf{R}$, the local density of s' at R' is, in view of Eq. (4.14),

$$\rho^\star_{s'}(\mathbf{R}') = Z^\star(N-1,\ V,\ T)^{-1} \int \hat{\rho}_{s'}(\mathbf{R}') e^{-\beta U^\star_{N-1}}\, d^3r_2 \cdots d^3r_N \tag{4.19a}$$

together with the proviso that, even if $s' = s$, the fixed particle $1s$, not being one of the $N-1$ particles of the inhomogeneous system, is not included in the sum in Eq. (4.13). The star on $\rho^\star_{s'}(\mathbf{R}')$ indicates that this term is the density field in an inhomogeneous system—the inhomogeneity due, of course, to the s particle at $\mathbf{R}$. Now in view of Eq. (4.16) we may rewrite Eq. (4.19a) in terms of U_N if only we require that particle 1, the one that makes the difference between $N-1$ and N, is located at $\mathbf{R}$:

$$\rho^\star_{s'}(\mathbf{R}') = Z^\star(N-1,\ V,\ T)^{-1} \int_V \delta(\mathbf{r}_{1s} - \mathbf{R})\, \hat{\rho}_{s'}(\mathbf{R}') e^{-\beta U_N}\, d^3r_1 \cdots d^3r_N \tag{4.19b}$$

$$= (N_{s'} - \delta_{ss'})\left(\frac{V}{Z(N,\ V,\ T)}\right) \int e^{-\beta U_N} \delta(\mathbf{r}_{1s} - \mathbf{R}) \delta(\mathbf{r}_{2s'} - \mathbf{R}')\, d^3r_1 \cdots d^3r_N \tag{4.19c}$$

In passing from Eq. (4.19b) to (4.19c) we use Eq. (4.18) and we apply again the analysis of the effect of $\hat{\rho}_s$ in passing from Eq. (4.14) to (4.15). In fact, in Eq. (4.19c) we can write $\delta(\mathbf{r}_{ms'} - \mathbf{R}')$, where m is the index of any of the particles of species s' except in the case in which species s' is the same as s, the species of the original fixed particle. Then the mth particle of species s can be any one but the original fixed particle. Taking $m = 2$ is one of many notational devices that would serve here.

The ratio of $\rho^\star_{s'}(\mathbf{R}')$ to the uniform density $\rho_{s'} = (N_{s'} - \delta_{ss'})/V$ we would have if the inhomogeneous field were removed is

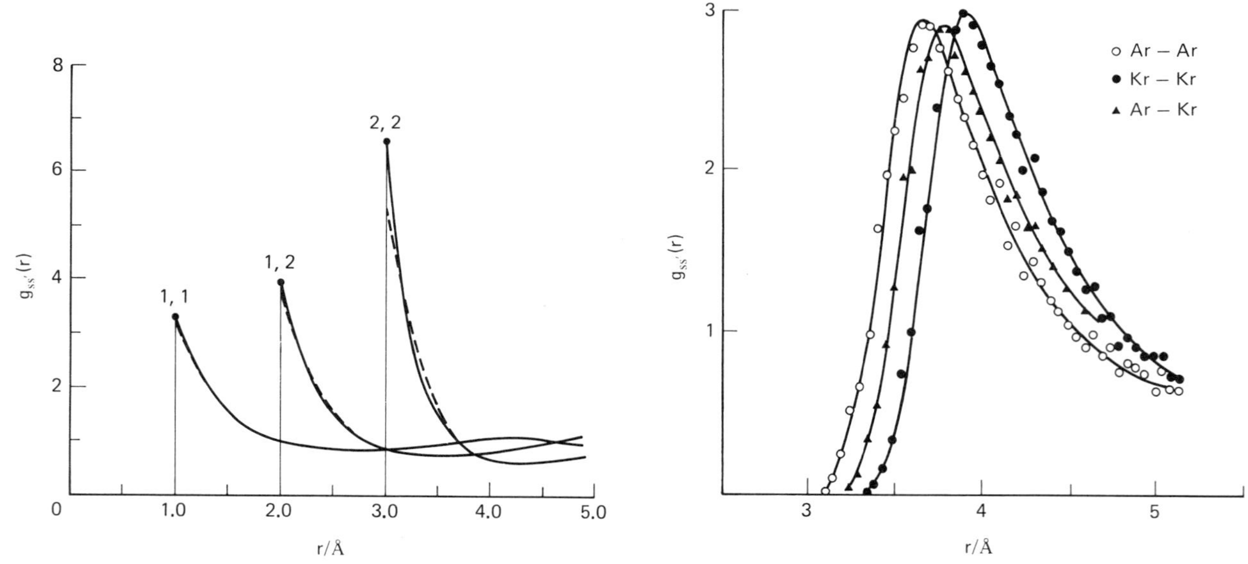
8
6
4
2
0
1.0
2.0
3.0
4.0
5.0
1, 1
1, 2
2, 2
$g_{ss'}(r)$
r/Å
3
2
1
3
4
5
$g_{ss'}(r)$
r/Å
○ Ar – Ar
● Kr – Kr
▲ Ar – Kr

(a)

(b)

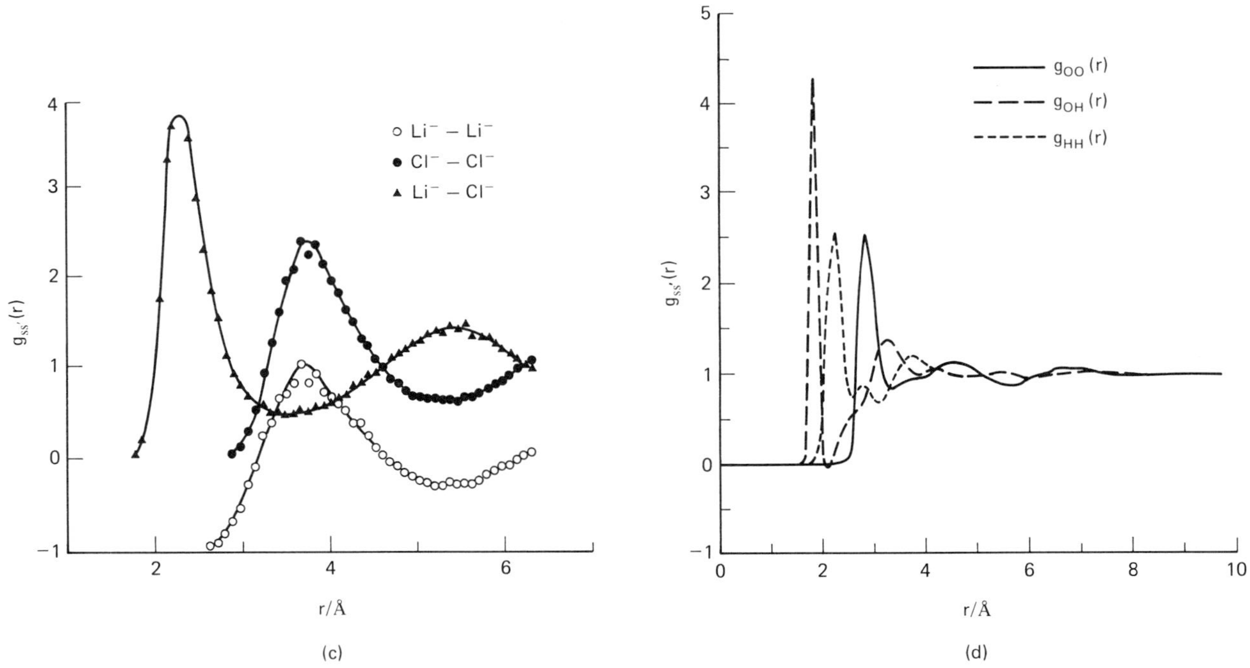

Figure 4.1 Examples of pair correlation functions. The methods of determination are described later. (a) Equimolar mixture of hard spheres with diameters $\sigma_1 = 1$ Å and $\sigma_2 = 3$ Å with packing fraction $\eta = (\pi/6)\ \Sigma_s\ \rho_s \sigma_s = 0.5806$. Dashed line, PY approximation; solid line, corrected PY (Verlet–Weis) generalized to mixtures; circles, MD simulation. (From Grundke and Henderson, 1972.) (b) Equimolar mixture of 6–12 model fluids corresponding to an Ar–Kr mixture at $T = 118$ K and $V = 33.3$ cm^3/mol. Data from MD simulation. (From McDonald and Singer, 1973.) (c) Charged soft-sphere model for liquid LiCl at $T = 883$ K, $V = 28.3$ cm^3/mol. Data from MC simulation. Notice that the peaks of g_{++} and g_{--} are at the same r, although the Li^+ and Cl^- diameters differ greatly. (From McDonald and Singer, 1973.) (d) The atom–atom correlation functions of liquid water at 25°C determined by neutron diffraction. The intramolecular peaks in g_{OH} (near 1.0 Å) and g_{HH} (near 1.5 Å) have been suppressed. (From Narten et al., 1982.)

$$\begin{aligned}\frac{\rho_s^\star(\mathbf{R}')}{\rho_{s'}} &= Z(N, V, T)^{-1} V^2 \int e^{-\beta U_N}\, \delta(\mathbf{r}_{1s} - \mathbf{R})\delta(\mathbf{r}_{2s'} - \mathbf{R}')\, d^3r_1 \cdots d^3r_N \\ &= V^2 \langle \delta(\mathbf{r}_{1s} - \mathbf{R})\delta(\mathbf{r}_{2s'} - \mathbf{R}')\rangle \\ &\equiv g_{ss'}(\mathbf{R}', \mathbf{R})\end{aligned} \tag{4.20}$$

which defines the pair correlation function $g_{ss'}$ in the canonical ensemble. It may be noticed that $g_{ss'}$ depends only on the magnitude of $\mathbf{R}' - \mathbf{R}$ if U_N does not depend on external fields; in this case $g_{ss'}(|\mathbf{R}' - \mathbf{R}|)$ is often called the *radial distribution function.* Some examples are given in Fig. 4.1. See if you can understand the main features of these curves. Some will be discussed later.

Many aspects of the theory of $g_{ss'}(r)$ are most readily developed in the grand ensemble. Thus we now denote by $g_{ss'}^{(\mathbf{N})}(\mathbf{R}, \mathbf{R}')$ the spatial pair correlation function defined in the canonical ensemble with variables $\mathbf{N}$, V, T and define the grand ensemble average of the *extensive* variable related to $g_{ss'}$, namely the average number of pairs of molecules such that an s molecule is at $\mathbf{R}$ and an s' molecule is at $\mathbf{R}'$,

$$\begin{aligned}&\langle N_s[N_{s'} - \delta_{ss'}] g_{ss'}^{(\mathbf{N})}(\mathbf{R}, \mathbf{R}')\rangle \\ &= \sum_{\mathbf{N}} \frac{\mathbf{z}^{\mathbf{N}} Z(\mathbf{N}, V, T)\, N}{\mathbf{N}!\, \Xi(\mathbf{z}, V, T)}{}_s[N_{s'} - \delta_{ss'}] g_{ss'}^{(\mathbf{N})}(\mathbf{R}, \mathbf{R}') \\ &= V^2 \sum_{\mathbf{N}} \frac{\mathbf{z}^{\mathbf{N}} N_s[N_{s'} - \delta_{ss'}]}{\mathbf{N}!\Xi} \int e^{-\beta U_N(\mathbf{r}_1, \ldots, \mathbf{r}_N)} \delta(\mathbf{R} - \mathbf{r}_{1s})\delta(\mathbf{R}' - \mathbf{r}_{2s'})\, d^3r_1 \cdots d^3r_N \\ &= V^2 z_s z_{s'} \sum_{\mathbf{L}} \frac{\mathbf{z}^{\mathbf{L}}}{\mathbf{L}!\Xi} \int e^{-\beta U_{s,s',\mathbf{L}}(\mathbf{R}, \mathbf{R}', \mathbf{r}_1, \ldots, \mathbf{r}_L)}\, d^3r_1 \cdots d^3r_L\end{aligned} \tag{4.21}$$

In these equations we employ the definitions

$$\mathbf{N} = N_1, \ldots, N_s, \ldots \tag{4.22a}$$

$$\mathbf{z}^{\mathbf{N}} = \prod_s z_s^{N_s} \tag{4.22b}$$

$$\mathbf{N}! = \prod_s N_s! \tag{4.22c}$$

$$\mathbf{L} = \mathbf{N} - 1_s - 2_{s'} \tag{4.22d}$$

and $U_{s,s',\mathbf{L}}(\mathbf{R}, \mathbf{R}', \mathbf{r}_1, \ldots, \mathbf{r}_L)$ is the potential of a configuration with an s particle at $\mathbf{R}$, an s' particle at $\mathbf{R}'$, and the remaining $L = L_1 + \cdots + L_N$ particles at $\mathbf{r}_1, \ldots, \mathbf{r}_N$. The pair correlation function in the grand ensemble is recovered by dividing Eq. (4.21) by $\langle N_s\rangle \langle N_{s'}\rangle$ to get

$$g_{ss'}(\mathbf{R}, \mathbf{R}') = \frac{z_s z_{s'}}{\rho_s \rho_{s'}} \frac{1}{\Xi} \sum_{\mathbf{L}} \frac{\mathbf{z}^{\mathbf{L}}}{\mathbf{L}!} \int e^{-\beta U_{s,s',\mathbf{L}}(\mathbf{R}, \mathbf{R}', \mathbf{r}_1, \ldots, \mathbf{r}_N)}\, d^3r_1 \cdots d^3r_N \tag{4.23}$$

where we use the notation $g_{ss'}$ just as for the canonical ensemble average.

DO PROBLEMS 4.5 TO 4.8.

4.3 INTERMOLECULAR FORCES

Some of the important results of statistical mechanics specify how measurable coefficients, like the pressure, or functions that provide a useful molecular-level description, like the pair correlation function, can be expressed as functionals of the Hamiltonian. So it is interesting to ask what we know about the Hamiltonian or, more to the point, about the term U_N that carries the effect of the interactions among the particles. To summarize some of the main features, we begin with the case $N = 2$ and write $u(\mathbf{r}_1, \mathbf{r}_2) = U_2(\mathbf{r}_1, \mathbf{r}_2)$.

For two uncharged monatomic particles the intermolecular force is derived from a potential with a shape such as curve b or c in Fig. 4.2. When the distance r between the particles is small, the potential is large and positive due to the action of the Pauli exclusion principle: When the electrons of the two particles are crowded into the same space, some of them must gain in kinetic energy in order to be in different quantum states than the others. If one suppresses the electronic coordinates and sees only the interparticle distance, the effect appears as a repulsion between the particles, called the *core repulsion.* For large enough r the core repulsion is dominated by an attractive effect which may be described in terms of the polarization of each atom by polar excited states of the other. The corresponding forces of attraction is sometimes called the *van der Waals force,* after the author of the first statistical-mechanical theory of its effect; the *London force,* after the author of the first quantum theory of its origin; or the *dispersion force,* a name arising from the relation of the theory of radiation to the theory of this force.

Perhaps the simplest mathematical function used to represent $u(r)$ is the *hard-sphere potential,*

$$u(r) = \begin{cases} \infty & \text{if} \quad r < \sigma \\ 0 & \text{if} \quad \sigma < r \end{cases} \tag{4.24}$$

It represents only the core repulsion but remarkably, as indicated by some of the data in Fig. 4.1 and shown more fully in Chapter 5, the model fluid governed by the hard-sphere pair potential exhibits the main structural features found for real liquids.

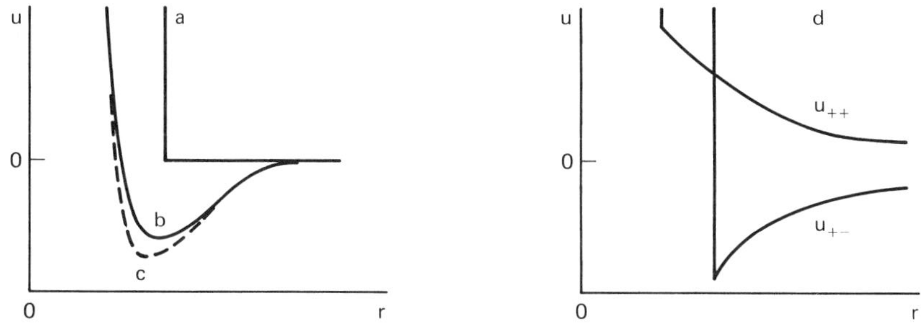

Figure 4.2 Pair potentials. (a) Hard-sphere model; (b) 6–12 model for Ar–Ar, $\epsilon/k_B = 120$ K, $\sigma = 3.2$ Å; (c) accurate potential for Ar–Ar (Barker et al., 1971); (d) primitive model for 1–1 aqueous electrolytes [Eq. (4.26)].

A more realistic function, called the Lennard-Jones or 6–12 potential (see Problem 3.11) is

$$u(r) = 4e_0\left[\left(\frac{\sigma}{r}\right)^{12} - \left(\frac{\sigma}{r}\right)^{6}\right] \tag{4.25a}$$

The r^{-6} term has the right functional form for the dispersion interaction in the range in which r is many molecular diameters. (For macroscopic r the correct form is r^{-7}, owing to a relativistic effect.) The r^{-12} form for the core repulsion is dictated by convenience. A better form on the basis of theory as well as beam-scattering experiments is shown by the core repulsion term in the Born–Mayer pair potential

$$u(r) = \begin{cases} \infty, & r < r_c \\ A \exp(-Br) - Cr^{-6}. & r > r_c \end{cases} \tag{4.25b}$$

where A, B, C, and r_c are positive temperature-independent parameters. The cutoff below r_c is necessary to avoid having $u(r)$ equal to $-\infty$ at $r = 0$, but there is a range of values of r_c in which it has no significant effect on the properties of the model fluid governed by Eq. (4.25b).

These model pair potentials for the interactions between molecules are qualitatively similar in shape to the Morse potential,

$$u(r) = A[1 - \exp(-a(r - r_0)]^2 - A \tag{4.25c}$$

a model potential that is remarkably accurate for representing the pair potential functions for chemically bonded interactions.

An important feature of a pair potential is the depth of the well, the potential at the minimum relative to the potential for the widely separated particles. It is convenient to express the well depth as the product of the Boltzmann constant and a temperature. In this form typical values are $k_B \times 200$ K for intermolecular interactions and $k_B \times 50{,}000$ K for strong chemical bonds. For hydrogen-bonded interactions, as in the HF dimer, a typical well depth is $k_b \times 3000$ K, an intermediate case. Of course, there is a continuous spectrum of well-depth energies; these are merely typical values.

However, there also is a qualitative difference between typical interactions among molecules and the interactions that hold the molecule together; namely, the saturation of interactions associated with chemical valence is a dominant feature only for chemically bonded interactions. The corresponding deviations from pairwise additivity for *intermolecular* interactions are comparatively minor, so that, as we shall see, pairwise additivity of intermolecular interactions often is a useful approximation.

Another model that is widely employed is appropriate for systems of charged particles. The *primitive model* for a pair of ions in an electrolyte solution is

$$u_{ab}(r) = \begin{cases} \infty & \text{if } r < \sigma_{ab} \\ \dfrac{e_a e_b}{\epsilon r} & \text{if } \sigma_{ab} < r \end{cases} \tag{4.26}$$

where e_a and e_b are electric charges and ϵ is the solvent dielectric constant. With $\epsilon = 1$ the primitive model is applied to fused salts; the "restricted" primitive model in either case is one in which all of the σ_{ab} and $|e_a|$ are the same.

These very simple model potentials are useful to fix our ideas, but even in the best case, namely the 6–12 potential applied to the Ar–Ar interaction, the model potential is not adequate to represent all of the experimental data.

Of course, most molecules of interest are not monatomic and then the intermolecular potential function is much more complicated. In the first place, it then depends on variables in addition to the center-to-center distance, namely variables fixing the relative orientations of the molecules, and also generally has contributions from the interactions of the electric moments (charges, dipoles, quadrupoles, etc.) of the molecules with each other and with the polarizabilities. We treat examples of some of these complications in Chapter 8.

It is most often assumed that U_N is *pairwise additive,* that is,

$$U_N(q_1 \cdots q_{3N}) = \sum_{\text{pairs}} u_{ss'}(r) \tag{4.27}$$

where, in each of the $N(N-1)/2$ terms on the right, one uses the value of the pair potential $u_{ss'}$ at the value of r given by the separation of that particular pair in the configuration $q_1 \cdots q_{3N}$.

To explain this, let us consider the case in which $N = 3$, say a set of particles of one species which we label *a, b,* and *c.* Then in general we may write

$$U_3(\mathbf{r}_a, \mathbf{r}_b, \mathbf{r}_c) = u(|\mathbf{r}_a - \mathbf{r}_b|) + u(|\mathbf{r}_b - \mathbf{r}_c|) + u(|\mathbf{r}_c - \mathbf{r}_a|) + u^{(3)}(\mathbf{r}_a, \mathbf{r}_b, \mathbf{r}_c)$$

where the first three terms on the right are pair potentials, like those discussed above, while the $u^{(3)}$ term is the defect or component potential obtained when one compares the three-point potential with the sum of pair potentials. For hard spheres, point charges, and certain other simple model particles, $u^{(3)}$ vanishes by construction. A very simple model for which the reader can verify that it does not vanish is one in which each particle is a point with a polarizability and a charge. Neither does it vanish for real particles interacting through London forces. However, it is generally agreed that many interesting problems can be treated within the approximation that U_N is pairwise additive. Furthermore, little is known about $u^{(3)}$, $u^{(4)}$, and higher component potentials at this time and, while much of the theory that we describe can be extended to include the effects of nonpairwise-additive terms in U_N, this area of study has not received much attention. Accordingly, we shall invoke pairwise additivity whenever it is convenient.

It is interesting to recall that the description of a given system can be formulated at several levels. We may call the Schrödinger level the one in which the particles are nuclei and electrons. The Schrödinger Hamiltonian is well known and is made up of one-particle kinetic terms and a pairwise-additive potential-energy function. Although deeper levels of description are known, they do not seem relevant to chemical systems.

In the Born–Oppenheimer level of description the explicit particles are atoms or molecules, the electronic coordinates being averaged away, and the potential function of the Hamiltonian is the Born–Oppenheimer potential surface of the nuclei, typically in the ground electronic state. The nonpairwise terms in the potential can be traced to suppression of the electronic coordinates as illustrated in the examples of nonpairwise contributions mentioned above.

In the McMillan–Mayer level of description, one or more of the species in a mixture are designated the solvent. Particles of the solute species are treated explicitly, and the potential term of the Hamiltonian carries the solvent-averaged forces among the solute particles (Chapter 10). The suppression of the coordinates of the solvent particles will contribute to deviations of this potential from pairwise additivity.

4.4 SOME PROPERTIES OF THE SPATIAL PAIR CORRELATION FUNCTION

The basic significance of the pair correlation function follows from its definition in Eq. (4.20); namely, $\rho_{s'} g_{ss'}(\mathbf{R}', \mathbf{R})$ is the local concentration $\rho_{s'}(\mathbf{R}')$ of particles of species s' near $\mathbf{R}'$ when there is a particle of species s at $\mathbf{R}$. In the absence of external fields this local concentration depends only on the distance $|\mathbf{R}' - \mathbf{R}|$. If we also have only one component, then $\rho g(r)$ is the local concentration of particles of the fluid near the surface of a sphere of radius r when one particle of the fluid is fixed at the center of the sphere. In either case we notice that the ratio of concentrations $\rho_{s'}(\mathbf{R}')/\rho_{s'} = g_{ss'}(r)$ is the equilibrium constant (mass action law constant) for the process

s' particles in bulk fluid $\rightleftharpoons$ s' particles a distance r from a particular s particle

These simple concepts can be applied to obtain insight into the following results.

The limiting behavior

$$g_{ss'}(r) \longrightarrow 1 \qquad \text{as } r \longrightarrow \infty \tag{4.28}$$

reflects the fact that the inhomogeneity due to one fixed particle in a fluid is limited to a finite region; if the fluid system is sufficiently large, there will be regions at a large distance r from the fixed particle where the local densities $\rho_{s'}(r)$ and the stoichiometric density $\rho_{s'}$ are the same and hence $g_{ss'}(r) = 1$. Strictly speaking, in the canonical ensemble Eq. (4.28) holds only to order $1/N$ in a finite system and becomes exact in the thermodynamic limit (Section 1.4).

We also have the *core condition*

$$g_{ss'}(r) = 0 \qquad \text{if } r < d_{ss'} \tag{4.29}$$

where $d_{ss'}$ is a "diameter" within which the short-range repulsive force between a particle of species s and one of species s' is so great that $\rho_{s'}(r)$ is quite negligible [Eq. (4.20)].

These considerations and many others can be elucidated by introducing the definition of the potential of average force $w_{ss'}(r)$:

$$g_{ss'}(r) = e^{-\beta w_{ss'}(r)} \tag{4.30}$$

whose negative gradient $-\partial w_{ss'}(r)/\partial r$ is the ensemble average force between a particle of species s and one of species s held at a separation r.

Two of the main features of $w_{ss'}(r)$, corresponding to Eqs. (4.28) and (4.29), respectively, are

$$w_{ss'}(r) \longrightarrow 0 \qquad \text{as } r \longrightarrow \infty \tag{4.31}$$

$$w_{ss'}(r) = \infty \qquad \text{for } r < d_{ss'} \tag{4.32}$$

Another important feature of the potential of average force follows from the concept of $g_{ss'}(r)$ as an equilibrium constant as noted above. Recalling the relation $\Delta G° = -RT \ln K$ of the standard free-energy change to the equilibrium constant, we see that $w_{ss'}(r) = -k_BT \ln g_{ss'}(r)$ is the free-energy change associated with the process for which $g_{ss'}$ is the equilibrium constant.

For a fluid of hard-sphere particles, for which the potential function is given in Eqs. (4.24) and (4.27), the correlation function has been accurately determined at several densities. In Fig. 4.3 some results in the range of liquid densities are given. It is interesting to estimate the potential of average force $w(r)$ for these cases and to consider the origins of the features in these curves. The *structure* of a liquid is represented by precisely these features in the case of the hard-sphere fluid. For example, it seems likely that the first peak in $g(r)$ is an aspect of what has been called the *cage effect* in the study of the chemical kinetic processes in solution. Other examples of these features may be found in Fig. 4.1.

When the potential U_N is pairwise additive [i.e., when Eq. (4.27) is exact],

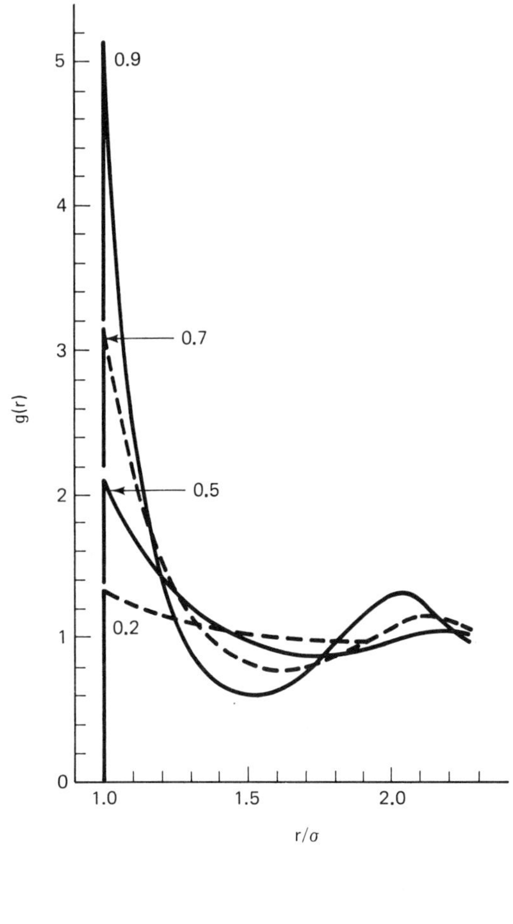

Figure 4.3 Pair correlation functions for the hard-sphere fluid at reduced densities $\rho\sigma^3$ equal to 0.2, 0.5, 0.7, and 0.9, where σ is the diameter of a sphere. These data were obtained by Monte Carlo simulation by Barker and Henderson (1971). At $\rho\sigma^3 = 0.94$ the hard-sphere fluid is in equilibrium with a crystalline phase with $\rho\sigma^3 = 1.04$, as discussed in the review by Barker and Henderson (1976). Thus at $\rho\sigma^3 = 0.9$ the hard-sphere fluid is in a liquid-like state, although this fluid does not exhibit a vapor–liquid phase transition. The large peak in $g(r)$ near the contact distance and the oscillatory behavior are typical features of the structure of liquids. The figure shows how these features are altered as the density is reduced.

many useful ensemble averages may be expressed as functionals of $g_{ss'}(r)$. For example, in this case Eq. (4.6b) may be modified as follows:

$$\begin{aligned}\bar{E}^{\text{ex}} &= \frac{\int d^{3N}q\, U_N(q_1, \ldots, q_{3N}) e^{-\beta U_N}}{Z} \\ &= \sum_{a=1}^{N} \sum_{b=a+1}^{N} \frac{\int d^{3N}q\, u_{s_a s_b}(|\mathbf{r}_a - \mathbf{r}_b|) e^{-\beta U_N}}{Z} \\ &= \frac{1}{2Z} \sum_{s} \sum_{s'} N_s[N_{s'} - \delta_{ss'}] \int d^3R \int d^3R' \int d^{3N}q \\ &\quad \times u_{ss'}(|\mathbf{R} - \mathbf{R}'|)\delta(\mathbf{R} - \mathbf{r}_{1s})\delta(\mathbf{R}' - \mathbf{r}_{2s'}) e^{-\beta U_N}\end{aligned} \tag{4.33}$$

The assumed pairwise additivity of U_N is exploited in the second equality, where the sums are over particle labels a and b while $\mathbf{r}_a$ and $\mathbf{r}_b$, the particle location vectors, are implicitly determined by some of the coordinates $q_1, \ldots, q_{3N}$. In the third equality the factor $\frac{1}{2}$ comes from the fact that there are $N_s(N_s - 1)/2$ pairs of species ss, while with the specified sum the number of unlike pairs is counted as both $N_s N_{s'}$ and $N_{s'} N_s$. Comparison of Eq. (4.33) with Eq. (4.20) leads to the result

$$\frac{\bar{E}^{\text{ex}}}{V} = \frac{1}{2} \sum_{s} \sum_{s'} \rho_s \rho_{s'} \int u_{ss'}(r) g_{ss'}(r)\, d^3r \tag{4.34}$$

The same procedure applied to Eqs. (4.11) and (3.38) gives, respectively,

$$\beta p = \sum_{s} \rho_s - \frac{\beta}{6} \sum_{s} \sum_{s'} \rho_s \rho_{s'} \int \frac{r\, \partial u_{ss'}}{\partial r} g_{ss'}(r)\, d^3r \tag{4.35}$$

$$\beta A = \beta A_{\text{classical}} + \frac{(\beta\hbar)^2 V}{48} \sum_{s} \sum_{s'} \frac{\rho_s \rho_{s'}}{\bar{m}_{ss'}} \int [\nabla_r^2 u_{ss'}(r)] g_{ss'}(r)\, d^3r \tag{4.36}$$

where

$$\bar{m}_{ss'} \equiv \frac{m_s m_{s'}}{m_s + m_{s'}}$$

$$\nabla_r^2 u \equiv \frac{\partial^2 u}{\partial r^2} + \frac{2}{r}\left(\frac{\partial u}{\partial r}\right)$$

These remarkably simple equations show how the thermodynamic properties are determined by the solution structure at the pair level for models in which U_N is pairwise additive. For real systems the pairwise additivity of U_N is an approximation which is quite accurate in some systems, such as fluids made up of rare-gas molecules, but not so good in others. However, for fluids in which the forces between the particles depend on the orientations as well as separation, equally simple equations can be recovered by a generalization of the notation in this chapter. In such cases one replaces the location vector $\mathbf{r}$ for each particle by the location vector *and* the set of angles specifying the particle's orientation, as discussed in Chapter 8. In any case, these equations illustrate the role of $g_{ss'}(r)$ as a probability density; $u(r)g(r)$ as the local energy density, etc.

Turning to another aspect, we notice that Eqs. (4.34) to (4.36) have been derived for the canonical ensemble. However, $g_{ss'}$ is the same in the canonical ensemble and in the grand ensemble in the thermodynamic limit (i.e., for a macroscopic system), so one may equally well use the grand ensemble $g_{ss'}$ in these equations. It follows that the equations themselves could also have been derived for the grand ensemble. The choice of ensemble is merely a matter of convenience. However, there is one more expression for thermodynamic coefficients as functionals of $g_{ss'}$ that should be listed with the preceding equations; and it can only be derived from the grand ensemble theory because it relates to fluctuations in density. Thus we integrate $\mathbf{R}'$ in Eq. (4.23) to find [see the passage from Eq. (4.19b) to (4.19c)]

$$\begin{aligned}\int g_{ss'}(\mathbf{R}, \mathbf{R}')\, d^3\mathbf{R}' &= \frac{z_s z_{s'}}{V\rho_s\rho_{s'}\Xi} \sum_{\mathbf{L}} \frac{\mathbf{z}^{\mathbf{L}}}{\mathbf{L}!} Z(\mathbf{L} + 1_s + 2_{s'}, V, T) \\ &= \frac{z_s z_{s'}}{V\rho_s\rho_{s'}\Xi}\left(\frac{\partial^2 \Xi}{\partial z_s \partial z_{s'}}\right) \\ &= (V\rho_s\rho_{s'})^{-1}\langle N_s(N_{s'} - \delta_{ss'})\rangle\end{aligned} \tag{4.37}$$

where we have used Table 1.5 and have assumed that the system is homogeneous. A more useful form is obtained by subtracting V from both sides of the equation and recalling that $\rho_s = \langle N_s\rangle / V$. So we find that

$$\begin{aligned}\int g_{ss'}(\mathbf{R}, \mathbf{R}') - 1]\, d^3\mathbf{R}' &= \int [g_{ss'}(r) - 1]\, d^3 r \\ &= (V\rho_s\rho_{s'})^{-1}[\langle N_s(N_{s'} - \delta_{ss'})\rangle - \langle N_s\rangle\langle N_{s'}\rangle]\end{aligned} \tag{4.38}$$

or, using the notation in Eq. (1.59), we have

$$\begin{aligned}\rho_s\delta_{ss'} + \rho_s\rho_{s'}\int (g_{ss'}(r) - 1)\, d^3 r &= \frac{M_{ss'}}{\beta} \\ &= \frac{\partial\rho_s/\partial\mu_{s'}}{\beta}\end{aligned} \tag{4.39}$$

If and only if there is but one component in the system, we have [Eq. (1.86)] the isothermal compressibility

$$\kappa_T = \frac{(\partial\rho/\partial\mu)_{V,T}}{\rho^2} \tag{4.40}$$

in the one-component version of Eq. (4.39), namely

$$1 + \rho\int [g(r) - 1]\, d^3 r = \frac{\kappa_T \rho}{\beta} \tag{4.41}$$

More generally, the theory of the generalized compressibilities $M_{\varsigma\varsigma'}$ is given in Eqs. (1.59) to (1.66). It should be noted that the theory of $M_{\varsigma\varsigma'}$, unlike Eqs. (4.34) to (4.36), does not depend on the pairwise-additivity assumption.

DO PROBLEMS 4.9 AND 4.10.

THE LITERATURE

Helpful accounts of the theory of spatial correlation functions may be found in many places, including Hill (1956) and Hansen and McDonald (1976). Basic aspects of the theory of intermolecular forces are presented by Karplus and Porter (1970), while recent results for highly accurate pair potentials for atomic particles, both like and unlike, are described by Scoles (1980).

PROBLEMS

4.1. Generalize Eq. (4.3) to apply to a mixture using Eqs. (1.21), (1.39), and (1.44).

4.2. Derive Eq. (4.5) from Eq. (1.22.1), the fundamental equation for A.

4.3. The last equality in Eq. (4.6a) implicitly assumes that U_N does not depend on temperature. Derive the following more general result:

$$E^{\text{ex}} = \left\langle \frac{\partial(\beta U_N)}{\partial \beta} \right\rangle \tag{4.42}$$

4.4. In view of Eqs. (4.42) and (4.5), show that a temperature-dependent $U_N(q_1, \ldots, q_{3N})$ is the reversible work, rather than the $\Delta \bar{E}$, for the process in which the N widely separated particles are brought to the specified configuration.

4.5. **(a)** Find the form for $g_{\varsigma,\varsigma',\varsigma''}(\mathbf{R}, \mathbf{R}', \mathbf{R}'')$, the three-particle spatial correlation function, that is similar to Eq. (4.20).

(b) First write down the grand ensemble equation for this three-particle correlation function just by analogy with Eq. (4.23). Then carry through the derivation in the three-particle case, in analogy to the derivation of Eq. (4.23).

(c) For the n-particle spatial correlation function

$$g_{\varsigma 1, \ldots, \varsigma n}(\mathbf{R}_1, \ldots, \mathbf{R}_n)$$

write down the equations analogous to Eqs. (4.20) and (4.23). Compare your result with Eq. (10.2c).

4.6. In Eq. (4.21) we find the grand ensemble average of $g_{\varsigma\varsigma'}$ by working with a related extensive quantity. To illustrate the reason for this, verify that in the grand ensemble (see Table 1.4) we have

$$\langle N \rangle = \sum_{N,j} NP(N, j) \qquad \text{and} \qquad \langle E \rangle = \sum_{N,j} E(N, j)P(N, j)$$

Then see if you can find any simple significance in the average

$$\sum_{N,j} \frac{E(N, j)}{N} P(N, j)$$

Even more simply compare the concept of average wind velocity with average rainfall, which has an obvious significance and which has a meaning only in terms of some definition.

4.7. Write out the first five terms of Eq. (4.23) for a mixture of two species, a and b.

4.8. As an application of the idea in Eq. (4.16) we define

$$U_{N-2}^{\star\star}(\mathbf{r}_3, \ldots, \mathbf{r}_N) = U_N(\mathbf{R}, \mathbf{R}', \mathbf{r}_3, \ldots, \mathbf{r}_N)$$

for the potential of a system of $N-2$ particles in the field of a particle fixed at $\mathbf{R}$ and one at $\mathbf{R}'$.

(a) Show that the ensemble average force on the particle fixed at $\mathbf{R}$ is

$$\langle \mathbf{F}_1(\mathbf{R}) \rangle^{\star\star} = \frac{\int -\nabla_1 U_{N-2}^{\star\star} \exp(-\beta U_{N-2}^{\star\star})\, d^3r_3 \cdots d^3r_N}{\int \exp(-\beta U_{N-2}^{\star\star})\, d^3r_3 \cdots d^3r_N} \tag{4.43a}$$

where $\nabla_1 = \partial/\partial \mathbf{R}$ is the gradient with respect to change of $\mathbf{R}$.

(b) Show that

$$\nabla_1 \ln g(\mathbf{R}, \mathbf{R}') = \beta \langle \mathbf{F}_1(\mathbf{R}) \rangle^{\star\star} \tag{4.43b}$$

(c) Show that

$$-\nabla_1 w(\mathbf{R}, \mathbf{R}') = \langle \mathbf{F}_1(\mathbf{R}) \rangle^{\star\star} \tag{4.43c}$$

so that w defined in Eq. (4.29) is indeed the potential of the average force.

(d) Show that $(\partial/\partial R)[k_B T \ln g_{s,s',s''}(\mathbf{R}, \mathbf{R}', \mathbf{R}'')]$ is the average force on a particle of species s at $\mathbf{R}$ if a particle of species s' is at $\mathbf{R}'$, one of species s'' at $\mathbf{R}''$, and the coordinates of all of the other particles in the system are averaged over.

4.9. For the one-component case, derive Eq. (4.36) for the first quantum correction to βA from Eqs. (3.38) and (4.27).

4.10. On the basis of Eqs. (4.23) and (4.30), show that we have

$$g_{ss'}(r) \to \exp[-\beta u_{ss'}(r)] \qquad \text{as} \quad \mathbf{z} \to 0 \tag{4.44a}$$

$$w_{ss'}(r) \to u_{ss'}(r) \qquad \text{as} \quad \mathbf{z} \to 0 \tag{4.44b}$$

where $\mathbf{z} = 0$ means $z_1 = 0$, $z_2 = 0$, and so on. [To do this carefully is difficult. Do not be surprised to find an extraneous factor of e^{-N} in $g(r)$. This difficulty is typical of the problems that motivate the theory in Chapter 6.]

5

Simulation and Diffraction

5.1 INTRODUCTION

In this chapter we consider the simulation methods for determining the thermodynamic coefficients and spatial correlation functions of Hamiltonian *model* systems. We also consider the theory of interpretation of diffraction experiments, the most nearly direct source of data for spatial correlation functions for *real* systems.

5.2 PERIODIC BOUNDARY CONDITIONS

Consider a model system specified by the Hamiltonian

$$H_N = \sum_{i=1}^{N} \frac{\mathbf{p}_i^2}{2m_i} + \sum_{\text{pairs}} u_{ij}(r_{ij}) \tag{5.1}$$

where $\mathbf{p}_i$ is the momentum of particle i and $u(r_{ij})$ the pair potential and $\mathbf{p}_i^2 \equiv \mathbf{p}_i \cdot \mathbf{p}_i$. To determine ensemble-average quantities by simulation, one constructs a sequence of states such that the potential energy, the pressure, the pair correlation function, and so on, can be expressed as averages over these states. In each such state the locations of all N particles are specified and to go from one step to the next one has to compute the forces or potentials connecting each of the $N(N-1)/2$ pairs of particles. This step is the bottleneck in the computation and greatly limits the number N of particles that can be accommodated. Most often $N \simeq 200$ is regarded as a practical compromise between so small a value that the results of the simulation are difficult to interpret and so large a value that the cost of computation is out of reach.

A system of 200 particles is a rather small body whose properties are all very sensitive to the boundary conditions. With sufficiently strong attractive forces the particles will tend to stay together in a body and it is possible to proceed with the simulation according to one algorithm or another and to construct ensemble averages which characterize the isolated body or "cluster" of N particles. The simulation may be thought of as a "computer experiment," the object of study being an isolated cluster of N particles governed by the model Hamiltonian. If the small cluster is the object of study, no further attention needs to be paid to the problem of boundary conditions, but if we seek the bulk properties, those that correspond to the system in the thermodynamic limit ($N \longrightarrow \infty$ at fixed N/V), we should notice that to deduce the numerical values of bulk properties from such an experiment is a subtle exercise that is comparable to deducing bulk properties from laboratory experiments on isolated clusters of small numbers of molecules. Even so, the results are often reasonably satisfactory.

To obtain bulk properties it is usually recommended to combine the model Hamiltonian with *periodic boundary conditions.* In straightforward cases this corresponds to simulating a system of $N_r \times N$ molecules in the limit $N_r = \infty$ and such that the system is a three-dimensional array of N_r cells, each housing a given configuration of N particles. Thus with periodic boundary conditions the object of study of the computer experiment is an infinite crystal whose unit cell contains a fluid-like configuration of N particles. This configuration changes, according to the algorithm of the simulation, as one goes from one state of the system to the next.

The way that periodic boundary conditions are imposed is illustrated in Fig. 5.1 for a two-dimensional system. One actually follows the N particles in one unit cell, say the central one in the figure. When a change of state carries a particle out through one face of the basic cell, the move also carries the corresponding particle through the opposite face into the basic cell. Thus N is conserved.

A particularly important system is the hard-sphere model, in which $u(r)$ is given by Eq. (4.24). In some other important models a soft "tail" is added to the hard-sphere potential. There are some special aspects in the simulation of such systems having infinite forces that would complicate the following discussion. Therefore, the rest of this chapter is explicitly limited to systems with continuous potentials, although our discussion of the literature at the end of the chapter is more complete.

Having chosen periodic boundary conditions, one has a further choice to make regarding the range of the forces. These may be discussed with respect to the size of the unit cell, here assumed to be a cube of edge L, as follows.

Spherical cutoff. For some $r_c \leq L/2$ one replaces $u(r)$ in Eq. (5.1) by $u(r) - u(r_c)$ for $r < r_c$ and by zero for $r \geq r_c$. Thus the direct interactions between particles separated by r_c or more are neglected. This procedure is relatively easy to justify for short-range interactions such as the 6–12 potential, but even seems to be satisfactory for models for water.

Minimum image. For each particle in the basic cell one counts its interaction with the $N - 1$ other particles in a new unit cell with it at the center. The distinction between minimum image on the one hand and spherical cutoff with $r_c = L/2$ on

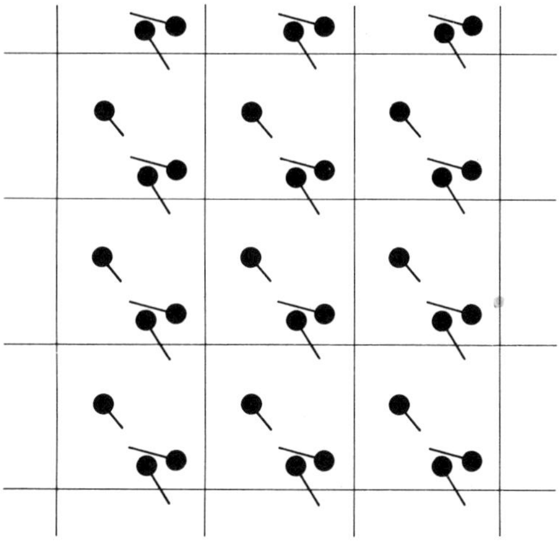

Figure 5.1 Periodic boundary conditions: a given state (filled circles) and the move to the next state (lines), represented in two dimensions.

the other is rather subtle (Fig. 5.2), but turns out to be significant in simulations of systems of polar molecules.

Ewald sum. Borrowing from the theory of the Madelung constant of ionic crystals, we can do the sums over the infinite lattice required to compute the energy per particle for each state of the system. Relatively little extra expense is required to do the lattice sums, but in the cases in which they are especially appropriate, namely systems with Coulomb interactions, the sums, now called *Ewald sums,* are only conditionally convergent and much care is required in interpreting the results.

Other methods, such as the combination of spherical cutoff with the Onsager reaction field, have been proposed for dynamical as well as Monte Carlo simulations.

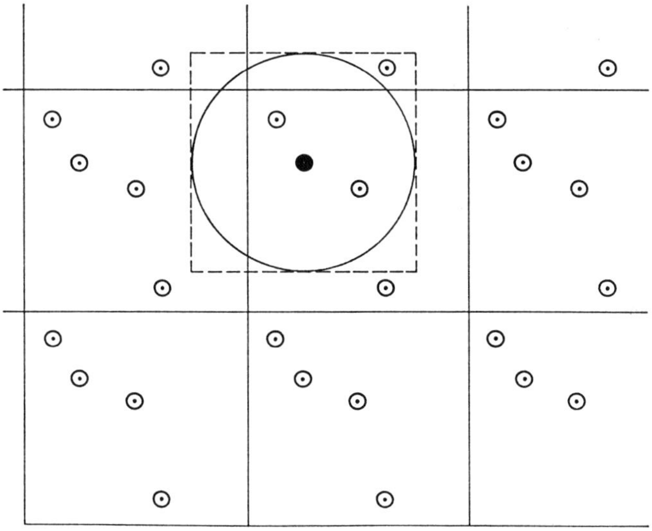

Figure 5.2 Representation in two dimensions of "minimum image" and spherical cutoff. In the former we count the interaction of the black particle with other particles within the dashed square. In the latter, if the cutoff distance is at $L/2$, we count the interactions of the black particle with other particles within the circle inscribed in the dashed square. As in this example, the two procedures sometimes count different numbers of interactions.

The problems concerning the appropriate truncation (or not) of the interactions have excited a lot of interest, as may be learned by consulting the literature section of this chapter.

Thus the interpretation of the results of a simulation method for generating ensemble averages is subject to some uncertainty, both associated with small N and the use of periodic boundary conditions and associated with the truncation of the interactions. In the absence of long-range forces the first uncertainty is minimal and the second is absent, but it seems wise in general to regard the simulation methods as approximate rather than as exact. However, it must be said that there is a widespread view that difference between the results of simulation methods for estimating ensemble averages and other methods must always be ascribed to the limitations of the latter.

5.3 MOLECULAR DYNAMICS

As will be apparent at once, this simulation method yields ensemble averages for the correlation functions and coefficients of rate processes as well as for equilibrium states. It seems convenient to consider molecular dynamics (MD) here with emphasis on the equilibrium-state information and to return to the subject for the rate information in Chapter 12.

For a system governed by classical mechanics with the Hamiltonian in Eq. (5.1) we may write the equation of motion

$$\begin{aligned} m_a \ddot{q}_a &\equiv F_a \\ &= -\frac{\partial}{\partial q_a} \sum_{\text{pairs}} u(r_{ij}), \qquad a = 1, 2, \ldots, 3N \end{aligned} \tag{5.2}$$

where, in the notion of Eq. (4.1), q_a is one of the $3N$ coordinates of a system of N particles. Also, m_a is the mass associated with the degree of freedom q_a and $\dot{q}_a \equiv dq_a/dt$, $\ddot{q}_a \equiv d^2q_a/dt^2$ and so on, where d/dt is the derivative with respect to time. In the molecular dynamics simulation method we generate a trajectory of the model system in configuration space (i.e., the space of $q_1, \ldots, q_{3N}$) by following the equations of motion. With some care we can generate a trajectory that represents the equilibrium fluctuations in the microcanonical ensemble. We now discuss various aspects of this procedure.

The microcanonical ensemble is a collection of replica systems of given N, V, and E. Thus each system is isolated; it exchanges neither molecules nor energy with the environment. The theory of ensemble averages in this ensemble is distinct from the theory of the two ensembles treated in Chapter 1. For example, in a microcanonical ensemble the temperature is an ensemble average which may be derived from the expression

$$\begin{aligned} \tfrac{3}{2} N k_B T &= \langle \text{K.E.} \rangle \\ &= \sum_a \tfrac{1}{2} m_a \langle \dot{q}_a^2 \rangle \end{aligned} \tag{5.3}$$

where $\langle \cdot \cdot \cdot \rangle$ means the microcanonical ensemble average. Thus in this ensemble the kinetic energy is a fluctuating quantity, owing to the interchange of energy between kinetic (K.E.) and potential (P.E.) forms, while E = K.E. + P.E. remains fixed.

For a numerical procedure we need to replace Eq. (5.2) by an equation with finite time steps, whose duration will be called $\hat{t}$. The most commonly used algorithm is due to Verlet (1967). We expand $q_a(t \pm \hat{t})$ around t:

$$q_a(t \pm \hat{t}) = q_a(t) \pm \hat{t}\dot{q}_a(t) + \frac{1}{2}\hat{t}^2\ddot{q}_a(t) \pm \frac{1}{3!}\hat{t}^3\dddot{q}_a(t) + \frac{1}{4!}\hat{t}^4\ddddot{q}_a(t) + \cdots \tag{5.4}$$

and add the forward and reverse forms of the equation to get the series

$$q_a(t + \hat{t}) = 2q_a(t) - q_a(t - \hat{t}) + \frac{\hat{t}^2 F_a(t)}{m_a} + \frac{1}{12}\hat{t}^4\ddddot{q}_a(t) + \cdots \tag{5.5}$$

which is truncated after the $\hat{t}^2$ term to get *Verlet's algorithm.* We have used Eq. (5.2) to replace $\ddot{q}_a$ by F_a/m_a. It should be noticed that with this algorithm we need not compute the velocities $\dot{q}_a$ explicitly. The full dynamics is obtained by following the system in the configuration space rather than in the full phase space.

To assess the effect of the truncation, we examine the ratio of the $\hat{t}^4$ to the $\hat{t}^2$ term, which is small provided that we have

$$\hat{t}^2 \ll 12\frac{\ddot{q}_a}{\ddddot{q}_a} \tag{5.6}$$

where the ratio can also be expressed in terms of derivatives of F_a, again using Eq. (5.2). Thus for a model for argon with the Lennard-Jones potential and in the ordinary temperature range we find that $\hat{t}$ = 0.01 ps (1 ps = 10^{-12} s) is satisfactory, but more often $\hat{t} \sim 10^{-15}$ s is required. This is a stringent requirement; it means that to follow the system for only 10 ps of its evolution requires 10,000 moves.

A system that is governed by Eq. (5.2) will conserve energy; E is constant. But a system that is governed by the approximate equation of motion given by the Verlet algorithm can be expected to have time-dependent E, although less so as $\hat{t}$ is made smaller. So the constancy of E may be used as a measure of whether $\hat{t}$ is small enough. Although one can keep more terms in Eq. (5.5) or construct other algorithms in order to allow larger time steps with no loss in accuracy, the price is inevitably more computations per time step to evaluate additional derivatives of the potential. Thus the more refined algorithms do not necessarily result in a lower cost of computation for, say, a 10-ps trajectory.

In order to establish the constancy of the total energy, it is necessary to know the velocities of the particles at each time t. A knowledge of the velocities will be even more important when considering the dynamical behavior of the fluid (see Chapter 13). Although Eq. (5.5) does not yield the particle velocities explicitly, it is easy to show that for the Verlet algorithm one has

$$\dot{q}_a(t) = \frac{q_a(t + \hat{t}) - q_a(t - \hat{t})}{2\hat{t}} \tag{5.7}$$

The Verlet algorithm propagates the system along a trajectory in location space. As each state is reached it is necessary to calculate the $3N$ forces F_a. This is the source of the larger part of the computational expense. Part of the cost is due to the need to calculate the $N(N-1)/2$ pair separations r_{ij}, while another part, which may be substantial for some models, is due to the need to calculate the pair forces from the appropriate algebraic expressions or to find them by interpolation in an appropriate table. It is in this part of the program that it is especially advantageous to give attention to the efficiency of the computation code for the particular model and the particular computing machine.

To start the simulation we need to choose initial values for $q_1, \ldots, q_{3N}$. This is often done by putting the N particles on a lattice within the chosen volume V. The starting configuration may be specified as the configuration at $t = 0$. To calculate the configuration at $t = \hat{t}$ (i.e., after one move), by Verlet's algorithm we obviously need a configuration at $t = -\hat{t}$. This may be constructed in such a way as to correspond to giving the molecules a Maxwell–Boltzmann velocity distribution for some chosen temperature T at $t = 0$. Thus from such a distribution we pick, at random, values of p_a/m_a for $a = 1, 2, \ldots, 3N$ and we specify

$$q_a(-\hat{t}) = q_a(0) - \frac{\hat{t}p_a}{m_a}, \qquad a = 1, \ldots, 3N \tag{5.8a}$$

which is just where the particles would have been had they traveled with fixed velocity in the interval from $t = -\hat{t}$ to $t = 0$. A way to pick values at random from a specified distribution is discussed in the following section.

Having initiated the simulation we compute our way from one state at t to the next at $t + \hat{t}$ by means of the move algorithm. It is usually found that the temperature as defined by Eq. (5.3) increases drastically for many moves after any sort of arbitrary initiation configuration, such as the one described above. This is a consequence of the trajectory finding its way to states of lower P.E. as the particles move under the influence of the intermolecular forces. To get near a chosen temperature, we scale the velocities, typically by calculating

$$q_a'(t-\hat{t}) = q_a(t) - x[q_a(t) - q_a(t-\hat{t})] \tag{5.8b}$$

and then by calculating $q_a(t+\hat{t})$ from $q_a(t)$ and $q_a'(t-\hat{t})$, all for $a = 1, \ldots, 3N$. This has the effect of changing the temperature at t from T to $T' = x^2T$.

It may take more than 1000 time steps, after initiation, and with more or less frequent scaling of the velocities as required, in order to be in a position to calculate a trajectory that represents the equilibrium fluctuations in the system. The procedure may be monitored by making a *control chart* showing how T or the potential energy per particle varies along the trajectory. When these averages seem to stop drifting, the trajectory is continued for at least another 1000 time steps to provide the data for calculating the ensemble averages of interest. Depending on the application, as well as on the capacity of the computing machine, we accumulate the averages of interest while this part of the trajectory is being generated, or we may store the locations of the particles at various time steps for later averaging. In either case it may be advantageous to notice that $\hat{t}$, having been chosen to be small enough so that the move algorithm is a good approximation to the equations of motion, is so small that the configurations change very little from one step to the next. Then it may be adequate to calculate the ensemble averages of interest from only every nth step, where n may be 10 or more. Another standard procedure is to calculate the ensemble averages of interest for each of three or four consecutive segments of the trajectory. Then comparing the segment averages enables an estimate of the statistical variation of the results. This is analogous to repeating a laboratory measurement several times to obtain an estimate of random errors.

There is no corresponding procedure for assessing the contributions of systematic errors. For example, the finite size of the N-particle system automatically suppresses correlations at any distance greater than the box length, even if such correlations are implied by the Hamiltonian. Another source of systematic error is the limited extent to which a simulation with finite cost can adequately sample the phase space accessible to the system. Neither of these problems is serious in the case of simulations of the 6–12 model in the range of states corresponding to the fluid, but one or both may be important if one looks at the onset of crystallization, at systems with long-range interactions, or at systems in which small complexes of particles are very stable, as when the well depth in the pair potential (see Fig. 4.2) exceeds $10k_BT$ and the density is well below the closed-packed value.

For a suitable property f we construct the time average [see Eqs. (1.1) and (1.5)] by averaging over the states i of the trajectory,

$$\langle f \rangle = \sum_{i=1}^{\nu} \frac{f_i}{\nu} \tag{5.9}$$

where f_i is the value in state i. For the potential energy per particle, for example, we take

$$f_i = \frac{1}{N}\left[\sum_{a=1}^{N}\sum_{b=a+1}^{N} u(r_{ab})\right]_i \tag{5.10}$$

With f_i = "T_i" = $(\text{K.E.})_i/(\frac{3}{2}Nk_B)$ as suggested by Eq. (5.3) we can calculate the heat capacity C_v by the microcanonical ensemble equation

$$\frac{\langle T^2 \rangle}{\langle T \rangle^2} - 1 = \frac{3}{2N}\left(1 - \frac{3Nk_B}{2C_v}\right) \tag{5.11}$$

However, it should be remarked that it is far more difficult to get good statistics from simulation results analyzed with fluctuation theory, of which this is an example, than from straightforward averages as in Eq. (5.9). So it is usually more satisfactory to obtain C_v from the temperature coefficient of the average potential energy obtained from simulations at more than one temperature, combined with the heat capacity associated with the kinetic energy, $\frac{3}{2}Nk_BT$.

With f_i defined as the number of pairs of particles with separation between r and $r + \Delta r$, the corresponding average is $\langle n(r) \rangle$ from which one can estimate the pair correlation function as

$$g(r) = \frac{V}{N}\left[\frac{\langle n(r) \rangle}{4\pi r^2 \Delta r}\right] \tag{5.12}$$

provided that Δr is small enough that the change of $g(r)$ in this range is small. The MD results in Fig. 5.3 were obtained in this way.

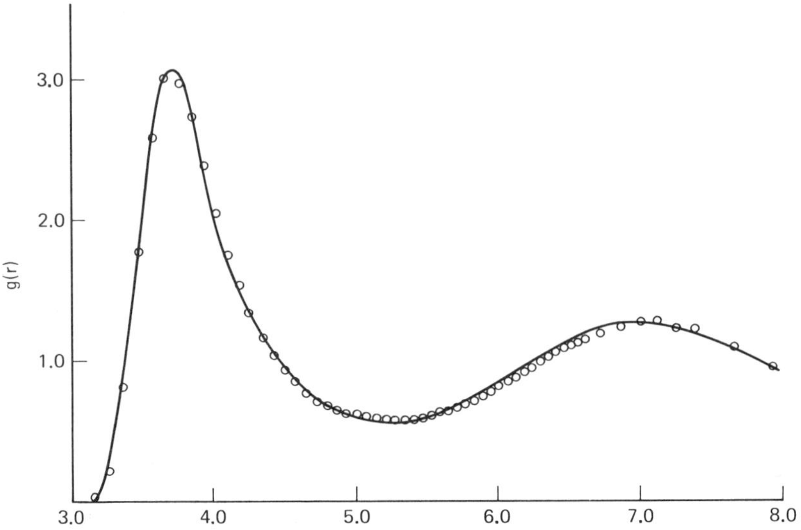

Figure 5.3 Pair correlation function for liquid argon near its triple point. ———— MD simulation for a 6–12 model [Eq. (4.25)] with parameters $\sigma = 3.405$ Å, $\epsilon/k_B = 119.8$ K that had been chosen to fit the thermodynamic data for argon. The results shown are for the state $T = 86.56$ K, $\rho = 0.02138$ Å^{-3}. ———— (indistinguishable from the line above): MC calculation for the Barker et al. (1971) model for argon (see Fig. 4.2), which includes three-body terms. o o o o o: Result of Fourier inversion of the experimental $S(k)$ data in Fig. 5.5 (From Yarnell et al., 1973.)

Finally, it is clear that the sensitivity of the ensemble averages to the boundary conditions depends inversely on N, the number of particles, while the cost per move tends to vary like $N(N - 1)/2$, as already noted. However, with larger N it takes fewer moves for given accuracy in ensemble averages, as one may see from Eqs. (5.9) and (5.10). The number of data points varies like $N\nu$ if, as in Eq. (5.3) all of the particles contribute directly to the average of interest. So the real cost per move may be more nearly proportional to N, up to a limit imposed by the storage capacity of the central processing unit of the computer.

The MD simulation method, described here only for a system of point particles, has been developed as well for systems of molecules. In models in which the molecular bond lengths and angles are flexible, we can take the atoms to be the particles of the system. In this case the potential functions from which the forces are derived include the intra- as well as the intermolecular interactions, but in other respects the equations described above can be applied without change. However, it is often interesting to simulate models in which the molecular geometry is rigid, either because the same models can be treated by one of the more analytical approximation methods described in Chapter 8 or because assemblies of rigid molecules may, in fact, be more like real molecular systems than assemblies of flexible molecules in which the quantum-mechanical aspects of the intramolecular motions are overlooked.

To simulate assemblies of rigid molecules, we may use "generalized coordinates" to include orientational coordinates and momenta in addition to the Cartesian coordinates and momenta of the centers of mass of the molecules. Sometimes quaternions are used instead of generalized coordinates to avoid the cross-terms in the move algorithm which are associated with the latter.

A recent development (Ryckaert et al., 1977) for MD simulations of assemblies of rigid molecules is to treat the atoms as the particles, just as for simple systems, but to incorporate the constraints imposed by fixed molecular bond lengths and angles by introducing the appropriate Langrangian multipliers. The resulting move algorithm may be similar to Verlet's algorithm except that in each move there is an auxilliary calculation that corresponds to determining the Lagrangian multipliers. It seems that this procedure, sometimes known as SHAKE, the name of a subroutine for doing the auxilliary calculation, is often more satisfactory than using generalized coordinates.

5.4 MONTE CARLO

In the MD simulation method we generate a sequence of states of the system by following the natural motion very accurately. In the Monte Carlo (MC) simulation method we generate a different sequence of states, often called the *Markov chain,* by a move algorithm that is constructed without reference to the equations of motion. Instead, it is designed to assure that averages over the Markov chain become identical to ensemble averages in the limit of infinite chain length. In either case the simulation leads to a sequence of states from which we can construct ensemble averages. In the MD case we calculate time averages and rely on Eq. (1.5) to identify these with ensemble averages. In the MC case we calculate ensemble averages very directly. We illustrate the procedure for the usual case in which we construct averages in the canonical ensemble.

Consider the move from state i to state j. As in the MD theory, a *state* means specified locations $q_1, \ldots, q_{3N}$ of N particles. We define $T(i \longrightarrow j)$ as the transition probability for the passage to state j from i. To generate a stationary Markov chain, it should be governed by Eqs. (5.13).

$$T(i \longrightarrow j) \geqslant 0 \tag{5.13a}$$

$$\sum_j T(i \longrightarrow j) = 1 \tag{5.13b}$$

In Eq. (5.13) we explicitly include the possibility that $j = i$ (i.e., that no transition will occur). These equations satisfy the conditions necessary for $T(i \longrightarrow j)$ to be a probability. Also in terms of P_i, the a priori probability of state i, we have

$$P_j = \sum_i P_i T(i \longrightarrow j) \tag{5.13c}$$

which assures stationarity of the process, which means that on the average the flow out of any state j is equal to the flow into it, and

$$P_i T(i \longrightarrow j) = P_j T(j \longrightarrow i) \tag{5.13d}$$

which is an expression of microscopic reversibility; on the average the flow into j from i is balanced by the flow into i from j.

It is easy, and completely allowable, to consider the stationary Markov chain in terms of a metaphor in which the states are chemical species, the P_i their concentrations, and the $T(i \longrightarrow j)$ the first-order rate constants for their interconversion. Then Eq. (5.13b) merely implies that we define an unconventional rate constant $T(i \longrightarrow i)$ (for a species i not to react in unit time) by the condition

$$T(i \longrightarrow i) \equiv 1 - \sum_{j \neq i} T(i \longrightarrow j) \tag{5.14}$$

All we need to add to be able to identify the *species average* of some quantity f,

$$\langle f \rangle = \sum_i f_i P_i$$

with the canonical ensemble average is to choose $T(i \longrightarrow j)$ in a way which implies that

$$P_i = \frac{\exp(-\beta U_i)}{\sum_i \exp(-\beta U_i)} \tag{5.15}$$

Equations (5.13c) and (5.13d) are the necessary conditions required to assure this consistency. Thus with the canonical ensemble choice for P_i we have the *equilibrium constant*

$$\frac{P_i}{P_j} = \exp(-\beta[U_i - U_j]) = \frac{T(j \longrightarrow i)}{T(i \longrightarrow j)} \tag{5.16}$$

where the second equality follows from (5.13d).

The algorithms for the Monte Carlo move from one state to another depend on random numbers, either integers or decimal numbers, which are generated by the computer. Actually, a sequence of numbers $N_1, N_2, \ldots, N_\nu$ which are very nearly uncorrelated with each other, for example,

$$n^{-1} \sum_{i=1}^{n} N_i N_{i+j} \simeq 0 \qquad \text{if } j \neq 0$$

and which conform to a desired distribution $W(N)$ can be computed in a deterministic way as described in Abramowitz and Stegun (1965 pp. 949ff.).

In the common Monte Carlo algorithm we begin a move by picking a random number from 1 to N, to indicate which of the N particles to try to move. Then we pick three random numbers, each between $-l/2$ and $l/2$, to see how far to try to move the chosen particle in the x, y, and z Cartesian coordinate directions. Thus the attempted move is from a state i to a new state j which is the same as i except that one particle is displaced to a new location within a cube of size l^3 centered on its location in the i state. It is convenient to introduce the *attempted* move probability

$$T^0(i \longrightarrow j) = \begin{cases} l^{-3} & \text{if } j \text{ is within the } l^3 \text{ cube centered on } i \\ 0 & \text{otherwise} \end{cases} \tag{5.17}$$

Next we have to calculate the energy U_j of the tentative new state:

$$U_j = U_i + \sum_{b \neq a} [u_{ba}(j) - u_{ba}(i)] \tag{5.18}$$

where a is the particle that is being moved, while b ranges over the $N - 1$ other particles. This is the costly part of the computation.

Now we can compute P_j and compare it with P_i. We accept the attempted move with probability W_{ij}:

$$W_{ij} = \text{Min}\left(1, \frac{P_j}{P_i}\right) \tag{5.19}$$

where Min(x, y) means "the smaller of x and y." Thus if $P_j > P_i$, the attempted move from i to j is accepted and state j is added to the growing end of the Markoff chain. Otherwise, P_j/P_i is in the range from 0 to 1. Then we generate a new random number in the range 0 to 1. If it is greater than P_j/P_i, the move is rejected and we add state i to the growing end of the Markov chain (i.e., we count it as a new state but with the original configuration). [In this procedure we generate $T(i \longrightarrow i)$ as specified by Eq. (5.13b).] If the random number is less than P_j/P_i, we accept the attempted move to state j and it is added to the Markov chain. In either case we are ready to begin the next move after recording various state properties that may be needed for calculating ensemble averages.

An important quantity in Monte Carlo work is the chain average $\langle W \rangle$ of the acceptance probability specified in Eq. (5.19). It can be varied by changing l. In dense systems, values of l that exceed a molecular diameter result in very small values of $\langle W \rangle$. Small $\langle W \rangle$ implies an inefficient simulation since most of the computing cost is in the calculation of $U_j - U_i$; it is required whether the attempted move ultimately is accepted or not. If l is chosen to be small compared to the average distance that a particle can move before colliding strongly with another, we may have $\langle W \rangle$ near unity, so very little computing is wasted on rejected attempts to move. However, the Markov chain samples the configuration space more and more slowly as l is reduced. This extreme is like choosing too small a value of $\hat{t}$ in an MD simulation. Most often, l is adjusted so that $\langle W \rangle \simeq 0.5$, although it is rather tedious to establish, in a given MC simulation, whether this is the optimum.

$\langle W \rangle$ is also sensitive to the number of particles ν_p we attempt to move in each step. Although we assumed that $\nu_p = 1$, the theory allows any $\nu_p \leq N$. However, it does not seem advantageous to increase ν_p when doing so implies reducing l to maintain $\langle W \rangle \sim 0.5$.

MC is quite similar to MD with respect to procedures for initiating a simulation and then following it for a while, reaching stationary values of some of the ensemble average quantities, such as the potential energy, and then beginning the accumulation of data for estimating ensemble averages.

Since we move only one particle at a time in MC, the number of moves required to attain equilibrium and to get ensemble averages with small noise is expected to be larger than in MD. So we may define a "pass" in an MC simulation as a sequence of N moves, in which we will, on the average, move every one of the N particles once. Thus the computing expense of an MD move is about the same as for an MC pass. But whereas for the typical $N \simeq 200$ system about 1000 moves may be satisfactory for MD ensemble averages, the corresponding MC simulation might well require 5000 or 10,000 passes to achieve the same quality. In some ways MD is more difficult than MC because of the disadvantages of working in the microcanonical ensemble and calculating the forces, and for more general systems, the torques. However, the choice between MC and MD in those cases in which we do not seek dynamical information is often not so clear until we compare them using the Hamiltonian model of interest.

It should also be mentioned that in some cases useful dynamical information can be extracted from MC simulations. This can happen when the system undergoes a process that is slow compared with the diffusive motion of a single particle and we can identify the ratio of these two rates in the MC simulation.

Finally, we note that the accumulation of ensemble averages is the same in MC as described for MD, except for the details affected by the differences in ensembles.

5.5 DIFFRACTION

Diffraction of radiation by a system is due to inhomogeneities. For diffraction of radiation with wavelength in the angstrom range, the relevant inhomogeneities have molecular scale, so these phenomena can be used to measure quantities closely related to $g_{ss'}(r)$, the spatial pair correlation function. Because the incident radiation most simply comprises a plane wave which may be represented as

$$e^{i(\mathbf{k}\cdot\mathbf{r}-\omega t)}$$

where $\mathbf{r}$ is spatial location, t is time, and $\mathbf{k}$ the "wave vector" conjugate to $\mathbf{r}$, it is not surprising that the theory of diffraction is conveniently expressed in terms of Fourier transforms of spatial functions, for example,

$$\tilde{f}(\mathbf{k}) \equiv \int f(\mathbf{r}) e^{i\mathbf{k}\cdot\mathbf{r}}\, d^3r \tag{5.20}$$

which defines the Fourier transformation $\tilde{f}(\mathbf{k})$ of the general function $f(\mathbf{r})$. Reference should be made to Appendix 1 for some of the important characteristics of Fourier transforms.

Beginning, as in Eq. (4.12), with the local one-particle density

$$\rho_s(\mathbf{r}) = \langle \hat{\rho}_s(\mathbf{r}) \rangle$$

we get the corresponding density in waves by Fourier transformation.

$$\tilde{\rho}_s(\mathbf{k}) = \langle \tilde{\hat{\rho}}_s(\mathbf{k}) \rangle \tag{5.21}$$

$$\tilde{\hat{\rho}}_s(\mathbf{k}) = \sum_{j=1_s}^{N_s} e^{-i\mathbf{k}\cdot\mathbf{r}_j} \tag{5.22}$$

For the pair correlation function defined in Eq. (4.20a),

$$g_{ss'}(\mathbf{R}' - \mathbf{R}) = V^2 \langle \delta(\mathbf{r}_{1_s} - \mathbf{R})\, \delta(\mathbf{r}_{2_{s'}} - \mathbf{R}') \rangle$$

the Fourier transform is

$$\int g_{ss'}(\mathbf{R}' - \mathbf{R}) e^{i\mathbf{k}\cdot(\mathbf{R}'-\mathbf{R})}\, d^3R\, d^3R' \equiv \tilde{g}_{ss'}(\mathbf{k}) = V \langle e^{i\mathbf{k}\cdot\mathbf{r}_{1s}} e^{-i\mathbf{k}\cdot\mathbf{r}_{2s'}} \rangle \tag{5.23}$$

This is not as nice as it looks because $\tilde{g}_{ss'}(\mathbf{k})$ "contains" a Dirac delta function. Thus $g_{ss'}(\mathbf{r})$ is very nearly equal to unity except for a small region in which r is less than several molecular diameters. Since we have [Eq. (3.18)]

$$\int_{-\infty}^{\infty} dx\, e^{ixk} = 2\pi\, \delta(k)$$

it follows that a three-dimensional Dirac delta function is generated by

$$\begin{aligned}\int d^3r\, e^{i\mathbf{k}\cdot\mathbf{r}} &= 2\pi\, \delta(k_x) \times 2\pi\, \delta(k_y) \times 2\pi\, \delta(k_z) \\ &= (2\pi)^3\, \delta(\mathbf{k})\end{aligned} \tag{5.24}$$

so we introduce the new spatial correlation function

$$h_{ss'}(r) \equiv g_{ss'}(r) - 1 \tag{5.25}$$

which is very nearly zero except in a very small region in which r is less than a few molecular diameters. (Although $h_{ss'}$ and $g_{ss'}$ are both very important in the statistical mechanics of fluids, there is no widely accepted nomenclature to distinguish between them except for the use of the symbols g and h.) It follows that $\tilde{h}_{ss'}(k)$, unlike $\tilde{g}_{ss'}(k)$, is a definite finite number for every k. Now we rewrite Eq. (5.23) as

$$V\langle e^{i\mathbf{k}\cdot\mathbf{r}_{1s}} e^{-i\mathbf{k}\cdot\mathbf{r}_{2s'}}\rangle = (2\pi)^3\, \delta(\mathbf{k}) + \tilde{h}_{ss'}(k) \tag{5.26}$$

A related quantity that is widely used is the partial structure factor $S_{ss'}$,

$$[N/V]S_{\;s'}(k) = \rho_s \delta_{ss'} + \rho_s \rho_{s'} \tilde{h}_{ss'}(k) \tag{5.27}$$

where N/V is the total density of particles in the system.

If plane-wave radiation with wave vector $\mathbf{k}_0$ bathes (Fig. 5.4) a collection of

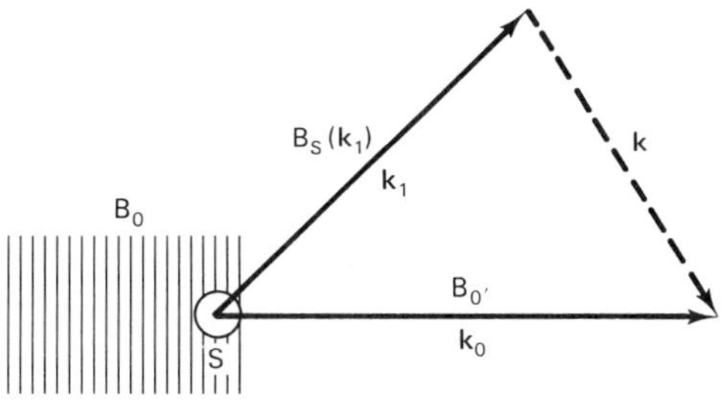

Figure 5.4 Experimental measurement of scattering of radiation by a material sample S. B_0 is the incoming beam of radiation along vector $\mathbf{k}_0$, $\mathrm{B}_{0'}$, the undiffracted beam, again along $\mathbf{k}_0$, and $B_S(\mathbf{k}_1)$ is the part of the scattered beam that lies along $\mathbf{k}_1$. For each beam i the vector momentum of the radiation is $\hbar\mathbf{k}_i$ so $\hbar k$ is the momentum lost (i.e., $|\hbar\mathbf{k}_0 - \hbar\mathbf{k}_1|$) by the beam $B_S(\mathbf{k}_1)$. In a diffraction experiment no energy resolution of $B_S(\mathbf{k}_1)$ is made, while in inelastic scattering experiments (Chapter 13) the energy distribution of $B_S(\mathbf{k}_1)$ is measured. This schematic diagram applies to all sorts of radiation (neutrons, electrons, x-rays or visible light, etc.), but the experimental details and the interpretation of the results do depend on the radiation.

particles (N_s particles of the species s), the wave function of the scattered radiation along wave vector $\mathbf{k}_0 - \mathbf{k}$ is approximately

$$A(\mathbf{k}) = \sum_s f_s(\mathbf{k}) \sum_{j=1_s}^{N_s} e^{-i\mathbf{k}\cdot\mathbf{r}_j} \tag{5.28}$$

where some factors that are only important for a quantitative interpretation of the intensities have been omitted so that we can focus on the part that depends on the scattering material. Here f_s is a measure of the scattering power of the atom. X-rays and electrons are scattered mainly by electrons while neutrons are scattered mainly by nuclei. In the case of x-rays, f_s is a combination of the scattering powers of the electrons associated with the atoms, so it increases roughly in proportion to the atomic number, and it is k dependent in the Å^{-1} range of k because the electrons that scatter the x-rays are found in regions about 1 Å in diameter, the atoms. For neutrons f_s is the coherent scattering length [f_s^{coh} in Eq. (13.48)], an average that depends on the nuclear levels of the scattering atom, so it is isotope dependent and not correlated with atomic number. Also it has no k dependence in the Å^{-1} range because the nucleus is so small.

The average observable intensity per unit volume of scattering material in the incident beam is proportional to

$$\begin{aligned}\frac{\langle A(\mathbf{k})A^*(\mathbf{k})\rangle}{V} &= \frac{1}{V}\sum_s \sum_{s'} f_s(\mathbf{k}) f_{s'}(\mathbf{k}) \sum_{j=1_s}^{N_s} \sum_{j'=1_{s'}}^{N_{s'}} \langle e^{i\mathbf{k}\cdot(\mathbf{r}_j - \mathbf{r}_{j'})}\rangle \\ &= \sum_s \sum_{s'} f_s f_{s'} \{\rho_s \delta_{ss'} + (\rho_s \rho_{s'})[(2\pi)^3 \delta(\mathbf{k}) + \tilde{h}_{ss'}(k)]\}\end{aligned} \tag{5.29}$$

where the A* signifies a complex conjugate, as in the usual quantum-mechanical expression for a probability density. Also, to vary $\mathbf{k}$ at fixed $\mathbf{k}_0$ and with elastic scattering (no energy change) as required here, we change the angle of observation, (i.e., the scattering angle) so that this equation implicitly gives the intensity of scattered radiation as a function of angle from the incident beam. The term in $\delta(\mathbf{k})$, called the forward-scattered beam, has no structural information and is practically negligible for $k > 10^{-4}\ \text{Å}^{-1}$, so it can be neglected in analyzing the experimental data.

The application of Eq. (5.29) to fluids with only one atomic component is straightforward (see Fig. 5.5) with x-rays or neutrons or even with electrons as the radiation. In more complicated systems there are great advantages in using neutrons together with substitution of isotopes in the material to exploit the fact that such substitution changes the f_s factors while not changing the structure significantly, except in the case of substitution of H by D in some systems. With enough substitutions and enough accuracy in the scattering measurements, we can determine all of the $\tilde{h}_{ss'}(k)$ functions. Of course, we should add that the diffraction experiment allows us to determine each $\tilde{h}_{ss'}(k)$ only over some range of k, depending on instrumental details, and that the physics of neutron diffraction limits us to the case in which the species s are all single-atom species. Thus N in Eq. (5.27) is the total number of atoms. Another diffraction method, EXAFS, can also be used to probe the $\tilde{h}_{ss'}(k)$ functions in fluids, but its use is somewhat less developed at this time.

It may be noted that the use of diffraction measurements to determine a spatial

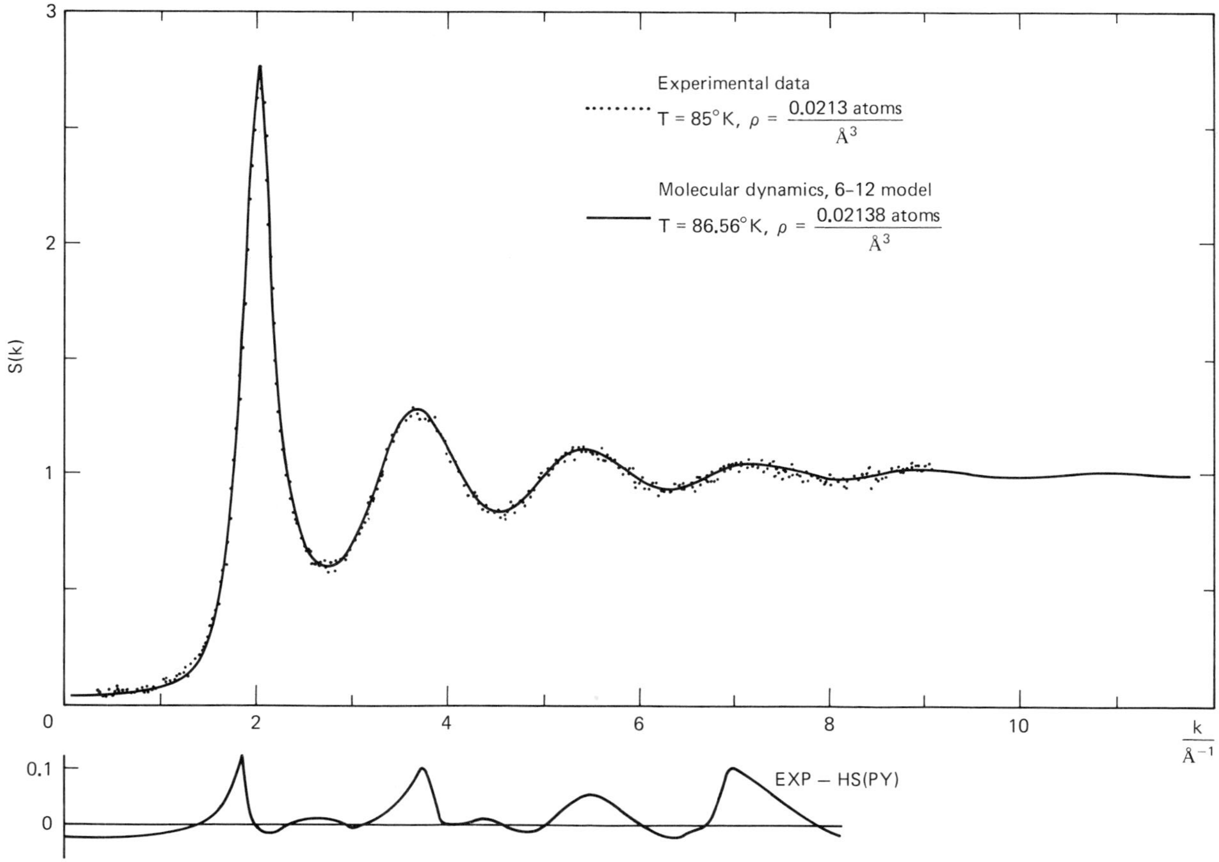

Friedman 5.5

Figure 5.5 Neutron diffraction data for liquid argon compared with the results of model calculations. The parameters of the 6–12 model are given in the caption of Fig. 5.3. The lower figure, which has a 2.4-fold expanded vertical scale compared to the upper, is the difference between $S(k)$ from neutron diffraction and from a calculation, via the PY approximation (Chapter 7), for a hard-sphere model with $\sigma = 3.590$ Å [Eq. (4.24)]. The rms deviation for the hard-sphere model is 0.033, compared to 0.022 for the 6–12, but apparently would be smaller if the exact hard-sphere $g(r)$ had been used, as indicated by some results of Verlet (1968). (Adapted from Yarnell et al., 1973.)

correlation function $g(r)$, even for a one-component system, is somewhat limited because of the difficulty of inverting the $\tilde{h}(k)$ to $h(r)$ when the former is not known accurately over the whole range from $k = 0$ to $k = \infty$. In the case of fluid structure determination a conservative way to proceed by using some of the exactly known conditions on $g(r)$ and $\tilde{h}(k)$ is illustrated by the work of Yarnell et al. (1973) on liquid argon. Even so, reference to Figs. 4.3, 5.3, and 5.5 shows that different Hamiltonians that imply noticeably different $g(r)$ functions may give almost indistinguishable $\tilde{h}(k)$ functions, again showing the stringent requirements for accuracy to determine structural features in fluids by diffraction.

THE LITERATURE

The MD method was developed by Alder and Wainright (1957) for assemblies of hard spheres and was extended to continuous potentials by Rahman (1964). In these early MD papers and those of Verlet (1967, 1968) there is discussion of many details that are not mentioned in more recent work. The MC method was introduced even earlier, by Metropolis et al. (1954). In Berne (1977a,b), there are detailed discussions of MC (Torrie, Valleau, Whittington) and MD (Berne, Erpenbeck, Kushik, Wood). The theory of the microcanonical ensemble averages is given by Lebowitz et al. (1967). See also Lado (1981). The simulation of polar and polarizable fluids has been reviewed by Alder and Pollock (1981). Some important subtleties in the applications of Ewald sums to the systems of polar molecules are investigated by deLeeuw et al. (1980). Water and aqueous solutions are among the more complicated systems for which models have been studied by simulation techniques. See, for example, Stillinger (1975) and Wood (1979). A comparison of the Metropolis MC algorithm described here with two recently proposed modifications is investigated in terms of a model for water by Mehrotra et al. (1983). For the theory of neutron and x-ray scattering, see also Hansen and McDonald (1976), Eglestaff (1967), and Blum and Narten (1976). For an account of XAFS, see Sandstrom and Lytle (1979).

PROBLEMS

5.1. Consider a hard-sphere system in a microcanonical ensemble. How does Eq. (5.3) apply; that is, does the temperature fluctuate?

5.2. From Eq. (5.6) estimate how small a step $\hat{t}$ is needed for MD for a 6–12 potential model system [Eq. (4.25)]. How does your estimate compare with the standard one,

$$\hat{t} \simeq 0.03 \left(\frac{\sigma^2 m}{48 e_0}\right)^{1/2}$$

which is about 0.01 ps in the case of Ar? (see Table 3.1.)

5.3. An investigator doing a molecular dynamics simulation fails to notice that his system, after "equilibration," has a center-of-mass velocity of 10^4 m/s. He is simulating KCl, near 500 K, with 200 ions of each sign. If he calculates T by the usual formula, how many degrees will he be off?

5.4. Write Verlet's algorithm [Eq. (5.5) truncated after the $\hat{t}^2$ term] in a form which shows that it applies to a three-dimensional system of, say, 200 particles.

5.5. Better results from MD simulations using a machine of finite percision may be expected if the Verlet algorithm is employed in the "summed form" (Swope et al., 1982):

$$q_a(t+\hat{t}) = q_a(t) + \Delta q_a(t) \tag{5.30}$$

$$\Delta q_a(t) = \Delta q_a(t-\hat{t}) + \frac{\hat{t}^2}{m} F_a(t) \tag{5.31}$$

where we define

$$\Delta q_a(t) = q_a(t+\hat{t}) - q_a(t) \tag{5.32}$$

(a) Show that introducing $\Delta q_a(t)$ into Verlet's algorithm gives Eq. (5.31).

(b) Show that in terms of the velocity defined in Eq. (5.7) the summed form of Verlet's algorithm is

$$q_a(t+\hat{t}) = q_a(t) + \hat{t}\dot{q}_a(t) + \frac{\hat{t}^2}{2m_a} F_a(t) \tag{5.33}$$

$$\dot{q}_a(t+\hat{t}) = \dot{q}_a(t) + \frac{\hat{t}}{2m_a} [F_a(t+\hat{t}) + F_a(t)] \tag{5.34}$$

5.6. Consider a one-dimensional system of particles interacting with the pair potential

$$u(x) = 4k_B T_0 \left[\left(\frac{x_0}{x}\right)^{12} - \left(\frac{x_0}{x}\right)^{6}\right]$$

where x is the distance between centers and T_0 and x_0 are parameters.

(a) Write a Fortran program to generate an MD simulation of the system, including the pair correlation function $g(x)$ defined so that $\rho g(x)$ is the local line concentration of particles at a distance x from a given particle if ρ is the average number of particles per unit length. Write the program for periodic boundary conditions along the line, with N_c particles in the unit cell.

(b) Write a program for the corresponding MC simulation.

(c) Run both programs with the data $\rho = \frac{3}{4}x_0$, $T = T_0$, and $N_c = 10$.

(d) Which simulation is cheapest as measured by the cost to fix the height of the first peak in $g(x)$ with an accuracy of 10%?

5.7. Assume that an atomic species s is available in two isotopes s_1 and s_2 and that the neutron scattering lengths are related by $f_{s_1} = -2f_{s_2}$. To what form does Eq. (5.29) simplify in an experiment in which the composition is adjusted so that $N_{s_2} = 2N_{s_1}$?

6

Cluster Expansions

6.1 THE VIRIAL EQUATION OF STATE

The equation of state of a fluid specifies how the pressure of the fluid depends on density (i.e., concentration) and temperature. The infinite series

$$\beta p = \rho + B_2\rho^2 + B_3\rho^3 + \cdots \tag{6.1}$$

written here for a one-component fluid, is called the *virial equation of state.* Two questions arise at once: Is the series convergent? How are the virial coefficients B_n related to the Hamiltonian? In this chapter we describe a systematic way to answer the second question. The first is much more difficult unless the answer is quite obviously "no" as it is in some important cases.

We also may make use of virial expansions of other quantities than the pressure. One such expansion,

$$\frac{-\beta A^{\text{ex}}}{V} = B_2^A\rho^2 + B_3^A\rho^3 + \cdots \tag{6.2}$$

is especially important. Here $A^{\text{ex}} = A - A^{\text{id}}$, the excess over the free energy of an ideal gas in the same N, V, T state, is usually called the *excess* or *configurational free energy* [see Eqs. (3.54) and (4.2)]. It is found that the coefficients B_n^A in Eq. (6.2) are related to the virial coefficients B_n by the equation

$$B_n = -(n-1)B_n^A \tag{6.3}$$

It may be noticed that these equations too have been written for a one-component system, but they are readily generalized.

The virial expansion of A^{ex}, Eq. (6.2), is especially important because other thermodynamics excess functions are derivatives of A^{ex} and, as we shall see, some structural functions are functional derivatives of A^{ex}.

6.2 THE SECOND VIRIAL COEFFICIENT, B_2

The expression for B_2 as a functional of the Hamiltonian is easily found without resorting to systematic methods. For example, we begin with Eq. (4.41):

$$1+\rho\int[g(r)-1]\,d^3r=\frac{\rho\kappa_T}{\beta} \tag{6.4}$$

and substitute the compressibility derived from Eq. (6.1) to get

$$1+\rho\int[g(r)-1]\,d^3r=(1+2B_2\rho+3B_3\rho^2+\cdots)^{-1} \tag{6.5}$$

In the limit $\rho = 0$ we have, in view of Eq. (4.44),

$$B_2=-B_2^A=-\tfrac{1}{2}\int[e^{-\beta u(r)}-1]\,d^3r \tag{6.6}$$

The derivation is similar for a mixture, but since the theory given later in this chapter is quite general, we merely cite the result here: For a mixture [see Eq. (6.75)]

$$\frac{-\beta A^{\text{ex}}}{V}=\sum_s\sum_{s'}\left(B^A_{ss'}\rho_s\rho_{s'}+\sum_{s''}B^A_{ss's''}\rho_s\rho_{s'}\rho_{s''}+\cdots\right) \tag{6.7}$$

with

$$B^A_{ss'}=\tfrac{1}{2}\int\{[\exp(-\beta u_{ss'}(r))]-1\}\,d^3r \tag{6.8}$$

It is interesting to compare these results for the leading effects of intermolecular forces on the thermodynamics of the gas with the corresponding results for a chemical association model. Thus suppose that we have the chemical equilibrium

$$2X=X_2,\qquad K=\frac{\rho_2}{\rho_1^2} \tag{6.9a}$$

where ρ_1 is the concentration N_1/V of monomers and $\rho_2 = N_2/V$ is the dimer concentration. In a chemical model we neglect *inter*molecular forces, so we write

$$\beta p=\rho_1+\rho_2 \qquad \text{(Dalton's law)} \tag{6.9b}$$

and we find, with the stoichiometric concentration of X given by $\rho = \rho_1 + 2\rho_2$,

$$\beta p=\rho-K\rho^2+O(\rho^3) \tag{6.9c}$$

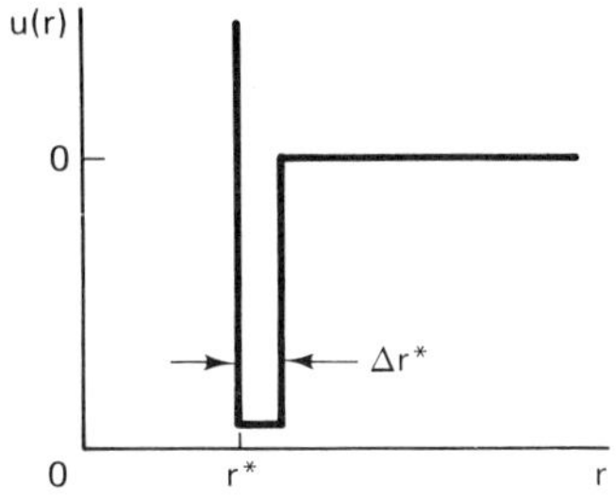

Figure 6.1 Pair potential employed in Eq. (6.11).

which may be compared with Eq. (6.1) to find

$$K = -B_2 = B_2^A \tag{6.10}$$

When the chemical model is apt, the pair potential $u(r)$ has a sharp and deep minimum, say at $r = r^\star$ and with width $\Delta r^\star$ near the bottom (Fig. 6.1) [see Eq. (6.13)]. Then from (6.6) and (6.10) we find

$$K \simeq \tfrac{1}{2}(4\pi r^{\star 2}\, \Delta r^\star) \exp\left[-\beta u(r^\star)\right] - \frac{2\pi}{3} r^{\star 3} \tag{6.11}$$

which is, except for the negative term, the same as one deduces from thermodynamics if $u(r^\star) - k_BT \ln (2\pi r^{\star 2}\, \Delta r^\star \rho_0)$ is identified as the standard free-energy change for the process in Eq. (6.9a). (Here ρ_0 is the reference concentration for the standard free-energy change.)

For a hard-sphere model [Eq. (4.24)] we find that

$$B_2 = \frac{2\pi\sigma^3}{3} \tag{6.12}$$

with no temperature dependence. For any realistic model B_2 can be evaluated only by numerical integration, but the square-well model

$$u(r) = \begin{cases} \infty & \text{if } r < \sigma \\ h & \text{if } \sigma < r < b\sigma \\ 0 & \text{if } b\sigma < r \end{cases} \tag{6.13}$$

may be used to crudely represent any potential with a strongly repulsive core and (if $h < 0$ and $b > 1$) an attractive tail, like the 6–12 potential [Eq. (4.25)]. In this case the second virial coefficient has a temperature dependence similar to that shown in Fig. 6.2. The negative values below the *Boyle temperature* show that at lower temperatures the attractive forces dominate B_2, leading to condensation as we know. The experimental data for B_2 of the rare gases, N_2, CH_4, and many other simple molecules can be fit to the corresponding graph for $B_2(T)$ for the 6–12 model, an aspect of the principle of corresponding states (Fig. 6.2).

DO PROBLEMS 6.1 TO 6.4.

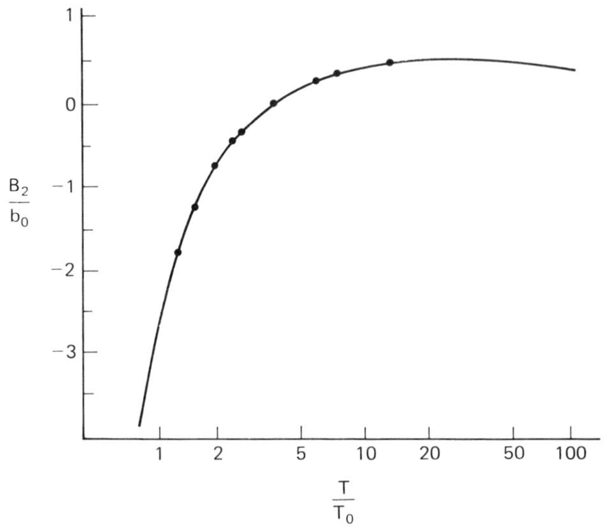

Figure 6.2 Reduced second virial coefficient B_2/b_0 as a function of reduced temperature T/T_0. The line is calculated for the 6–12 potential in which case $b_0 = \frac{2}{3}\pi\sigma^3$ and $T_0 = \epsilon/k_B$. A similar curve is given if one uses the model in Eq. (6.13). The points are for argon, in which case b_0 and T_0 are given in Table 3.1. The fit is equally good for CH_4, N_2, Ne, and other rare gases except He. For He and H_2 the fit is equally good only after correction for quantal effects. (From Hirschfelder et al., 1954, p. 164.)

6.3 FUNCTIONAL DERIVATIVES

Here we introduce a mathematical tool which is especially useful for relating the theory of thermodynamic coefficients to the theory of the spatial correlation functions [e.g., for obtaining the density expansion of $g_{ss'}(r)$ from Eq. (6.2)]. Although functional differentiation is straightforward in the ordinary mathematical notation in which we present it first, it becomes extremely simple in the graph theory notation that we come to next.

We notice that Eq. (6.6) expresses B_2 as a function of β, and as a "functional" of $u(r)$. That is, B_2 depends on the numerical value of β, while it depends on the numerical value of $u(r)$ at every r. Here we will use the notation $B_2[u(r)]$ or $B_2[u]$ to say that B_2 is a function*al* of u.

More generally, we consider $p[q(x)]$, p as some functional of an arbitrary function $q(x)$, and consider that a small variation δq in the function q gives a corresponding variation δp in the functional:

$$\delta p = \int_a^b \frac{\delta p}{\delta q(x)} \delta q(x)\, dx \tag{6.14}$$

where $a \leqslant x \leqslant b$ is the range of x in which p depends on $q(x)$. For example, in Eq. (6.6), B_2 depends on u in $0 \leqslant r \leqslant \infty$. The quantity $\delta p/\delta q$ is called the *functional derivative.* Notice that it has the same units as p/qx.

An example, a somewhat generalized case that is typical for our purpose, is one in which we have

$$p = \int Q(\mathbf{r}_1, \ldots, \mathbf{r}_n)\, d^3r_1 \cdots d^3r_n \tag{6.15}$$

with the abbreviation

$$Q = q(\mathbf{r}_1)q(\mathbf{r}_2) \cdots q(r_n)F(\mathbf{r}_1 \cdots \mathbf{r}_n) \tag{6.16}$$

where F does not depend on q. Then the variation in p due to a small variation in q has the form[1]

$$\begin{aligned} p + \delta p &= \int [q(\mathbf{r}_1) + \delta q(\mathbf{r}_1)] \cdots [q(\mathbf{r}_n) + \delta q(\mathbf{r}_n)] F \, d^3r_1 \cdots d^3r_n \\ &= p + \sum_{j=1}^{n} \int \frac{Q}{q(\mathbf{r}_j)} \delta q(\mathbf{r}_j) \, d^3r_1 \cdots d^3r_n + O([\delta q]^2) \end{aligned} \tag{6.17}$$

where, as in the theory of ordinary derivatives, we neglect the terms that are more than first order in the differentials. We can rewrite this equation in a form that looks more like (6.14), namely

$$\delta p = \int \left[\sum_{j=1}^{n} \int \frac{Q}{q(\mathbf{r}_j)} \delta(\mathbf{r}_j - \mathbf{R}) \, d^3r_1 \cdots d^3r_n \right] \delta q(\mathbf{R}) \, d^3R \tag{6.18}$$

It may be verified that (6.18) reduces to (6.17) by doing the integral over $\mathbf{R}$ before the others. Comparing (6.18) with (6.14) shows that we have

$$\begin{aligned} \frac{\delta p}{\delta q(\mathbf{R})} &= \sum_{j=1}^{n} \int \frac{Q}{q(\mathbf{r}_j)} \delta(\mathbf{r}_j - \mathbf{R}) \, d^3r_1 \cdots d^3r_n \\ &= n \int \frac{Q}{q(\mathbf{r}_1)} \delta(\mathbf{r}_1 - \mathbf{R}) \, d^3r_1 \cdots d^3r_n \end{aligned} \tag{6.19}$$

where the last step follows because all n terms in the sum give the same integral. We also notice that we have

$$\frac{\delta p}{\delta q(\mathbf{R})} = \frac{n}{q(\mathbf{R})} \int Q(\mathbf{R}, \mathbf{r}_2, \mathbf{r}_3, \ldots, \mathbf{r}_n) \, d^3\mathbf{r}_2 \cdots d^3\mathbf{r}_n \tag{6.20}$$

which shows explicitly how functional differentiation has the effect of undoing integration; compare Eq. (6.20) with (6.15).

The result in Eq. (6.19), which follows from Eq. (6.16), becomes, when Q is a general functional of the q functions,

$$\frac{\delta p}{\delta q(\mathbf{R})} = \int \sum_j \left[\frac{\delta Q}{\delta q(\mathbf{r}_j)} \right] \left[\frac{\delta q(\mathbf{r}_j)}{\delta q(\mathbf{R})} \right] d^3r_1 \cdots d^3r_n \tag{6.21}$$

where the first bracket is evaluated simply as an ordinary partial derivative and the second bracket is

$$\frac{\delta q(\mathbf{r}_j)}{\delta q(\mathbf{R})} = \delta(\mathbf{r}_j - \mathbf{R}) \tag{6.22}$$

[1] See footnote to Eq. 3.37.

6.4 FUNCTIONAL DERIVATIVE FORMULAS FOR PAIR CORRELATION FUNCTIONS

We may extend the external field results that were obtained in Section 4.2 as follows. We write

$$U_N^\star(1, \ldots, N) = U_N(1, \ldots, N) + \sum_{i=1}^{N} \phi(\mathbf{r}_i) \tag{6.23}$$

where $\phi(\mathbf{r})$ is the potential of interaction of the external field with a particle at $\mathbf{r}$. In this field the grand partition function is

$$\begin{aligned} \Xi^\star &= \sum_{N=0}^{\infty} \frac{z^N}{N!} \int e^{-\beta U_N^\star (1,\ldots,N)}\, d(1) \cdots d(N) \\ &= \sum_{N=0}^{\infty} \frac{1}{N!} \int e^{-\beta U_N(1,\ldots,N)} z^\star(1) z^\star(2) \cdots z^\star(N) d(1) \cdots d(N) \end{aligned} \tag{6.24}$$

where we introduce the *local activity* by the definition

$$z^\star(i) \equiv z^\star(\mathbf{r}_i) \equiv z e^{-\beta\phi(\mathbf{r}_i)} \tag{6.25}$$

Also in these equations we use $U_N(1, \ldots, N)$ as an abbreviation for $U_N(\mathbf{r}_1, \ldots, \mathbf{r}_N)$ and $d(1) \cdots d(N)$ for $d^3r_1 \cdots d^3r_N$.

Now we evaluate

$$\frac{\delta \Xi^\star}{\delta z^\star(1')} = \sum_{N=0}^{\infty} \frac{N}{N!} \int e^{-\beta U_N} z^\star(2) \cdots z^\star(N) \delta(1-1') d(1) \cdots d(N) \tag{6.26}$$

which is valid at any external field ϕ. Here $\delta(1-1')$ is the same as $\delta(\mathbf{r}_1 - \mathbf{r}_1')$. In view of the equation for N in Table 1.5 it is interesting to look at

$$\begin{aligned} &\frac{z^\star(1')}{\Xi^\star} \frac{\delta \Xi^\star}{\delta z^\star(1')} \\ &\quad = \sum_{N=0}^{\infty} \frac{N}{N!\Xi^\star} \int e^{-\beta U_N} z^\star(1) \cdots z^\star(N) \delta(1-1') d(1) \cdots d(N) \\ &\quad \equiv \rho^\star(1') \end{aligned} \tag{6.27}$$

In the limit of vanishing external field this becomes

$$\frac{z}{\Xi} \left(\frac{\delta \Xi^\star}{\delta z^\star(1')} \right)_{\phi=0} = \rho = \frac{N}{V} \tag{6.28}$$

The appearance of the $1/V$ factor in the $\phi = 0$ limit in the sum in Eq. (6.27) is related to the appearance of this same factor in Eqs. (4.15) and (4.18), where it has been discussed. Proceeding in the same way we find that

$$\begin{aligned} &\frac{\delta^2 \Xi^\star}{\delta z^\star(1') \delta z^\star(2')} \\ &\quad = \sum_{N=0}^{\infty} \frac{N(N-1)}{N!} \int e^{-\beta U_N} z^\star(3) \cdots z^\star(N) \delta(1-1') \delta(2-2') d(1) \cdots d(N) \end{aligned}$$

and in the $\phi = 0$ limit

$$\left(\frac{\delta^2 \Xi^\star}{\delta z^\star(1')\delta z^\star(2')}\right)_{\phi=0} = \sum_{N=0}^{\infty} \frac{z^{N-2}N(N-1)}{N!} \int e^{-\beta U_N(1',2',3,4,\ldots,N)} d(3) \cdots d(N) \tag{6.29}$$

Comparing this result with Eq. (4.23) (specialized to one component) shows that we have

$$\rho^2 g(1'\ 2') = \frac{z^2}{\Xi}\left(\frac{\delta^2 \Xi^\star}{\delta z^\star(1')\delta z^\star(2')}\right)_{\phi=0} \tag{6.30}$$

We have derived these results for a one-component system to keep the notation as simple as possible. More generally, we have

$$\rho_s = \frac{z_s}{\Xi}\left(\frac{\delta \Xi^\star}{\delta z_s^\star(1')}\right)_{\phi=0} \tag{6.31}$$

$$\rho_s \rho_{s'} g_{ss'}(1'\ 2') = \frac{z_s z_{s'}}{\Xi}\left(\frac{\delta^2 \Xi^\star}{\delta z_s^\star(1')\delta z_{s'}^\star(2')}\right)_{\phi=0} \tag{6.32}$$

We also may express the pair correlation function as a functional derivative of A^{ex}. In this case we do not apply an external field. To begin we rewrite Eq. (3.27),

$$Z(N, V, T) = \int_V W_N(1, \cdots, N) d(1) \cdots d(N) \tag{6.33}$$

where we define the Boltzmann factor

$$W_N(1, \ldots, N) = \exp[-\beta U_N(1, \ldots, N)] = \prod_{\text{pairs}} e(i\ j) \tag{6.34}$$

where we have assumed that U_N is pairwise additive and introduced the function

$$e(i\ j) = e^{-\beta u(i\ j)} \tag{6.35}$$

In the product in Eq. (6.34) there are $N(N-1)/2$ factors, one for each pair of the N particles. Now we find

$$Z + \delta Z = \int \Pi[e(i\ j) + \delta e(i\ j)] d(1) \cdots d(N) = Z + \frac{N(N-1)}{2} \int \frac{W_N}{e(1\ 2)} \delta e(1\ 2) d(1) \cdots d(N) + O((\delta e)^2) \tag{6.36}$$

and the functional derivative can be identified as

$$\frac{\delta Z}{\delta e(1'\ 2')} = \frac{N(N-1)}{2e(1'\ 2')} \int W_N(1, \ldots, N) \delta(1-1')\delta(2-2') d(1) \cdots d(N) \tag{6.37}$$

where the Dirac delta functions have been inserted because in (6.36) the locations 1 and 2 in W_N match those in $e(1\ 2)$, while in Eq. (6.37) these locations have been labeled 1′ and 2′ for clarity. Comparing (6.37) with Eq. (4.20), which for a one-component system reads

$$g(1'\ 2') = V^2 \langle \delta(1-1')\delta(2-2') \rangle$$

we find, for $N \gg 1$,

$$Z^{-1} \frac{\delta Z}{\delta e(1'\ 2')} = \frac{\rho^2}{2e(1'\ 2')} g(1'\ 2') \tag{6.38}$$

The left side of this equation can also be written $\delta(-\beta A^{\text{ex}})/\delta e(1\ 2)$ [see Eq. (4.2)]. The multicomponent form is as easily derived:

$$g_{ss'}(1'\ 2') = \frac{2e_{ss'}(1'\ 2')}{\rho_s \rho_{s'}} \frac{\delta(-\beta A^{\text{ex}})}{\delta e_{ss'}(1'\ 2')} \tag{6.39}$$

This is a canonical ensemble version of Eq. (6.32). Another analogy to the latter, involving A^{ex} in an external field, will be developed later [Eq. (6.86)].

DO PROBLEM 6.5.

6.5 GRAPHS, DIAGRAMS, CLUSTER EXPANSIONS

Certain statistical mechanical functions can be expressed in terms of multicenter integrals which may be associated with graphs or diagrams consisting of circles, lines, and perhaps other features.

As an example, consider the cluster function (also called the Mayer f function)

$$f(1\ 2) \equiv e(1\ 2) - 1 \equiv e^{-\beta u(1\ 2)} - 1 \tag{6.40}$$

which is introduced in Eq. (6.24) to get

$$\begin{aligned} \Xi^\star = \sum_{N=0}^{\infty} \frac{1}{N!} \int \Big\{ 1 &+ \sum_{\text{pairs}} f(i\ j) \\ &+ \sum_{\text{triples}} [f(i\ j)f(j\ k) + f(i\ j)f(j\ k)f(k\ i)] \\ &+ \cdots \Big\} z^\star(1) \cdots z^\star(N)\, d(1) \cdots d(N) \end{aligned} \tag{6.41}$$

The terms on the right may be arranged as follows:

$N = 0$:

$$\text{term} = 1 \tag{6.42a}$$

$N = 1$:

$$\text{term} = \bullet \equiv \int z^\star(1) d(1) \tag{6.42b}$$

$N = 2$:

terms =  (6.42c)

 $\equiv \frac{1}{2!}\int z^\star(1)d(1) \times \int z^\star(2)d(2) = \frac{1}{2!}\left[\int z^\star(1)d(1)\right]^2$ (6.42d)

 $\equiv \frac{1}{2!}\int f(1\ 2)z^\star(1)z^\star(2)d(1)d(2)$ (6.42e)

$N = 3$:

terms =  (6.42f)

 $\equiv \frac{1}{3!}\left[\int z^\star(1)d(1)\right]^3$ (6.42g)

 $\equiv \frac{3}{3!}\int z^\star(1)d(1)\int f(2\ 3)z^\star(2)z^\star(3)\ d(2)d(3)$ (6.42h)

 $\equiv \frac{3}{3!}\int f(1\ 2)f(2\ 3)z^\star(1)z^\star(2)z^\star(3)\ d(1)d(2)d(3)$ (6.42i)

 $\equiv \frac{1}{3!}\int f(1\ 2)f(2\ 3)f(3\ 1)z^\star(1)z^\star(2)z^\star(3)\ d(1)d(2)d(3)$ (6.42j)

A transparent shorthand way to represent these eight terms for $N = 3$ is

terms =  + "3"  + "3"  +  (6.42k)

We use this shorthand to represent the 64 terms of $N = 4$.

$N = 4$:

terms =  + "6"  + "3"  + "12"  + "4"  + "12"  + "4" 
+ "12"  + "3"  + "6" 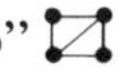 +  (6.42l)

"6"  $\equiv \frac{6}{4!}\left[\int z^\star(1)d(1)\right]^2\int f(3\ 4)z^\star(3)z^\star(4)d(3)d(4)$ (6.42m)

"3"  $\equiv \frac{3}{4!}\int f(1\ 2)f(2\ 3)f(3\ 4)f(4\ 1)z^\star(1)z^\star(2)z^\star(3)z^\star(4)\,d(1)d(2)d(3)d(4)$

(6.42n)

and so on.

Thus each cluster function $f(i\ j)$ is associated with a line or "bond" in the graph and each particle i is associated with a circle whose "weight" is $z^\star(i)$ in these examples. A black circle for particle i means that $\int_V d(i)$ appears in the equation for the value of the graph. A white circle represents a particle whose coordinates are not integrated, as in the following graph with an f bond on a white $z^\star(1)$ circle and a black $z^\star$ circle.

$$= z^\star(1)\int f(1\ 2)z^\star(2)d(2)$$

Graphs with white circles will be encountered shortly, so they are included in the following general remarks.

For each graph in Eq. (6.42) the number in quotation marks, call it σ, has an obvious significance in terms of the way these graphs are generated by Eq. (6.41). For some graphs σ is troublesome to determine and it is fortunate that the "symmetry number" $S = N!/\sigma$ is much easier to calculate in general.

As we can verify from the examples given and those that follow, the symmetry number of a graph with m black circles is the number of different ways that m labels can be assigned to the black circles without affecting the *topology,* the pattern of connections, of the labeled circles. For example, consider the graph of f bonds on $z^\star$ circles,

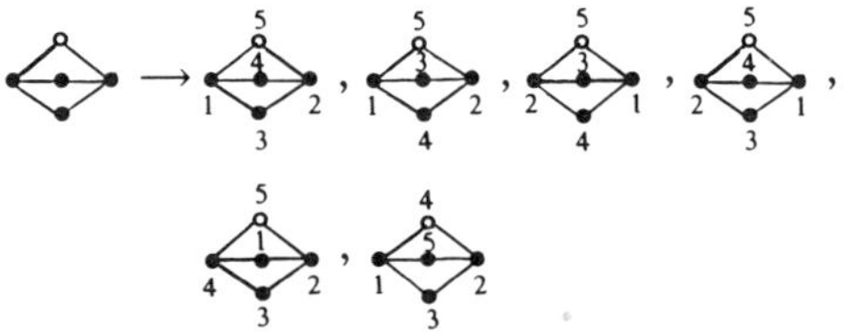

All of the labeled graphs in the first line imply the integral

$$\int f(1\ 4)f(4\ 2)f(1\ 3)f(3\ 2)f(1\ 5)f(2\ 5)$$
$$\times z^\star(1)z^\star(2)z^\star(3)z^\star(4)z^\star(5)d(1)d(2)d(3)d(4)$$

while those in the second line and any others that are obtained by permuting the labels correspond to different integrals. In this case $S = 4$. [In these considerations we must notice the symmetry $f(1\ 2) = f(2\ 1)$ implied by the definitions.]

Now we make the following definition of the value of a graph, which applies to any graph with only pairwise bonds, as in the examples above: In any statistical-mechanical expression a graph has a numerical value given by the equation

$$\text{value(graph)} = \frac{1}{S\boldsymbol{\nu}!} \times [\text{integration and summation implied by graph}] \tag{6.43}$$

where S is the symmetry number of the graph and $\boldsymbol{\nu}! \equiv \nu_1!\ \nu_2! \cdots \nu_i! \cdots$, where ν_i is the number of identical bonds connecting the ith pair of circles of the diagram. For examples of how this works, see Eq. (6.42) and the following additional examples. Besides graphs with white circles, this definition, with one small change, also covers nonsimple graphs (i.e., with multiple bonds) and graphs for multicomponent systems which will be encountered later. Additional examples; f bonds on z circles:

 $$= \frac{1}{2}\int f(1\ 4) f(1\ 3) f(3\ 2) f(4\ 2) d(3) d(4) \cdot z^4, \qquad S = 2 \tag{6.44a}$$

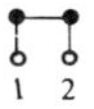 $$= \int f(1\ 3) f(3\ 4) f(4\ 2) d(3) d(4) \cdot z^4, \qquad S = 1 \tag{6.44b}$$

 $$= \frac{1}{3!}\int f(1\ 2) f(1\ 3) f(3\ 2)^3 d(3) \cdot z^3, \qquad S = 1 \tag{6.44c}$$

 $$= \frac{1}{2}\int f(1\ 2) d(2) f(2\ 3) f(1\ 3) f(3\ 4) f(3\ 5) d(3) d(4) d(5) \cdot z^5, \qquad S = 2 \tag{6.44d}$$

We may now write Eq. (6.41) in the language of graph theory:

$$\Xi^\star(z, \beta, V) = 1 + \text{sum of all simple graphs of one or more black } z^\star \text{ circles and zero or more } f \text{ bonds} \tag{6.45}$$

Equation (6.45) is written in the usual language of graph theory in which any mathematical operations on a graph means the mathematical operation on the *value* of the graph. In this case the sum of the graphs means the sum of their values.

This expansion has a very annoying characteristic, namely that for any given N the largest term, the one with no bonds, grows with V like V^N, while the fully connected graph grows only like V, and the other graphs are in between. It is a serious problem because we expect that bulk properties are obtained from a z, V, T ensemble in the thermodynamic limit: $V \longrightarrow \infty$ at fixed z, and we are particularly interested in how those properties are controlled by the interparticle forces which, in Eq. (6.45), are manifest in the f bonds. The medicine for this disease is *topological reduction.* It is convenient to prepare for topological reduction by introducing some definitions and theorems related to graphs.

DO PROBLEMS 6.6 TO 6.8.

6.6 GRAPH THEORY

J. E. Mayer (1942) used linear graphs as notational abbreviations for the integrals in expansions of the configurational integral and distribution functions for various model systems. The parallel developments in the use of graphs in combinatorial theory indicated how the use of graphs could also avoid the combinatorial problems involved in treating many-body statistical mechanical systems. Morita and Hiroike (1961) and DeDominicis (1962) independently developed a graphical notation for expressing entire integrals such that the usual combinatorial difficulties do not appear when the graphs are manipulated in prescribed ways. Here we exploit these advances with the help of Stell's (1967) notation for the graphs. For the sake of completeness in this section some earlier definitions are repeated.

A *graph* is a collection of circles, between certain pairs of which are lines, among certain triples of which are triangular faces, and in general, among certain r-tuples of which there are regular polyhedra termed r faces or r bonds. In most important cases and in all of the examples in this chapter we deal only with the case $r = 2$; hence with pairwise bonds and "linear" graphs as in Eq. (6.43). In the correspondence between graphs and integrals, a symmetric function of r coordinates $F_r(1, \ldots, r)$ is represented by an "r face" labeled F_r. The circles themselves represent functions of one variable: A z circle represents a function z at some coordinate; a 1 circle represents the function unity. Moreover, there are two types of circles. Black circles, also called *field points*, are not labeled, and are associated with the dummy variables of integration, while white circles, also called *root points*, are labeled by coordinates and correspond to free variables. The *value* of a graph is given by

$$\text{value}(\text{graph}) = \frac{I(\text{graph})}{S \prod_i \prod_r q_{ir}!} \tag{6.46}$$

where I(graph) is the integral that corresponds to the graph, where S is the number of permutations of the black circles (which for this purpose are considered as labeled) which leave all connections of the graph invariant, and q_{ir} is the number of faces of a given type i connecting a particular set of r circles. The specification of type allows various functions of r coordinates to be represented. The product is over all such functions appearing in the graph, and over all distinct groups of r circles.

An example is given here for the graph in Fig. 6.3, where the solid lines represent F_2, the dashed line G_2, and the shaded triangle F_3. We have

$$\text{value}(\text{graph}) = \frac{\iiint z(1)z(2)z(3)z(4)F_3(2\ \ 3\ \ 4)F_2^2(1\ \ 2)G_2(3\ \ 4)d(2)d(3)d(4)}{2 \cdot 1! \cdot 1! \cdot 2!}$$

Now some additional important nomenclature in graph theory is described. A *simple graph* has at most one bond of a given type joining any given set of circles, so every $q_{ir}!$ factor is unity. While the graph in Fig. 6.3 is not a simple graph, the representation of $F_3(1\ \ 2\ \ 3)F_3(2\ \ 3\ \ 4)$, for example, would be a simple graph because the two F_3 faces do not link the same set of three circles.

In a *connected graph* we can pass between any two elements of the graph on a path through other circles and bonds of the graph.

An *articulation circle* in a connected graph is a white or black circle whose removal leaves two or more unconnected pieces, at least one of which contains no white circles and at least one black circle. Connected graphs that are free of articulation circles are called *irreducible*.

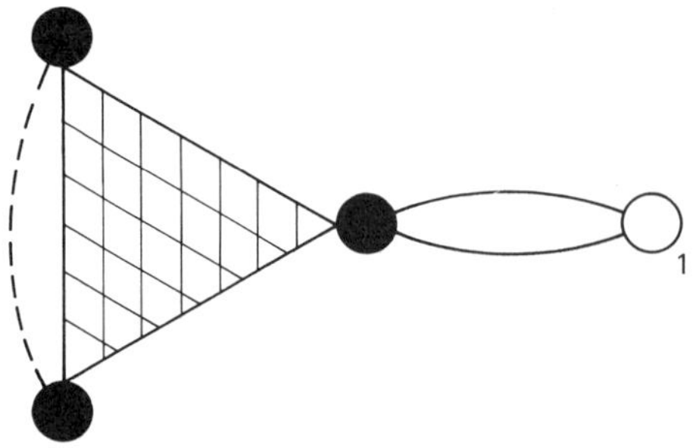

Figure 6.3 A nonsimple graph on $z(\mathbf{r})$ circles.

A *cutting circle* in an irreducible graph with two or more white circles is a circle whose removal leaves two or more unconnected pieces, at least two of which each contain a white circle. For graphs having only two white circles, a cutting circle is a black circle that is intersected by all of the paths connecting the white circles.

A *reducible pair* of circles in an irreducible graph is a pair of circles, white or black or one of each, whose removal leaves two or more unconnected pieces, at least one of which has no white circles. In the special case that any of the pieces with no white circles has at least one black circle the reducible pair is also called an *articulation pair* of circles.

A *residual* of a reducible pair of circles is a graph obtained by attaching a white circle to each end of any of the pieces containing no white circles that remain when the reducible pair of circles is removed. Reducible pairs and residuals are illustrated and used in Section 9.6.

A *product graph* is one whose value is the product of the values of two or more simpler graphs. In Eq. (6.42) every disconnected graph is a product graph. Considering the two connected graphs with white 1-circles ●–○–● and ○–●–○, the first is a product graph but not the second.

A *topological reduction* is a conversion of a series of graphs into another series with fewer terms, even though both may be infinite. The weights on the circles, the bond functions, or both may change in the process. Topological reduction is more formally defined in Section 9.5. It is illustrated in the procedure in Section 6.7 for converting Eq. (6.41) to a form that is more useful for many purposes.

Now we come to the three lemmas of DeDominicis, Morita, and Hiroike, which so greatly facilitate the use of graphs. Each lemma is an equation involving the values of graphs; following the usual convention we suppress any particular notations to specify "value," it being understood that "graph" in a mathematical expression means "value of the graph." At first sight each of the lemmas seems quite obvious, but to prove them one must delve into the symmetry numbers.

Lemma 1. Let *G be a collection of distinct graphs* $g_1, g_2, \ldots, g_k$ with each g_i a connected graph on some number $n \geqslant 0$ of white circles and any number $m_i \geqslant 0$ of black circles. Also, it is specified that $k \leq \infty$ and that no g_i is a product graph.

Let $H = G +$ sum of all product graphs formed from G. Then

$$H = e^G - 1 \tag{6.47a}$$

Proof. Let

$$H = \sum_{l_1=0}^{\infty} \sum_{l_2=0}^{\infty} \cdots \sum_{l_k=0}^{\infty} h(l_1, \ldots, l_k) - 1 \tag{6.47b}$$

where $h(l_1, \ldots, l_k)$ is a product of l_1 graphs g_1, l_2 graphs g_2, and so on. Thus

$$G = h(1, 0, 0, \ldots, 0) + h(0, 1, 0, \ldots, 0) + \cdots + h(0, 0, 0, \ldots, 1)$$

The -1 appears in Eq. (6.47a) because $h(0, 0, \ldots, 0)$ is not a term in H: According to the definition of H at least one of the l_i's must be nonzero in each term of (6.47b).

Using the notation in Eq. (6.46) we write I_i for the integral in the equation for the value of g_i. Then

$$h(l_1, \ldots, l_k) = \frac{I_1^{l_1} \times \cdots \times I_\nu^{l_k}}{S(l_1, \ldots, l_k)} \tag{6.48}$$

expresses the fact that each h is formed as a product of g_i's, but the combinatorial factor has to be calculated anew. According to the definition following Eq. (6.46) it is

$$S(l_1, \ldots, l_k) = \prod_{i=1}^{k} l_i!S_i^{l_i} \tag{6.49}$$

the product of the combinatorial factors of the individual g_i and the $l_i!$ factors to count the number of interchanges of g_i that change h if all of the black circles are labeled but not otherwise. So we find, since $I_i/S_i = g_i$, that

$$h(l_1, \ldots, l_k) = \frac{g_1^{l_1}}{l_1!} \times \cdots \times \frac{g_k^{l_k}}{l_k!} \tag{6.50}$$

and

$$\begin{aligned} H &= e^{g_1} \times e^{g_2} \times \cdots \times e^{g_k} - 1 \\ &= e^G - 1 \end{aligned} \tag{6.51}$$

Lemma 2. Let g be a graph with m black $\gamma(\mathbf{r})$ circles. The numbers (≥ 0) of bonds and of white circles are arbitrary. Then

$$\frac{\delta g}{\delta \gamma(\mathbf{R})} = \text{sum of all distinct diagrams obtained from } g \text{ by replacing one black } \gamma(\mathbf{r}) \text{ circle by a white 1 circle at } \mathbf{R} \tag{6.52}$$

Proof. Write

$$\begin{aligned} g &= \frac{I}{S} \\ &= \frac{1}{m!}[I_1 + I_2 + \cdots + I_\sigma] \end{aligned} \tag{6.53}$$

where the I_j are distinct graphs if the m black circles are labeled [see the discussion following Eq. (6.42)]. In view of Eq. (6.22), we have

$$\frac{\delta g}{\delta \gamma(\mathbf{R})} = \frac{1}{m!} \sum_{k=1}^{\sigma} \sum_{j=1}^{m} I_k(\mathbf{r}_j : \mathbf{R}) \tag{6.54}$$

where $I_k(\mathbf{r}_j : \mathbf{R})$ is obtained from I_k by replacing the black circle labeled j by a white 1-circle at $\mathbf{R}$. When the labels on the remaining black circles are removed, there are $m \times \sigma_1$ distinct graphs $I_k(\mathbf{r}_j : \mathbf{R})$ of whatever k and j that become topologically indistinguishable, perhaps another $m \times \sigma_2$ such graphs that also become topologically indistinguishable from each other but remain distinct from the first set, and so on, until we have $\sigma_1 + \sigma_2 + \cdots = \sigma$. So we finally have

$$\frac{\delta g}{\delta\gamma(\mathbf{R})} = \frac{1}{(m-1)!}[\sigma_1 I_{(1)}(\mathbf{R}) + \sigma_2 I_{(2)}(\mathbf{R}) + \cdots]$$
$$= \frac{I_{(1)}(\mathbf{R})}{S_1} + \frac{I_{(2)}(\mathbf{R})}{S_2} + \cdots \qquad (6.55)$$

where each $I_{(J)}(\mathbf{R})$ is derived from I by replacing a $\gamma(\mathbf{r})$ factor by $\delta(\mathbf{r} - \mathbf{R})$. It follows that $I_{(j)}/S_j$ is a graph obtained from $g = I/S$ by replacing a black $\gamma(\mathbf{r})$ circle by a white 1-circle.

Illustration:

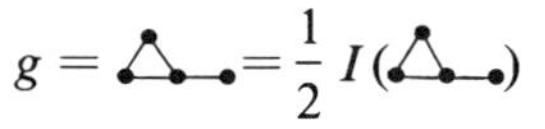
(6.56)

where I(graph) means the integral; here

 $= \int f(1\;2)f(2\;3)f(3\;1)f(3\;4)\gamma(1)\gamma(2)\gamma(3)\gamma(4)d(1)d(2)d(3)d(4)$ (6.57)

We find directly from Eq. (6.55) that

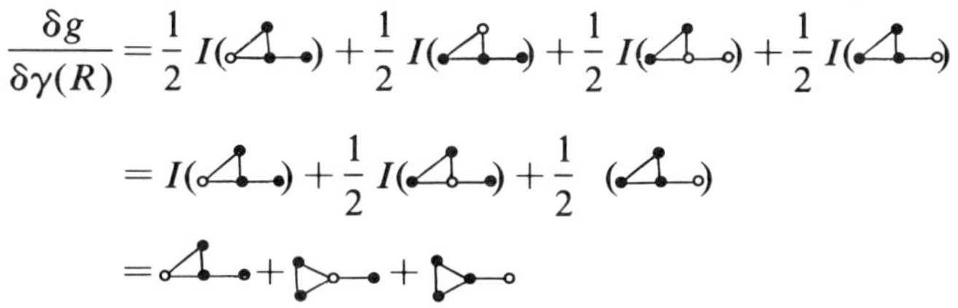
(6.58a)

in agreement with Lemma 2. It is also instructive to follow 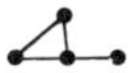 in detail through the proof of Lemma 2 to see how the factor m can indeed be extracted from the sum in Eq. (6.54).

A slightly modified version of Lemma 2 that is useful is

Lemma 2′. Let g be a simple graph of black γ circles and b bonds. Then

$\dfrac{\delta g}{\delta b(\mathbf{R}, \mathbf{R}')} = \frac{1}{2} \times$ the sum of all distinct graphs obtained from g by removing one b bond in all possible ways and replacing the two black γ circles, to which it was connected, by white γ circles at $\mathbf{R}$ and $\mathbf{R}'$, respectively (6.58b)

The method of proof follows Lemma 2. As illustrations we have

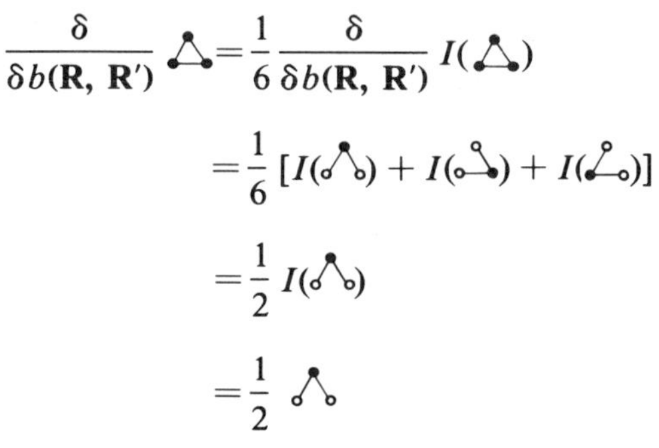
(6.59a)

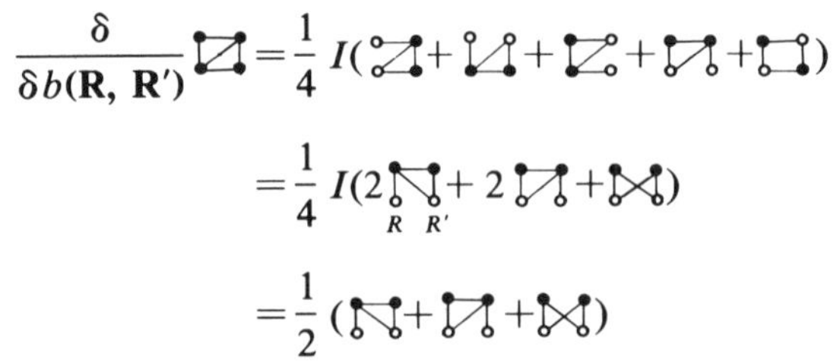

(6.59b)

Lemma 3. The third lemma concerns the reduction of graphs with black articulation circles. First we consider the relatively trivial reduction of white articulation circles, as in this example, which involves white 1-circles.

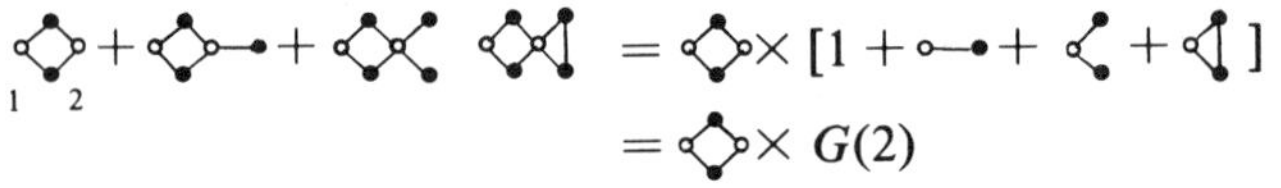

(6.60a)

(6.60b)

where $G(2)$ is the sum of the "hair" that can be attached to the white circle labeled 2. To be more explicit, $G(2)$ is the sum

$$G(2) = w_1(2) + w_2(2) + w_3(2) + w_4(2) \tag{6.60c}$$

$$w_1(2) = 1 \tag{6.60d}$$

$w_2(2) =$ (6.60e)

$w_3(2) =$ (6.60f)

$w_4(2) =$ (6.60g)

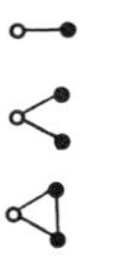

of the values of the hair graphs. The hair graphs themselves may incorporate black articulation circles, although in this example none do. Finally, the $w_\alpha(2)$ and $G(2)$ are functions of the location 2, although in the absence of an external field they reduce to constant-weight factors on circle 2.

It is noteworthy that each graph in Eq. (6.60a) has a symmetry number that includes a factor 2 from the two equivalent black circles of the hairless graph, the first graph on the left. The remaining part of the symmetry number for each graph (either 1 or 2 in these examples) is incorporated in the value w_α of the hair graph.

The corresponding reduction for graphs with black articulation circles is illustrated by the following example, where the w_α weights are the same as in Eqs. (6.60d)

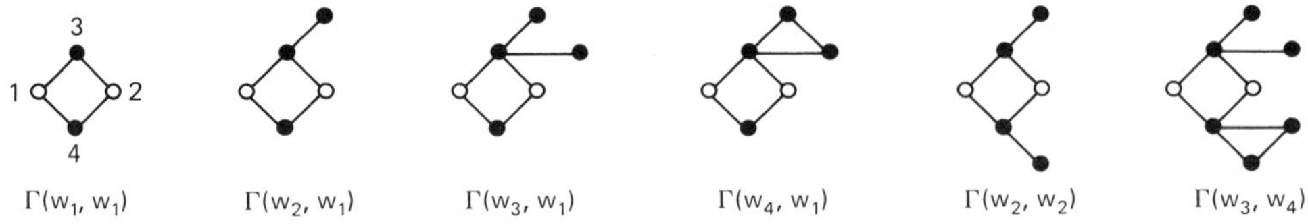

Figure 6.4 Graphs with black articulation circles.

to (6.60g) and where we employ the notation for the composite graphs given in Fig. 6.4. These definitions correspond to the following equations, where we use $\hat{\Gamma}(3, 4)$ to represent the product of bonds $f(1\ 3)f(3\ 2)f(1\ 4)f(4\ 2)$ for the hairless graph with all white circles of unit weight.

$$\Gamma(w_1, w_1) = \frac{1}{2}\int d(3)d(4)\hat{\Gamma}(3, 4)w_1(3)w_1(4) \tag{6.61a}$$

$$\Gamma(w_2, w_1) = \int d(3)d(4)\hat{\Gamma}(3, 4)w_2(3)w_1(4) \tag{6.61b}$$

$$\Gamma(w_3, w_1) = \int d(3)d(4)\hat{\Gamma}(3, 4)w_3(3)w_1(4) \tag{6.61c}$$

$$\Gamma(w_4, w_1) = \int d(3)d(4)\hat{\Gamma}(3, 4)w_4(3)w_1(4) \tag{6.61d}$$

$$\Gamma(w_2, w_2) = \frac{1}{2}\int d(3)d(4)\hat{\Gamma}(3, 4)w_2(3)w_2(4) \tag{6.61e}$$

$$\Gamma(w_3, w_4) = \int d(3)d(4)\hat{\Gamma}(3, 4)w_3(3)w_4(4) \tag{6.61f}$$

Of course, $\hat{\Gamma}(3, 4)$ depends on the coordinates 1 and 2, although this dependence is not explicit in the notation.

At first sight the sum of these graphs, even together with the others that may be formed with the same w_α, appears not to factor in a way that is analogous to Eq. (6.60). The reason is that the symmetry number $S = 2$ is lost when the hair attached to circle 3 is not the same as that attached to circle 4, as illustrated in Eq. (6.61c). However, if we look more closely into this question, we notice that the symmetry numbers S are generated in procedures in which we sum over all of the *distinct* graphs with some specified characterstics. An illustration of this fact may be found in the passage from Eq. (6.40) to (6.43). Therefore, we examine the following sum:

$$\sum_{\alpha,\alpha'}^{d} \Gamma(w_\alpha, w_{\alpha'}) = \sum_{\alpha,\alpha'}^{d} \frac{1}{S(w_\alpha, w_{\alpha'})}\int d(3)d(4)\hat{\Gamma}(3, 4)w_\alpha(3)w_{\alpha'}(4) \tag{6.61g}$$

Here we sum over all sets α, α' of two indices of the hair functions in Eq. (6.60) except that we suppress those terms that are not graphically *distinct* from those already counted. For example, we do not count both $\Gamma(w_1, w_2)$ and $\Gamma(w_2, w_1)$ because these graphs are indistinguishable because the black circles are not labeled. However, we also notice that we have

$$\sum_{\alpha,\alpha'}^{d} \Gamma(w_\alpha, w_{\alpha'}) = \frac{1}{2}\sum_{\alpha=1}^{4}\sum_{\alpha'=1}^{4}\int d(3)d(4)\hat{\Gamma}(3,4)w_\alpha(3)w_{\alpha'}(4) \tag{6.61h}$$

because if $\alpha = \alpha'$, the terms appear on the right with the symmetry number 2, while if $\alpha \neq \alpha'$, two equal terms appear on the right, giving the correct effective symmetry number $2/2 = 1$.

In view of Eq. (6.60c), the result in (6.61h) can be written in the compact form

$$\sum_{\alpha,\alpha'}^{d} \Gamma(w_\alpha, w_{\alpha'}) = \frac{1}{S(\alpha, \alpha)} \int d(3)d(4)\hat{\Gamma}(3, 4)G(3)G(4) \tag{6.61i}$$

where $S(\alpha, \alpha)$ is the symmetry number of any of the graphs in which the black circles have equal weights [i.e., $S(\alpha, \alpha) = 2$ in our example]. This example indicates that the effect of summing all of the graphs which differ in the hair on the black circles is to change the weight of each black circle in the hairless graph to a value that is associated with the sum over all of the hair graphs. This result is indeed parallel to the one we found with far less effort for hair on white articulation circles.

Now we turn to Lemma 3, which is a generalization of the result that we have obtained for a particular hairless graph and a particular set of hair graphs.

We again define the sum of the hair graphs as

$$G(\mathbf{R}) = w_1(\mathbf{R}) + \cdots + w_\alpha(\mathbf{R}) + \cdots + w_k(\mathbf{R}), \qquad 1 \leqslant k \leqslant \infty \tag{6.62a}$$

where $w_1 = 1$ and each remaining w_α is a distinct connected graph with one white circle of unit weight and one or more black circles. We also define

$\Gamma(w_\alpha, w_{\alpha'}, \ldots, w_{\alpha(m)})$

= simple connected graph with no articulation circles, with zero or more white circles, and with m black circles with weights $w_\alpha(\mathbf{R})$, $w_{\alpha'}(\mathbf{R})$, . . . , $w_{\alpha(m)}(\mathbf{R})$ respectively

$$= \frac{1}{S(\alpha, \alpha', \ldots, \alpha^{(m)})} \int d(1) \cdots d(m)\hat{\Gamma}(1, \ldots, m)w_\alpha(1) \cdots w_{\alpha(m)}(m) \tag{6.62b}$$

where $\hat{\Gamma}$ is the combination of bond functions specified by the graph and S is the symmetry number. Then we have Lemma 3:

$$\sum_{\boldsymbol{\alpha}}^{d} \Gamma(w_\alpha, w_{\alpha'}, \ldots, w_{\alpha(m)}) = \Gamma(G, G, \ldots, G) \tag{6.62c}$$

where the sum is over the list of indices $\boldsymbol{\alpha} = \alpha, \alpha', \ldots, \alpha^{(m)}$ of the hair graphs on the articulation circles and where the superscript d on Σ^d means to sum over all sets $\boldsymbol{\alpha}$ that give topologically distinct hairy graphs. Thus if two lists $\boldsymbol{\alpha}$ and $\boldsymbol{\alpha}'$ differ only in the interchange of w_n and w_m between equivalent black articulation circles of Γ, only one of them is to be counted in the sum.

To prove Lemma 3 we analyze the following sequence of equations:

$$\sum_{\boldsymbol{\alpha}}^{d} \Gamma(w_\alpha, w_{\alpha'}, \ldots, w_{\alpha(m)})$$

$$\equiv \sum_{\boldsymbol{\alpha}}^{d} \frac{1}{S(\alpha, \alpha', \ldots, \alpha^{(m)})} \int d(1) \cdots d(m)\hat{\Gamma}(1, \ldots, m)w_\alpha(1) \cdots w_{\alpha(m)}(m) \tag{6.63a}$$

$$= \sum_{\boldsymbol{\alpha}} \frac{1}{S(\alpha, \ldots, \alpha)} \int d(1) \cdots d(m)\hat{\Gamma}(1, \ldots, m)w_\alpha(1) \cdots w_{\alpha(m)}(m) \tag{6.63b}$$

$$= \frac{1}{S(\alpha, \ldots, \alpha)} \int d(1) \cdots d(m) \hat{\Gamma}(1, \ldots, m) \sum_{\alpha} w_{\alpha}(1) \cdots \sum_{\alpha} w_{\alpha}(m) \tag{6.63c}$$

$$= \Gamma(G, \ldots, G) \tag{6.63d}$$

The only difficult step is the passage from Eq. (6.63a) to (6.63b). Consider first the special case in which the black circles in the hairless graph are topologically equivalent so that we will have $S(\alpha, \ldots, \alpha) = m!$. In this case the coefficient σ defined by the equation $\sigma/m! = 1/S(\alpha, \alpha', \ldots, \alpha^{(m)})$ is just like the σ coefficient that figures in the discussion of Eq. (6.42); it is the number of ways of putting the bonds (here the weights) on the graph that become indistinguishable if the labels are removed from the black circles. So if we substitute $\sigma/m!$ for $1/S(\alpha, \alpha', \ldots, \alpha^{(m)})$ in Eq. (6.63a), then in passing to (6.63b) we generate a number σ of terms due to the unrestricted sum in the latter.

This analysis may be extended to the case in which the m black circles in the hairless graph may be partitioned into several subsets, within each of which the black circles are topologically identical. If the number of such subsets is larger than 1, then $S(\alpha, \ldots, \alpha)$ will differ from $m!$ Indeed, it will be the product of $m_i!$ over the subsets, where m_i is the number of black circles in the ith subset. If every $m_i = 1$, we have the other limit from that first considered in which all of the black circles are equivalent. If every $m_i = 1$, the passage from (6.63a) to (6.63b) is trivial. In the intermediate cases one has to consider the contributions of each subset to $S(\alpha, \ldots, \alpha)$ and the corresponding σ factors.

DO PROBLEMS 6.9 TO 6.11.

6.7 THE VIRIAL EXPANSION OF A^{ex}

Now we proceed to reduce Eq. (6.45) to a more useful form. In graphical notation Eq. (6.45) is

$\Xi^{\star} = 1 +$ 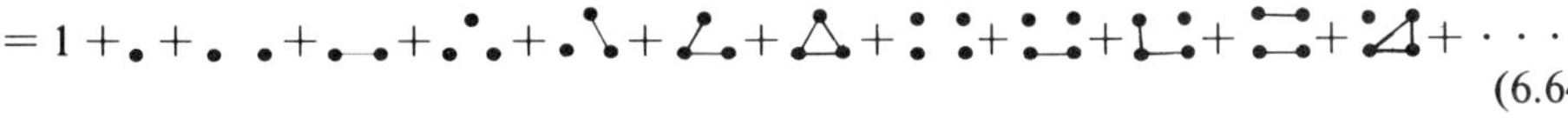 (6.64)

where we have f bonds on $z^{\star}$ circles. Now we apply Lemma 1, identifying $\Xi^{\star}$ as $1 + H$ and find for G, here identified as $\ln \Xi^{\star}$, the following graphical representation:

$\ln \Xi^{\star} =$ 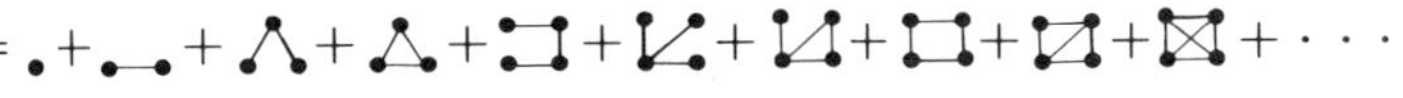 (6.65)

where again we have f bonds on $z^{\star}$ circles. If we turn off the external potential, we would recover the expansion of βp in powers of the activity z. Instead, we apply Lemma 2 to get [see Eq. (6.27)]

$\rho^{\star}(\mathbf{r}) = z^{\star}(\mathbf{r}) \dfrac{\delta \ln \Xi^{\star}}{\delta z^{\star}(\mathbf{r})} =$ 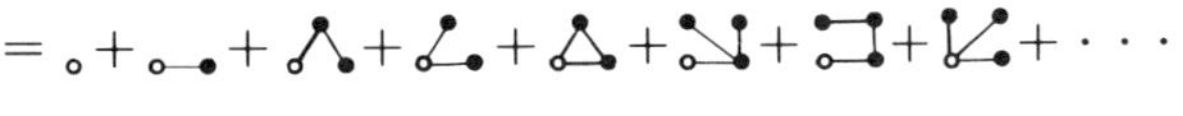 (6.66)

where we again have f bonds on $z^{\star}$ circles, of which one is white and the rest black. Here if we turn off the external potential, we recover another classical result:

the density as a power series in activity. Instead, we divide by $z^\star(\mathbf{r})$ and apply Lemma 1, identifying $1 + H = \rho^\star(\mathbf{r})/z^\star(\mathbf{r})$, to get

$$\ln \frac{\rho^\star(\mathbf{r})}{z^\star(\mathbf{r})} =$$ 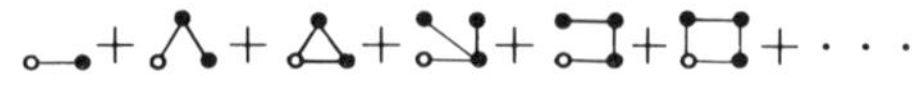 (6.67)

again with f bonds on black $z^\star$ circles, but now, because of the division by $z^\star(r)$, the white circle has unit weight.

Now we notice that the graphs in Eq. (6.67) are either irreducible or they are irreducible except for "hair" growing from articulation circles, the "hair" being derived from terms in the expansion of $\rho^\star(\mathbf{r})/z^\star(\mathbf{r})$ in Eq. (6.66). So we apply Lemma 3, identifying $G(\mathbf{r})$ as the set of graphs in Eq. (6.66), but now with unit weight white circles, and identifying Γ as, in turn, each of the irreducible graphs in Eq. (6.67) and so obtain Eq. (6.68),

$$\ln \frac{\rho^\star(\mathbf{r})}{z^\star(\mathbf{r})} =$$ 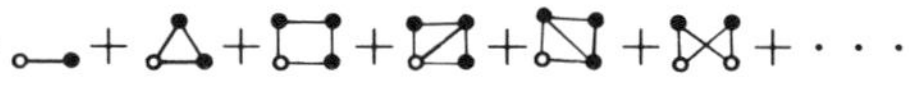 (6.68)

where now we have f bonds on one white 1-circle and black $\rho^\star$ circles and where now all of the diagrams are irreducible. If we now let the external field vanish, $\rho^\star \to \rho$ [see Eq. (6.27) and (6.28)] and we get $\ln(\rho/z)$ given by the expansion in Eq. (6.68), now with f bonds on one white circle and black ρ circles. So with the external field off we have

$$\ln z = \ln \rho - \rho\beta_1 - \rho^2\beta_2 - \rho^3\beta_3 - \cdots \tag{6.69}$$

with

$$\beta_n = \text{sum of all distinct irreducible graphs of } f \text{ bonds on one white 1-circle and } n \text{ black 1-circles} \tag{6.70}$$

For example,

$$\beta_3 =$$ 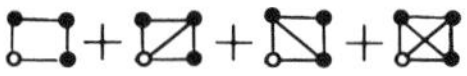

$$= \int d(2)d(3)d(4)f(1\;2)f(2\;3)f(3\;4)f(4\;1)\left[\frac{1}{2} + \frac{1}{2}f(1\;3) + \frac{1}{2}f(2\;4) + \frac{1}{3!}f(1\;3)f(2\;4)\right]. \tag{6.71}$$

Now we can express the coefficients in Eq. (6.2) as functionals of the Hamiltonian. Thus we have

$$\begin{aligned}\beta\mu^{\text{ex}} &= \beta\frac{\partial A^{\text{ex}}}{\partial N} = \frac{\partial(\beta A^{\text{ex}}/V)}{\partial(N/V)} \\ &= -2\rho B_2^A - 3\rho B_3^A - 4\rho B_4^A - \cdots \\ &= \ln\frac{z}{\rho}\end{aligned} \tag{6.72}$$

where the last equality follows from thermodynamics: In terms of the ideal chemical potential μ^{id} and reference state (hyp $\rho = 1$) chemical potential μ^0, we have

$$\begin{aligned}\beta\mu^{ex} &= \beta(\mu - \mu^{id}) \\ &= \beta(\mu - \mu^0) - \beta(\mu^{id} - \mu^0) \\ &= \ln z - \ln \rho\end{aligned} \tag{6.73}$$

So we find, in general,

$$\begin{aligned}B_n^A &= \frac{1}{n}\beta_{n-1} \\ &= \frac{1}{V}\text{(sum of all distinct irreducible graphs of } f \text{ bonds on } n \text{ black 1-circles)}\end{aligned} \tag{6.74}$$

since integrating over the white circle in a β_{n-1} graph raises the symmetry number by a factor n.

The first few terms of Eq. (6.2) are given here (f bonds on black ρ circles):

$\rho^2 V \times B_2^A =$

$\rho^3 V \times B_3^A =$ 

$\rho^4 V \times B_4^A =$ 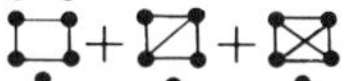 (6.75a)

$\rho^5 V \times B_5^A =$ 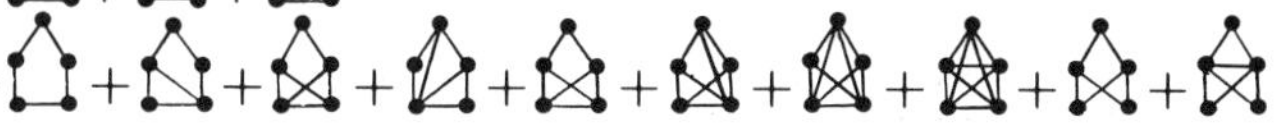

Thus we have

$$-\beta A^{ex} = \text{sum over all simple distinct irreducible graphs of } f \text{ bonds on two or more black } \rho \text{ circles} \tag{6.75b}$$

Since the number of distinct graphs increases very rapidly with increasing number of circles and since the highly connected graphs correspond to integrals that are very difficult to evaluate, even numerically, it is not surprising that Eq. (6.75) is not often used for practical calculations except when the B_2^A and B_3^A terms suffice (i.e., at low ρ). But these results, first obtained by J. E. Mayer, are useful for other theoretical developments.

Equations (6.1), (6.2), and (6.69), with the equations relating the coefficients B_n, B_n^A, and β_n, are three equivalent expressions of the virial expansion of the thermodynamic properties. The series in Eq. (6.2) has the advantage that the full symmetry of the problem is reflected in the coefficients; this advantage extends to Eq. (6.75b).

For a multicomponent system the derivation from Eq. (6.45) to (6.75b) goes through if only a rather simple change is made: A black circle corresponds to summation over species as well as integration over coordinates. Thus if $\rho_s = N_s/V$ and if s_l is the species at circle l, we have, for example,

 $$= \frac{1}{3!}\sum_{s_1}\sum_{s_2}\sum_{s_3}\int d(1)d(2)d(3)f_{s_1s_2}(1\ 2)f_{s_2s_3}(2\ 3)f_{s_3s_1}(3\ 1)\rho_{s_1}\rho_{s_2}\rho_{s_3} \tag{6.76}$$

The derivation given here is equally indifferent to the change from monatomic particles, where

$$(1) \simeq \mathbf{r}_1 = x_1, y_1, z_1 \tag{6.77}$$

to particles with internal coordinates where

$$\begin{aligned}(1) &\simeq \mathbf{r}_1,\ \Omega_1,\ \mathbf{I}_1 \\ &= x_1,\ y_1,\ z_1,\ \alpha_1,\ \beta_1,\ \gamma_1,\ \mathbf{I}_1\end{aligned} \tag{6.78}$$

where α, β, and γ are Eulerian angles fixing the orientation of a particle and $\mathbf{I}_1$ is something like a set of vibrational coordinates. See Section 3.3.

DO PROBLEMS 6.12 TO 6.15.

6.8 VIRIAL EXPANSION OF THE PAIR CORRELATION FUNCTION

If we apply Eq. (6.39) to Eqs. (6.2) and (6.75) with the aid of Eq. (6.58b), we obtain the cluster expansion for the pair correlation function:

$$g(1\ 2) = e(1\ 2)[1 +$$ 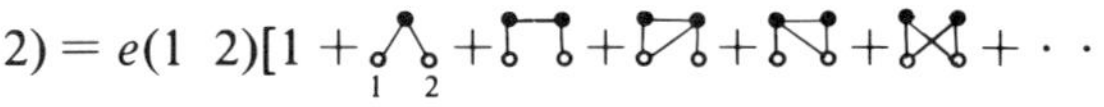 $$+ \cdots] \tag{6.79}$$

where the bracketed sum includes all distinct simple irreducible graphs of f bonds on white 1 circles, labeled 1 and 2, and one or more black ρ circles, and no bond directly connecting the white circles.

This expression, with its immediate applicability to multicomponent systems and systems of particles with internal coordinates, is the source of many of the developments in equilibrium statistical mechanics that we describe in Chapters 7 to 10. It is also easily extended to systems in which pairwise additivity fails.

6.9 VIRIAL EXPANSION FOR HARD-SPHERE FLUIDS

The first few B_n^A are given in Eq. (6.75) in graphical notation. The virial coefficients of the pressure B_n, which usually are discussed, are given in terms of the B_n^A in Eq. (6.3).

For the hard-sphere fluid one has pairwise additivity [Eq. (4.27)] with the pair potential given in Eq. (4.24). It follows that we have

$$\begin{aligned}f(r) &= e^{-\beta u(r)} - 1 \\ &= -\Theta(\sigma - r)\end{aligned} \tag{6.80.1}$$

where $\Theta(x)$ is the unit step function; $\Theta(x) = 1$ if $x > 0$ and $\Theta(x) = 0$ otherwise. We find that

$$\begin{aligned}B_2 &= -B_2^A \\ &= \frac{1}{2}\int_0^\infty 4\pi r^2 dr\, \Theta(\sigma - r) \\ &= \frac{2\pi\sigma^3}{3}\end{aligned} \tag{6.80.2}$$

and

$$\begin{aligned} B_3 &= -2B_3^A \\ &= \frac{1}{3}\int d^3r \int d^3R\ \Theta(\sigma - r)\Theta(\sigma - R)\Theta(\sigma - |\mathbf{r} - \mathbf{R}|) \end{aligned} \tag{6.80.3}$$

but the complete evaluation is difficult, as is true for the higher B_n, the more so the bigger n. However, the problem of evaluating these coefficients has attracted a lot of attention, with the results given in Table 6.1.

TABLE 6.1 VIRIAL COEFFICIENTS OF THE HARD-SPHERE FLUID

k	2	3	4	5	6	7
B_k/B_2^{k-1}	1	5/8	0.28695	0.1103 ±0.0003	0.0386 ±0.0004	0.0138 ±0.0004
$(4/B_2)^{k-1}B_k$	4	10	18.36	28.24	39.5	56.5

Source: Adapted from Table IV of Hansen and McDonald (1976).

It is apparent that convergence is not very rapid, but this aspect is clearer if we write the virial expansion (6.1) in terms of the reduced concentration η:

$$\eta \equiv \frac{\pi}{6}\rho\sigma^3 \tag{6.81.1}$$

which is $N(4\pi/3)(\sigma/2)^3/V$, hence the fraction of the space occupied by the N spheres. Often η is called the *packing fraction.* When $\eta \sim 0.5$ the hard-sphere fluid corresponds closely to a simple liquid like Ar near its triple point (Fig. 5.5).

Since, for the hard-sphere fluid,

$$\rho B_2 = 4\eta \tag{6.81.2}$$

Eq. (6.1) may be written

$$\begin{aligned} \frac{\beta p}{\rho} &= 1 + \sum_{k=2}^{6}\left(\frac{4}{B_2}\right)^{k-1} B_k \eta^{k-1} \\ &= 1 + 4\eta + 10\eta^2 + 18.36\eta^3 + 28.23\eta^4 + 39.5\eta^5 + 56.5\eta^6 + \cdots \end{aligned} \tag{6.81.3}$$

At $\eta = \frac{1}{2}$ none of the exhibited terms is less than 0.8. It was noticed by Carnahan and Starling (1969) that the same series is very nearly given by

$$\frac{\beta p}{\rho} = 1 + \sum_{k=1}^{6}(k^2 + 3k)\eta^k \tag{6.81.4}$$

Assuming that the coefficients are given by the same formula for $k > 6$ leads to an infinite series whose sum is given below in Eq. (7.56c), which turns out to be very accurate.

The hard-sphere fluid is important because it is the simplest model system that exhibits many of the characteristic properties of real fluids as illustrated by Figs. 4.1 and 5.5. We return to other equilibrium aspects in Section 7.3. At this point we see how slowly the density expansion converges and how difficult it is to evaluate many virial coefficients even for a simple model potential. Fortunately, there are other approaches to the calculation of the thermodynamics properties of model fluids.

THE LITERATURE

Guggenheim (1960) discusses various aspects of the relation of the second virial coefficient and the mass action constant for dimerization. For still other aspects related to Eq. (6.84), see Wood et al. (1978) and Bartsch and Ebeling (1971). Functional derivatives were introduced in statistical mechanics in a famous article by Percus (1964). The account in Hansen and McDonald (1976) is more pedogogical. Hirschfelder et al. (1954) give a complete account of the theory of the second virial coefficient. Many new data are in the book by Dymond and Smith (1969).

The theory of graphical expansions is given in many places, for example, Hansen and McDonald (1976), Andersen (1977), and Friedman and Dale (1977). For the graphical expansions in cases in which pairwise additivity, Eq. (4.27), is not assumed, see Friedman and Dale (1977), Stell (1967), and Friedman (1962). Perhaps the most general and complete exposition, by Stell (1975), extends to mixtures, to nonpairwise potentials, and to the full phase space.

PROBLEMS

6.1. Show that Eq. (1.22.4) can also be written

$$p = \rho^2 \left(\frac{\partial(A/N)}{\partial \rho} \right)_T \tag{6.82}$$

which is convenient for deriving Eq. (6.3) from Eqs. (6.1) and (6.2).

6.2. **(a)** For a model fluid with the pair potential given by the square-well function [Eq. (6.13)], express B_2 as a function of the parameters σ, b, and h. Then express the Boyle temperature T_B in terms of σ, b, and h.

(b) In the low-density range in which $\beta p = \rho + \rho^2 B_2$, what is the corresponding approximation to $g(r)$ for the square-well fluid [see (Eq. 6.5)]? That is, for this case write $g(r)$ as a function of σ, h, and b.

6.3. Choose the parameters so that the square-well model represents CH_4 as well as you can (see Table 3.1). Then use the square-well model to estimate the following for methane: T_B, B_2 at 300 K, the dimerization constant for $2CH_4 \rightarrow (CH_4)_2$ at 100 K. Estimate the dimer concentration at 100 K when $\rho = 1 \times 10^{-4}$ Å^{-3} (i.e., 0.16 M). Compare with the dimer concentration estimated as

$$\rho_d = \frac{1}{2} \rho^2 \int_\sigma^{b\sigma} g(r) 4\pi r^2 \, dr$$

6.4. Consider the model potential

$$u(r) = \begin{cases} \infty & \text{if } r < \sigma \\ \dfrac{-A}{r^x} & \text{if } r > \sigma \end{cases} \tag{6.83}$$

Prove that for this potential B_2 is finite only if $x > 3$. This potential with $x \leqslant 3$ is an example of a *long-range* potential.

6.5. Using the first equality in Eq. (6.24), evaluate

$$\frac{\delta \Xi^{\star}}{\delta \phi(1')}$$

and compare the result with Eq. (6.26). See that $z^{\star}$ is introduced merely to make the notation more convenient.

6.6. For the potential of Eq. (6.83), how does $f(r)$ [defined in Eq. (6.40)] vary at large r? For any physically possible pair potential, what is $f(r)$ at $r \longrightarrow 0$? Make a graph of $f(r)$ for the 6–12 potential with the parameters for methane (Table 3.1).

6.7. Prove that each of the three graphs in Eq. (6.42i) gives the same integral, a result that is indicated there. Prove that in the $\phi = 0$ limit this integral is proportional to the square of the integral in Eq. (6.42e).

6.8. In the graphs in $\Xi(z, V, T)$ each $\mathbf{r}_i$ is integrated over the volume V. If the pair potential is short range, how does the graph in Eq. (6.42i) depend on V as $V \longrightarrow \infty$? How about with the potential in (6.83) with $x = 1$ (Coulomb potential)?

6.9. Among the graphs in Eq. (6.42l):

(a) Give an example of a graph with an articulation circle, indicating this circle by an arrow.

(b) Give an example of a graph whose value grows like V^2 as $V \longrightarrow \infty$, if the pair potential is short ranged.

(c) Give an example of a graph whose symmetry number is 3.

(d) Which are irreducible?

6.10. Identify the cutting circles, the articulation circles, the articulation pairs of circles, and the product graphs in the graphs in Eq. (6.44).

6.11. Identify each pair of reducible circles and its residual in Eqs. (6.44b) and (6.44c).

6.12. From the $\phi = 0$ limit of Eq. (6.65) find the graphical prescription for the "reducible cluster integrals" b_n that appear in the expansion of the pressure in powers of the activity

$$\beta p = z + b_2 z^2 + b_3 z^3 + \cdots \tag{6.84}$$

Here the equation is written for a one-component system. Write the first few terms for a mixture of components a and b.

6.13. Consider a system of molecules with orientational and internal degrees of freedom as in Eq. (6.78). Assume that these can all be treated classically. How must Eq. (4.1) be modified to accommodate this case? Discuss *any* other modifications that are necessary to derive the coefficients B_n^A of the density expansion of A^{ex} for this case?

6.14. If we do not let the external field vanish after Eq. (6.68), we can derive

$$\left(\frac{-\beta A^{\text{ex}}}{V}\right)^{\star} = \sum_{n \geqslant 2} (\rho^{\star})^n B_n^A \tag{6.85}$$

where we know that the function on the left reduces to $-\beta A^{\text{ex}}/V$ and $\rho^\star \longrightarrow \rho$ when the external field vanishes. With the field on, the right side is the sum of simple distinct irreducible graphs of f bonds on two or more black $\rho^\star$ circles. Now using Lemma 2, find the graphical prescription for

$$c(1\ 2) \equiv \left[\frac{\delta^2(-\beta A^{\text{ex}})^\star}{\delta\rho^\star(1)\,\delta\rho^\star(2)}\right]_{\phi=0} \tag{6.86}$$

and compare it with (6.79). We shall find another way to generate the $c(1\ 2)$ expansion in Section 6.7.

6.15. It should be possible to derive Eq. (6.79) directly from Eq. (4.23) by expressing the latter in graphical form and then reducing the graphs using the techniques in Section 6.6. Try it.

7

Pair Correlation Function Theory

7.1 THE DIRECT CORRELATION FUNCTION AND THE ORNSTEIN–ZERNICKE EQUATION

Many of the most successful approximations for calculating the measurable properties of Hamiltonian model systems are based on the theory of the pair correlation functions, a fact that may be rationalized as follows. First we notice that, even though the pair correlation function is expressed in terms of the Hamiltonian of the system by the cluster expansion series in Eq. (6.79), we have no reason to expect that truncating this series after the first few terms will be a more successful approximation than the corresponding truncation of the virial expansion of the pressure which we have just examined. On the other hand, as we contemplate the variety of graphs that appear in the cluster expansion series of g(1 2), we soon notice hints of partial summation schemes based on the fact that deleting a given pair of black circles in such a graph often generates residuals (see Section 6.6) that are themselves graphs in g(1 2). This degree of self-similarity in turn leads to the possibility of topological reduction processes of various kinds. These general statements will be illustrated in a variety of ways in the developments in this and the following chapters.

In pair correlation function theory it is convenient to introduce another notation for the pair correlation function, namely [see Eq. (5.25)]

$$g(1\ 2) = 1 + h(1\ 2) \tag{7.1}$$

which is like [see Eqs. (6.35) and (6.40)]

$$e(1\ 2) = 1 + f(1\ 2) \tag{7.2}$$

The graphical expansion of $h(1\ 2)$ in terms of $f(1\ 2)$ bonds is easily derived from Eq. (6.29). It is

$h(1\ 2) =$ sum of all h-allowed graphs of f bonds on two white 1-circles labeled 1 and 2 and zero or more black ρ circles (7.3)

where by h-allowed graphs we specify all of the distinct, simple, irreducible graphs that can be made with the specified bonds and circles. A number of other important functions can be identified with subsets of the h-allowed graphs. For example, the function $y(1\ 2)$, which is defined as the bracketed sum of graphs in Eq. (6.79), is

$y(1\ 2) = 1 +$ sum of all h-allowed graphs of f bonds on two white 1-circles labeled 1 and 2 and one or more black ρ circles, with no direct bond between the white circles (7.4)

$= 1 +$ 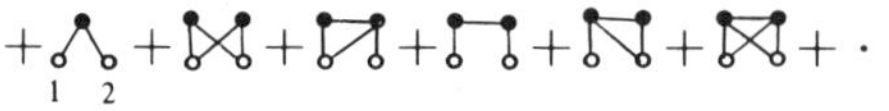

where all of the graphs with fewer than three black circles are shown.

Another important example is (see Problem 7.1)

$-\beta w(1\ 2) = -\beta u(1\ 2) +$ sum of all h-allowed graphs of f bonds on two white 1-circles labeled 1 and 2 and one or more black ρ circles, and with product graphs excluded (7.5)

$= -\beta u(1\ 2) +$ 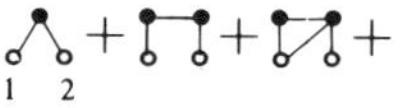

For a final example the answer to Problem 6.14, called the *direct correlation function,* is

$c(1\ 2) =$ sum of all h-allowed graphs of f bonds on two white 1-circles labeled 1 and 2 with zero or more black ρ circles, and with no cutting circles (7.6)

$=$ 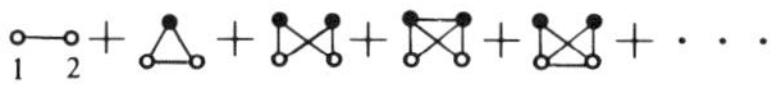

Now we shall exploit the fact that graphs with cutting circles can be constructed by combining graphs without cutting circles by a generalized sort of multiplication called *convolution* (see Appendix 1). At the simplest level we have, for f bonds, white 1-circles, and black ρ circles,

 $= \int f(1\ 3) d(3) \rho f(3\ 2) = f * f$

 $= \int f(1\ 3) d(3) \rho f(3\ 4) d(4) \rho f(4\ 2) = f * f * f$ (7.7)

where we also introduce a shorthand notation for convolutions.

Comparing Eq. (7.6) with (7.3) leads to the conclusion that we have

$$\begin{aligned} h &= c + c * c + c * c * c + \cdots \\ &= c + c * h \end{aligned} \tag{7.8a}$$

or

$$h(1\ 2) = c(1\ 2) + \int c(1\ 3) d(3) \rho h(3\ 2) \tag{7.8b}$$

Equation (7.8) is known as the *Ornstein–Zernicke equation.* It may equally well be considered as an integral equation that defines $c(1\ 2)$ in terms of $h(1\ 2)$. Generally, an integral equation is an equation for an unknown function in which that function appears in the integrand of some integral. This integral equation, which is of the convolution type, may be solved by means of Fourier transforms.

Another important equation for $c(1\ 2)$ follows from the definition of the sum $B(1\ 2)$ of *bridge graphs,* which we introduce as follows. We write

$$-\beta w(1\ 2) = -\beta u(1\ 2) + t(1\ 2) + B(1\ 2) \tag{7.9}$$

where $t(1\ 2) + B(1\ 2)$ is given as a sum of graphs in Eq. (7.5). Now we define $B(1\ 2)$ as the sum of those graphs with no cutting circles while each graph in $t(1\ 2)$ has at least one. We notice that an h-allowed graph with a cutting circle is not a product graph, so it appears in $t(1\ 2)$. Moreover, the sum of the t graphs is just $c * h$, the sum of all h-allowed graphs with at least one cutting circle. So we have

$$t = c * h = h - c \tag{7.10}$$

where we have used the Ornstein–Zernicke equation. Then with (7.9) we find an exact result for the direct correlation function.

$$c(1\ 2) = h(1\ 2) - \ln\,[1 + h(1\ 2)] - \beta u(1\ 2) + B(1\ 2) \tag{7.11}$$

For large r_{12} we have $|h(1\ 2)| << 1$ and we can expand the logarithm

$$\ln|(1 + h) = h - \tfrac{1}{2}h^2 + \tfrac{1}{3}h^3 - \cdots \tag{7.12}$$

to get

$$c(1\ 2) = \beta u(1\ 2) + B(1\ 2) + \tfrac{1}{2}h(1\ 2)^2 - \tfrac{1}{3}h(1\ 2)^3 + \cdots \tag{7.13}$$

By construction, the terms of $B(1\ 2)$ are h-allowed graphs with no cutting points and with product graphs excluded:

$B(1\ 2) =$ 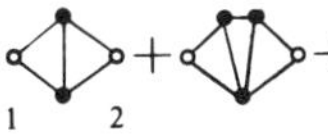 $+ \cdots$ (7.14)

where we have f bonds on black ρ circles and white 1-circles labeled 1 and 2. However, there are other terms in $B(1\ 2)$ which are like the terms in Eq. (7.14) except that each f bond connecting two circles is replaced by all h-allowed graphs connecting the same two circles. The conclusion that $B(1\ 2)$ is given by the graphical expansion in Eq. (7.14), but with h bonds instead of f bonds (as well as with some other qualifications that are not relevant to the terms shown), implies that $B(1\ 2)$ is no longer in range than $[h(1\ 2)]^2$. In all known cases in fluids (except at critical points), $h(1\ 2)$ decays at least as fast as $u(1\ 2)$ as $r \to \infty$ and then we have

$$c(1\ 2) \to -\beta u(1\ 2) \qquad \text{as } r_{12} \to \infty \tag{7.15}$$

by which we mean precisely that $c(1\ 2)/[-\beta u(1\ 2)] = 1$ to any desired accuracy if r_{12} is big enough.

The status of Eq. (7.15) is very interesting. It is often assumed to be correct. Theoretical deductions from it agree with results obtained in other ways. But it is not regarded as rigorously established because of uncertainty in the behavior of $B(1\ 2)$

at large r; the discussion above is merely a term-by-term analysis and is conceivably misleading for the infinite set of terms (see "The Literature" at the end of the chapter).

Next we consider several approximations to Eq. (7.11) that are widely used to estimate the correlation functions. Although this approach gives a unified view of these approximation methods, it should be noted that none of them was first formulated in this way.

DO PROBLEMS 7.1 TO 7.7.

7.2 THE LINEARIZED DEBYE–HÜCKEL APPROXIMATION

In this approximation we assume that the direct correlation function is given by its asymptotic form

$$c(1\ 2) = -\beta u(1\ 2), \qquad \text{large } r_{12} \tag{7.16}$$

over the whole range of coordinates [see Eq. (7.15)]. In the most important application we consider a system of ions, some positive and some negative, so we have the general equations

$$c_{ss'}(r) = -\beta u_{ss'}(r), \qquad \text{large } r \tag{7.17}$$

$$u_{ss'}(r) = e_s e_{s'}/\epsilon r, \qquad \text{large } r \tag{7.18}$$

where e_s is the ionic charge and the dielectric constant $\epsilon = 1$ for a simple ionic system (i.e., a plasma). We carry ϵ in the equations for later reference. The Debye–Hückel approximation for the ionic system is therefore

$$c_{ss'}(r) = -\frac{\beta e_s e_{s'}}{\epsilon r}, \qquad 0 < r < \infty \tag{7.19}$$

From Eqs. (A1.10) and (A1.11) in the $\alpha = 0$ limit we deduce that

$$\tilde{c}_{ss'}(k) = -\frac{4\pi\beta e_s e_{s'}}{\epsilon k^2} \tag{7.20}$$

This result needs justification because the integral following Eq. (A1.10) does not give a definite value if $\alpha = 0$. One way to elucidate this point is to notice that we can replace Eq. (7.18) with the more fundamental statements:

$$u_{ss'}(r) = \psi_s(r) e_{s'}, \qquad \text{large } r \tag{7.21}$$

$$\nabla^2 \psi_s(|\mathbf{r}|) = -\frac{4\pi}{\epsilon} e_s \delta(\mathbf{r}) \tag{7.22}$$

where $\psi_s(r) = \psi_s(|\mathbf{r}|)$ is the electrical potential at $\mathbf{r}$ due to an ion of species s at the origin and where this potential obeys Poisson's equation. Fourier-transforming both sides of (7.22), we find that

$$\tilde{\psi}_s(k) = \frac{4\pi e_s}{\epsilon k^2} \tag{7.23}$$

which is consistent with Eq. (7.20).

Having established that Eq. (7.20) indeed represents the Debye–Hückel approximation, we now calculate the pair correlation functions and thermodynamic coefficients of a hypothetical system that is governed by this equation. For this calculation, as for others for systems of many components (see Section 1.13 and Problems 7.6 and 7.7), it is convenient to use the tools of matrix algebra.

For ν ionic species the $\tilde{\mathbf{C}}$ matrix [Eq. (7.70)] is a $\nu \times \nu$ matrix with elements

$$\tilde{C}_{ss'} = -\frac{(\rho_s\rho_{s'})^{1/2} e_s e_{s'} 4\pi\beta}{\epsilon k^2} \tag{7.24}$$

and the ss' element of the product $\tilde{\mathbf{C}} \cdot \tilde{\mathbf{C}}$ is

$$(s|\tilde{\mathbf{C}} \cdot \tilde{\mathbf{C}}|s') = \sum_t \tilde{C}_{st}\tilde{C}_{ts'} = \frac{-\tilde{C}_{ss'}\kappa^2}{k^2} \tag{7.25}$$

where we introduce the "Debye kappa"

$$\kappa^2 \equiv \frac{4\pi\beta}{\epsilon} \sum_{t=1}^{\nu} \rho_t e_t^2 \tag{7.26}$$

It follows that we also have

$$(s|\tilde{\mathbf{C}} \cdot \tilde{\mathbf{C}} \cdot \tilde{\mathbf{C}}|s') = \tilde{C}_{ss'}\left(\frac{\kappa^2}{k^2}\right)^2 \tag{7.27}$$

and so on, from which we deduce that Eq. (7.73) becomes

$$\begin{aligned} \tilde{H}_{ss'} &= \tilde{C}_{ss'}\left[1 - \frac{\kappa^2}{k^2} + \left(\frac{\kappa^2}{k^2}\right)^2 - \left(\frac{\kappa^2}{k^2}\right)^3 + \cdots\right] \\ &= \frac{\tilde{C}_{ss'}}{1 + \kappa^2/k^2} \end{aligned} \tag{7.28}$$

or

$$\tilde{h}_{ss'}(k) = -\frac{4\pi\beta e_s e_{s'}}{\epsilon(k^2 + \kappa^2)} \tag{7.29}$$

and, with Eq. (A1.11),

$$h_{ss'}(r) = -\frac{\beta e_s e_{s'} e^{-\kappa r}}{\epsilon r} \tag{7.30}$$

This result for $h_{ss'} = g_{ss'} - 1$ is the famous Debye–Hückel pair correlation function; the Coulomb potential is *shielded* by the factor $e^{-\kappa r}$.

To extract the thermodynamic properties, we may use the virial equation [Eq. (4.35)], which in this case reads

$$\beta p = \sum_s \rho_s + \frac{1}{6}\beta \sum_s \sum_{s'} \rho_s \rho_{s'} \frac{e_s e_{s'}}{\epsilon}\left(-\frac{\beta e_s e_{s'}}{\epsilon}\right) \int d^3r \frac{1}{r}\frac{e^{-\kappa r}}{r} \tag{7.31}$$
$$= \sum_s \rho_s - \frac{\kappa^3}{24\pi}$$

The $\Sigma_s \rho_s$ term is of course βp^{id}, the ideal gas term, while $-\kappa^3/24\pi$ is the Debye–Hückel limiting law for $\beta(p - p^{\text{id}})$. For electrolyte solutions p is the osmotic pressure rather than the total pressure and $\beta p = \Sigma_s \rho_s$ is the van't Hoff equation for the osmotic pressure (Chapter 10). To appreciate the full significance of the Debye–Hückel limiting law, we notice that the virial equation of state [Eq. (6.1)], which is relevant to systems of molecules without long-range interactions [i.e., when the integral in Eq. (6.6) exists], has the form

$$\beta p = \rho + B_2\rho^2 + \cdots \tag{7.32}$$

while Eq. (7.31) has the form

$$\beta p = \rho + a\rho^{3/2} + \cdots \tag{7.33}$$

where $\rho = \Sigma_s \rho_s$ and where a does not depend on ρ. The difference is associated with the fact that Eq. (7.32) applies when the first correction to ideality is due to molecules interacting in pairs (see Section 6.2), while the $\rho^{3/2}$ term in Eq. (7.33) comes from interactions that are so long range that all the molecules interact collectively. This interpretation is consistent with the observation that each term in $[\cdot\cdot\cdot]$ in Eq. (7.28) diverges more strongly than the one before as $k \to 0$, as expected from the results in Problem 7.2. We get a finite result at $k = 0$ as in Eq. (7.29) only because all of the divergences, coming from all lengths of chains of ionic interactions, mutually cancel.

We recall [Eq. (4.29)] that it is realistic to expect

$$g_{ss'}(r) = 0 \qquad \text{for } r < \sigma_{ss'} \tag{7.34}$$

called the *core condition,* while Eq. (7.30) implies that

$$g_{ss'}(r) = h_{ss'}(r) + 1 \to \pm\infty \qquad \text{as } r \to 0 \tag{7.35}$$

This unrealistic behavior in the core repulsion region may be repaired by beginning with a better approximation than Eq. (7.19), but it is also customary to formulate an "extended Debye–Hückel theory" by merely writing

$$g_{ss'}(r) = \theta(r - \sigma_{ss'}) \times \left(1 - \frac{\beta e_s e_{s'} e^{-\kappa r}}{\epsilon r}\right) \tag{7.36}$$

where $\theta(x)$ is the unit step function [$\theta(x) = 0$ if $x < 0$, $\theta(x) = 1$ if $x > 0$]. A more elaborate extension, sometimes called the Debye–Hückel EXP approximation, is given by

$$g_{ss'}(r) = g_{ss'}^{\text{ref}}(r) \exp\left(-\frac{\beta e_s e_{s'} e^{-\kappa r}}{\epsilon r}\right) \tag{7.37}$$

where $g_{ss'}^{\text{ref}}$ is the pair correlation function for some nonionic reference system. Approximations like these are often accurate enough to be useful, obscure though the theoretical justification may be.

The $e^{-\kappa r}$ shielding in the pair correlation function and the dependence of $p - p^{\text{id}}$ on κ^3, which are consequences of the long range of the Coulomb potential, are only approximately valid for real ionic systems at finite concentrations, or even for simple models that are faithful to the core condition. This is to be expected in view of the severe approximation made in Eq. (7.19). However, it is remarkable that there are other simple consequences of the long range of the Coulomb potential, namely the Stillinger–Lovett moment conditions, that follow even when the short-range terms in $c_{ss'}$ are not neglected! We digress here to outline the theory of these moments.

7.3 THE STILLINGER–LOVETT MOMENT CONDITIONS

In place of Eq. (7.24), we may write the exact expression

$$(s|\tilde{\mathbf{C}}|s') = -\frac{x_s x_{s'}}{k^2} + D_{ss'}(k) \tag{7.38}$$

where we define

$$x_s = \left(\frac{4\pi\beta\rho_s}{\epsilon}\right)^{1/2} e_s \tag{7.39}$$

and where $D_{ss'}(k)$ derives from short-range contributions to the exact $c_{ss'}(r)$; $D_{ss'}(0)$ is finite. We also introduce the vector $\mathbf{x} = x_1, x_2, \ldots, x_s, \ldots$. We notice that $\mathbf{x} \cdot \mathbf{x} = \kappa^2$ [see Eq. (7.26)] and that the dyad $\mathbf{xx}$ can also be represented by a matrix whose s, s' element is $x_s x_{s'}$. Then the Ornstein–Zernicke equation (7.71) gives

$$\begin{aligned} \mathbf{1} + \tilde{\mathbf{H}} &= [\mathbf{1} - \tilde{\mathbf{C}}]^{-1} \\ &= \left[1 - \left(\mathbf{1} - \frac{\mathbf{xx}}{k^2 + \kappa^2}\right) \cdot \mathbf{D}\right]^{-1} \cdot \left(1 - \frac{\mathbf{xx}}{k^2 + \kappa^2}\right) \end{aligned} \tag{7.40}$$

where we have used the matrix identities

$$[\mathbf{1} + \mathbf{A} + \mathbf{B}]^{-1} = [\mathbf{1} + [\mathbf{1} + \mathbf{A}]^{-1} \cdot \mathbf{B}]^{-1} [\mathbf{1} + \mathbf{A}]^{-1} \tag{7.41}$$

$$\left(\mathbf{1} + \frac{\mathbf{xx}}{k^2}\right)\left(\mathbf{1} - \frac{\mathbf{xx}}{k^2 + \mathbf{x} \cdot \mathbf{x}}\right) = \mathbf{1} \tag{7.42}$$

We seek expressions for the moments

$$M_n = \sum_{ss'} \rho_s e_s \rho_{s'} e_{s'} \int d^3 r \, h_{ss'}(r) r^n \tag{7.43}$$

We notice that the even moments are coefficients of k^{2n} in the function

$$\begin{aligned} \hat{M} &\equiv \mathbf{x} \cdot \tilde{\mathbf{H}} \cdot \mathbf{x} \\ &= \frac{4\pi\beta}{\epsilon} \sum_n \frac{(ik)^{2n} M_{2n}}{(2n+1)!} \end{aligned} \tag{7.44}$$

where we have used the expansion of $\sin(kr)/kr$ in powers of kr. From Eq. (7.40) we find, after a little algebra,

$$\hat{M}=\frac{-\kappa^4}{k^2+\kappa^2}+\left[\frac{k^2}{k^2+\kappa^2}\right]^2\sum_{n=0}^{\infty}\mathbf{x}\cdot\mathbf{D}\cdot\left[\left(\mathbf{1}-\frac{\mathbf{xx}}{k^2+\kappa^2}\right)\cdot\mathbf{D}\right]^n\cdot\mathbf{x} \tag{7.45}$$

Equating the coefficients of k^0 and then of k^2 in Eqs. (7.44) and (7.45) gives

$$M_0=-\sum_s \rho_s e_s^2 \tag{7.46a}$$

$$M_2=-\frac{3\epsilon}{2\pi\beta} \tag{7.46b}$$

The higher moments depend on the short-range interactions through $\mathbf{D}$ and are less useful. The zeroth moment M_0 can also be derived from the well-known local electroneutrality condition

$$e_s+\sum_{s'} e_{s'}\rho_{s'}\tilde{h}_{ss'}(0)=0 \tag{7.46c}$$

DO PROBLEMS 7.8 TO 7.14.

7.4 THE PERCUS–YEVICK APPROXIMATION AND THE HARD-SPHERE FLUID

In the *Percus–Yevick* (PY) *approximation* the direct correlation function is given by

$$c(r)=\frac{f(r)g(r)}{e(r)} \tag{7.47a}$$

$$=f(r)y(r) \tag{7.47b}$$

$$=g(r)-y(r) \tag{7.47c}$$

where we write the condition for a one-component system with central forces only. This approximation, unlike the preceding one, is consistent with the exact core condition. The PY integral equation is obtained by using Eq. (7.47a) to reduce the Ornstein–Zernicke equation to one unknown:

$$h=c+c*h \tag{7.48a}$$

$$c=(1-e^{\beta u})(1+h) \tag{7.48b}$$

The resulting integral equation can be solved numerically as described in Section 7.7.

It is interesting to generate the graphical expansion of $h(1\ 2)$ in the PY approximation. We begin with the PY approximation in the form [see Eqs. (7.10), (7.47b), and (7.47c)]

$$t = c * c + c * c * c + c * c * c * c + \cdots \tag{7.49a}$$

$$c = (1 + t)f \tag{7.49b}$$

$$h = t + c \tag{7.49c}$$

and construct an iterative solution according to the scheme

$$t_{n+1} = c_n * c_n + c_n * c_n * c_n + \cdots$$

$$c_{n+1} = (1 + t_{n+1})f$$

$$h_n = t_n + c_n$$

where n is the index of the iteration. Here the first steps along these lines begin with $t_0 = 0$:

$$t_0 = 0, \qquad c_0 = f, \qquad h_0 = f \tag{7.49d}$$

$$t_1 = f * f + f * f * f + \cdots, \qquad c_1 = f + t_1 f, \qquad h_1 = f + t_1 + t_1 f \tag{7.49e}$$

$$t_2 = c_1 * c_1 + c_1 * c_1 * c_1 + \cdots, \qquad c_2 = f + t_2 f, \qquad h_2 = f + t_2 + t_2 f \tag{7.49f}$$

These equations are represented graphically in Fig. 7.1. By proceeding in this way we can establish a general prescription for the graphs of $h(1\ 2)$ in the PY approximation in terms of decorated polygons as follows.

We start with every polygon having three or more sides. Each side is covered by an f-bond (i.e., the two vertices connected by the side are connected by an f bond). Also, two adjacent vertices in each polygon are each covered by white circles labeled, respectively, 1 and 2. The rest of the vertices are covered by black ρ circles. Now for every polygon we make all of the distinct graphs which can be made by connecting nonadjacent circles by f bonds, subject to the restriction that no f bonds cross each other. Finally, these graphs are supplemented by a second set just like the first except that there is no f bond directly connecting the two white circles.

It may be verified that all of the graphs in Fig. 7.1 are among the graphs generated by this procedure except for the graph which is simply $f(1\ 2)$, and it is not hard to see how to continue the iterations beyond those in Fig. 7.1 to generate the rest, so we have

$$h^{\mathrm{PY}}(1\ 2) = f(1\ 2) + \text{the sum of all the graphs generated from polygons as described above} \tag{7.49g}$$

We also notice that the graphical expansion of h^{PY} can be obtained from the graphical expansion of h^{exact} given in Eq. (7.3) by omitting two classes of graphs:

1. All product graphs except for products of $f(1\ 2)$ with another graph
2. Bridge graphs and graphs with any residuals that are bridge graphs

The first few terms of $h^{\mathrm{exact}} - h^{\mathrm{PY}}$ are shown in Fig. 7.1. We return to these results in Section 7.6.

In the case of a hard-sphere fluid [Eq. (4.24)] the PY approximation can be written in a particularly symmetric form:

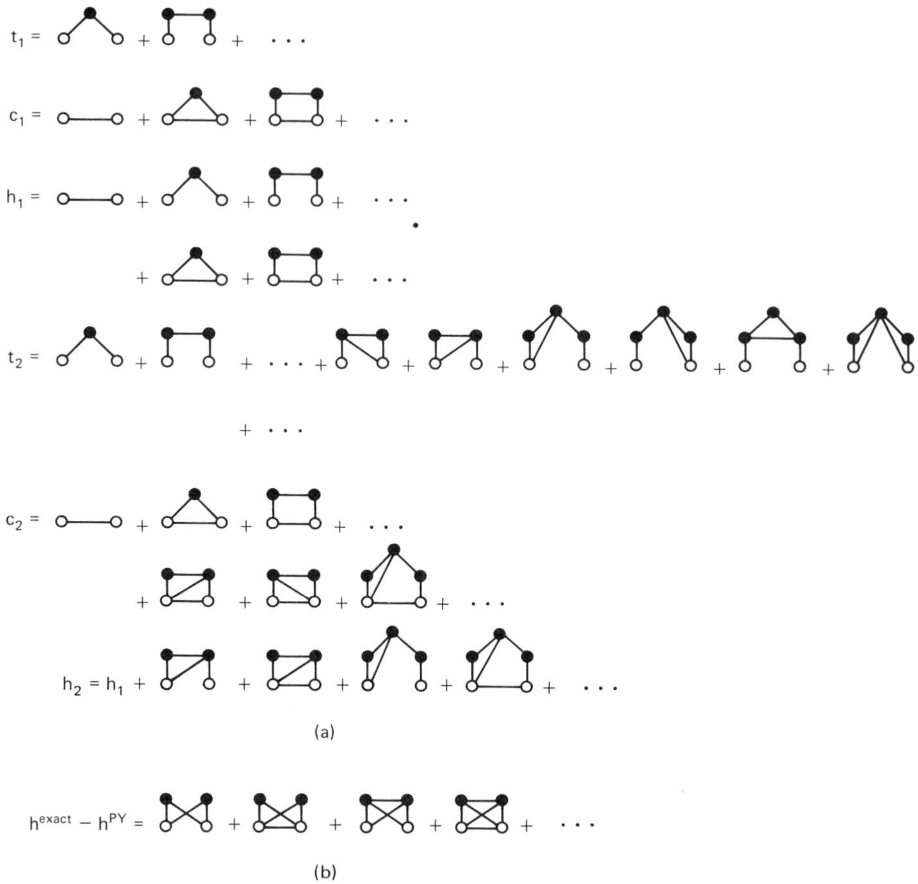

Figure 7.1
(a) Graphical analysis of the iterative solution of the PY approximation.
(b) Graphical representation of $h^{\text{exact}} - h^{\text{PY}}$.
In both cases the graphs show f bonds on black ρ circles and white 1-circles of unit weight.

$$g(r) = 0 \qquad \text{if } r < \sigma \tag{7.50a}$$

$$c(r) = 0 \qquad \text{if } \sigma < r \tag{7.50b}$$

since $g = y$ when $u = 0$. It is remarkable that this approximation for the hard-sphere fluid, namely the Ornstein–Zernicke equation and Eq. (7.50), can be solved analytically (Wertheim, 1963; Thiele, 1963). The result in reduced variables $x \equiv r/\sigma$, $\eta = (\pi/6)\rho\sigma^3$ is given by

$$c(\sigma x) = -\theta(1 - x)[(1 + \tfrac{1}{2}\eta x^3)\lambda_1 + 6\eta x\lambda_2] \tag{7.51a}$$

$$\lambda_1 = \frac{(1 + 2\eta)^2}{(1 - \eta)^4} \tag{7.51b}$$

$$\lambda_2 = -\frac{(1 + \frac{1}{2}\eta)^2}{(1 - \eta)^4} \tag{7.51c}$$

A formula for $g(r)$, which turns out to be analytic in each of the ranges $(n-1)\sigma < r < n\sigma$, where $n = 1, 2, 3, \ldots$, can be deduced from this equation for $c(r)$, as shown by Smith and Henderson (1970). For mixed hard spheres of various sizes the analytical solution to the PY approximation is also known.

The compressibility and hence the pressure and free energy can be calculated from $c(r)$ and Eq. (7.75). Also, for a hard-sphere fluid in the PY approximation, the virial equation can be expressed in terms of $c(\sigma)$. In general, Eq. (4.35) can be written

$$\begin{aligned} \frac{\beta p}{\rho} &= 1 + \frac{1}{6}\rho \int d^3r\, r \frac{\partial \ln e(r)}{\partial r}\, e(r)\, y(r) \\ &= 1 + \frac{1}{6}\rho \int d^3r\, r \frac{\partial e}{\partial r} y(r) \end{aligned} \tag{7.52}$$

In the special case of the hard-sphere fluid, we have [see (7.36) for notation]

$$e(r) = \theta(r - \sigma) \tag{7.53a}$$

In general,

$$\frac{\partial \theta(r-\sigma)}{\partial r} = \delta(r - \sigma) \tag{7.53b}$$

so we find, for the hard-sphere fluid,

$$\frac{\beta p}{\rho} = 1 + \frac{2\pi}{3}\rho\sigma^3 y(\sigma) \tag{7.54}$$

For r even infinitesimally smaller than σ, we have $y = -c$ in the PY approximation, and for very small positive ϵ we have $y(\sigma) \cong y(\sigma - \epsilon)$, since $y(r)$ is a continuous function of r, even around $r = d$. This is most easily seen from Eq. (7.4), where there are no direct bonds between the white circles. Therefore, we have, for the hard-sphere fluid in the PY approximation,

$$\frac{\beta p}{\rho} = 1 - \frac{2\pi}{3}\rho\sigma^3 c(\sigma) \tag{7.55}$$

where $c(\sigma)$ is evaluated as the limit of $c(x\sigma)$ as x approaches 1 from below.

The results for the equation of state of the hard-sphere fluid are given here.

PY, compressibility equation, (7.75):

$$\frac{\beta p}{\rho} = \frac{1 + \eta + \eta^2}{(1-\eta)^3} \tag{7.56a}$$

PY, virial equation, (7.55):

$$\frac{\beta p}{\rho} = \frac{1 + 2\eta + 3\eta^2}{(1-\eta)^2} \tag{7.56b}$$

Carnahan–Starling:

$$\frac{\beta p}{\rho} = \frac{1 + \eta + \eta^2 - \eta^3}{(1 - \eta)^3} \tag{7.56c}$$

The last of these, derived in another way as described in Section 6.9, is in amazingly good agreement with MD results for the hard-sphere fluid over the entire range of densities.

The comparison of Eqs. (7.56a) and (7.56b), often called a *thermodynamic consistency test,* provides a measure of the accuracy of $c(r)$ for the hard-sphere fluid in the PY approximation. At $\eta = 0.5$, corresponding to the liquid state near the triple point (i.e., near the freezing point at ordinary low pressures) the PY (compressibility) pressure is roughly 30% greater than the PY (virial) pressure, while the Carnahan–Starling value is roughly halfway between (Fig. 7.2). At the same time the PY pair correlation function $g(r)$ is almost indistinguishable from the accurate MD result except at r/σ just above unity (Fig. 7.3).

We have emphasized the application of the PY approximations to the hard-sphere fluid because of the importance of the exact solution of the PY integral equation. For many other models, requiring numerical solution of the PY integral equation, this approximation still gives useful results. However, for models for which the potential has long range [i.e., the integral in Eq. (6.6) does not converge], the PY approximation is not consistent with the correct asymptotic form of $c(r)$ as given in Eq. (7.15).

7.5 THE MEAN SPHERICAL APPROXIMATION

The *mean spherical approximation* (MSA) is applicable to model potentials with hard cores as in

$$u(r) = \infty \qquad \text{if } r < \sigma \tag{7.57}$$

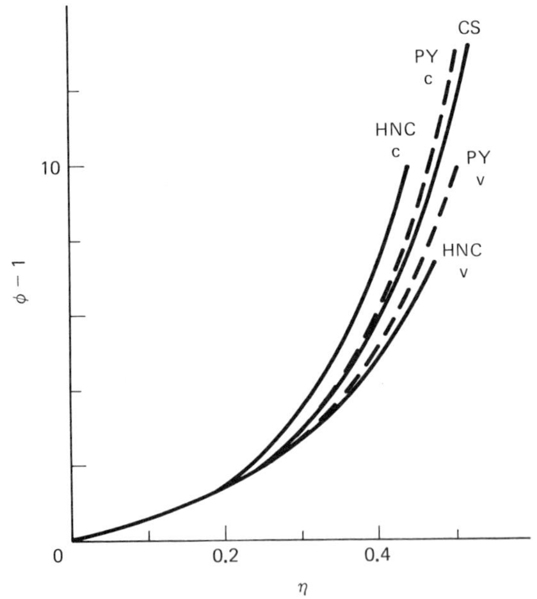

Figure 7.2 $\phi - 1$ as a function of η for the hard-sphere fluid. $\phi = \beta p/\rho$. CS is the Carnahan–Starling equation of state. The difference between Carnahan–Starling and the "exact" function (from MD simulation) is not visible on the scale of this graph. The other curves show ϕ calculated from the compressibility equation (c) and virial equation (v) for the HNC and PY approximations. (After Hansen and McDonald, 1976.)

where σ is the core diameter. The exact core condition

$$g(r) = 0 \qquad \text{if } r < \sigma \tag{7.58a}$$

together with the approximation

$$c(r) = -\beta u(r) \qquad \text{if } \sigma < r \tag{7.58b}$$

and the Ornstein-Zernicke equation constitute the MSA [see Eq. (7.49)].

In the case of hard-sphere models the MSA evidently is the same as the PY approximation, which, as described in Section 7.4, has been solved analytically. The MSA also has been solved analytically in some other cases, including charged hard spheres, which will be discussed here, and dipolar hard spheres, which will be discussed in detail in Chapter 8.

We consider a model for ionic fluids in which the potential U_N is pairwise additive [Eq. (4.27) together with Eq. (4.26)]:

$$u_{ss'}(r) = \begin{cases} \infty & \text{if } r < \sigma_{ss'} \\ \dfrac{e_s e_{s'}}{\epsilon r} & \text{if } \sigma_{ss'} < r \end{cases} \tag{7.59}$$

[see Fig. 4.1, Eqs. (7.16) and (7.31)]. With $\epsilon = 1$ this potential characterizes a model for a system of charged hard spheres. With $\epsilon = \epsilon_0$, the dielectric constant of a solvent, this potential U_N characterizes the *primitive model* for ionic solutions; it is even simpler than a model for hard spheres with central charges in a continuum solvent, which would require other terms such as a term proportional to $1/r^4$ to represent

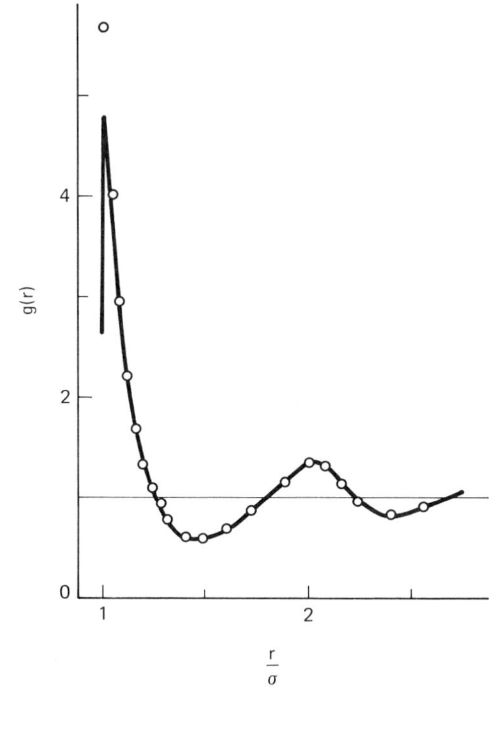

Figure 7.3 Pair correlation function for the hard-sphere fluid at $\eta = 0.49$. ○: simulation results; —: PY approximation. (After Hansen and McDonald, 1976.)

the interaction of the charge in one ion with the dipole it induces in a second ion when the interiors of the ions do not have the same dielectric constant as the solvent.

For the restricted primitive model (two species, $e_1 + e_2 = 0$, $\sigma_{11} = \sigma_{12} = \sigma_{22}$) Waisman and Lebowitz (1972) solved the MSA analytically with the following results:

$$c_{ss'}(r) = c_{ss'}^{\mathrm{PY}}(r) - \frac{1}{T_r}\frac{\sigma}{r}B\left(2 - \frac{Br}{\sigma}\right) \qquad \text{if } r < \sigma \tag{7.60}$$

$$T_r = \frac{\epsilon\sigma}{\beta e_1^2} \tag{7.61a}$$

$$B = \frac{1 + \kappa\sigma - (1 + 2\kappa\sigma)^{1/2}}{\kappa\sigma} \tag{7.61b}$$

where κ is the Debye kappa [Eq. (7.26)] and c^{PY} is the PY approximation to $c(r)$ for the uncharged hard-sphere fluid with the same diameters and density. The analytic solution of the MSA for quite general primitive models (various ion sizes and charges, mixed electrolytes) has been obtained by Blum (1980).

The thermodynamic properties can be obtained from the set of $c_{ss'}(r)$ via the compressibility or virial equations, but the best results are obtained by calculating the excess energy from the direct correlation functions and then calculating other thermodynamic functions from E^{ex} as a function of ρ_{ion} and T. For example,

$$\frac{A^{\mathrm{ex}}(T)}{T} = -\int_0^{1/T} E^{\mathrm{ex}}(t)\, d\left(\frac{1}{t}\right) \tag{7.61c}$$

where we take advantage of the fact that in an analytical theory $E(T)$ is known for all T.

In Fig. 7.4 we show comparisons of thermodynamic properties for various primitive model electrolytes calculated by the MSA and by other approximations. Evidently, the MSA is quite acceptable for primitive models for aqueous 1–1 electrolytes ($T_r \simeq 0.5$, $\eta \lesssim 0.1$) but not useful for models for aqueous 2–2 electrolytes ($T_r \sim 0.15$, $\eta \lesssim 0.1$). As might be expected because of the merely linear dependence of the direct correlation function on the potential in MSA [see Eq. (7.58b)], this approximation also fails for smaller values of T_r.

In the case of the Debye–Hückel theory, it has been well known since the report of Bjerrum's theory of ion pairing in 1925 that the useful range of the theory could be greatly extended by introducing a chemical model $M^+ + X^- \rightleftharpoons MX$ for ion pairing and treating the system as neutral MX particles mixed with unassociated M^+ and X^- ions governed by the Debye–Hückel theory. A similar procedure, with MSA in place of the Debye–Hückel approximation, is even more effective as reported by Ebeling and Grigo (1980).

7.6 THE HYPERNETTED CHAIN APPROXIMATION

In the HNC approximation we combine the Ornstein–Zernicke equation with

$$c(r) = h(r) - \ln\left[1 + h(r)\right] - \beta u(r) \tag{7.62}$$

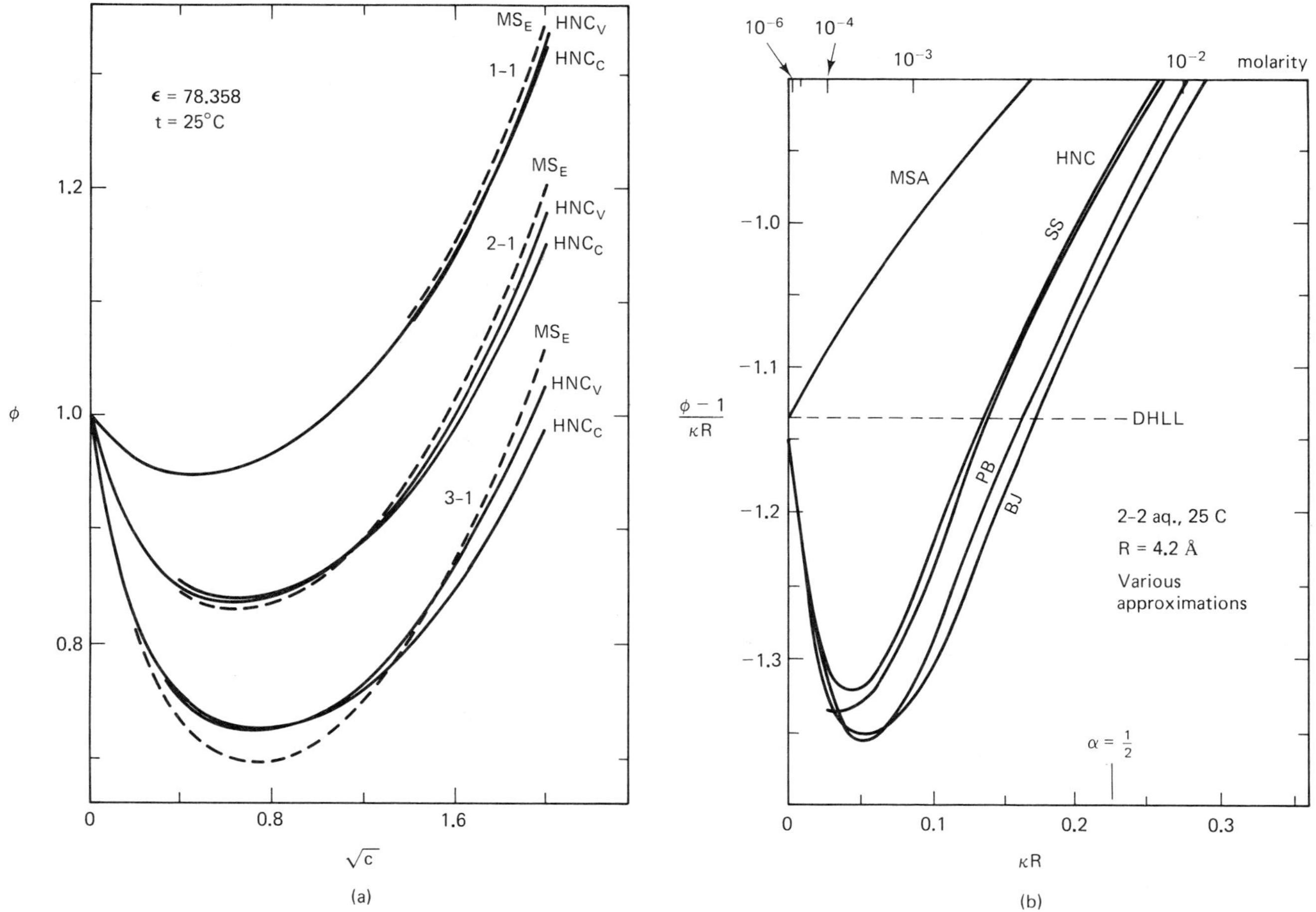

Figure 7.4 (a) Osmotic coefficients for restricted primitive model electrolytes of various charge types treated by various approximations, $\sigma = 4.2$ Å. MS_E is the MSA approximation with the osmotic coefficient calculated from the energy. HNC_v (HNC_c) is the HNC approximation with the osmotic coefficient calculated from the virial (compressibility) equation. (After Rasaiah, 1972.) (b) Various approximations for restricted primitive model for 2–2 aqueous electrolyte, except that SS is the HNC result for a similar model based on charged soft spheres. PB is the unlinearized Poisson–Boltzmann approximation of Debye and Hückel, while BJ is Bjerrum's approximation and α is the function of ions paired in this approximation. (After Rossky and Friedman, 1980.)

which is the same as the exact Eq. (7.11) but with the sum of bridge graphs omitted. It follows that $h^{\text{exact}} - h^{\text{HNC}}$ comprises all of the graphs of Eq. (7.3) which are bridge graphs or which have any residual which is a bridge graph. Although this is a smaller set of graphs than $h^{\text{exact}} - h^{\text{PY}}$, it turns out that the PY approximation often is more accurate than the HNC approximation in applications to simple fluids, as in the example in Fig. 7.2.

A typical—maybe even dominant—contribution to this situation is the graph, , which is counted in HNC but not in PY, but which may be largely canceled by 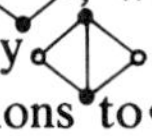 in the exact theory. Consider the case of hard spheres. Important contributions to  from configurations in which the two black circles are close together will tend to be canceled by 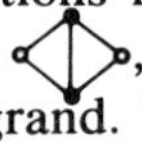, where these same configurations have an additional -1 factor in the integrand. Thus one may be better off suppressing 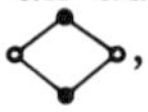, as PY does, rather than including it without 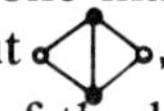, as HNC does.

On the other hand, the discussion of the large-r behavior of Eq. (7.11) applies equally to Eq. (7.62), so we find that Eq. (7.15) applies in the HNC approximation, which therefore is suitable for applying to models with long-range forces. It also follows that for ionic systems the HNC approximation is exactly consistent with the zeroth- and second-moment conditions. For models for electrolytes the HNC approximation is quite accurate for $T_r \gtrsim 0.15$ and $\eta < 0.2$ (Fig. 7.4), but it also is reasonably accurate for primitive models for fused salts (Fig. 7.5).

As long as $u(1\ 2) = u(r)$, the HNC approximation is quite straightforward to solve numerically. For example, the algorithm of Rossky and Dale is given next.

DO PROBLEMS 7.15 AND 7.16.

7.7 THE NUMERICAL SOLUTION OF INTEGRAL EQUATIONS

On the basis of Eqs. (7.9) to (7.11), one may formulate a class of approximate integral equations for the pair correlation function as follows:

$$h = c + c * h \tag{7.63a}$$

$$c = h - \ln(1 + h) - \beta u + B' \tag{7.63b}$$

where $B'(1\ 2)$ is either some known functional of h or some known function of r_{12}. The integral equation per se is obtained by eliminating c between these equations. Then h and c (as well as $t = h - c$) obtained by solving the integral equation will be approximations to the exact h and c (and $t = h - c$) for the given u, β, and ρ.

An equivalent pair of approximate equations,

$$c = (1 + f)e^{t + B'} - t - 1 \tag{7.64a}$$

$$t = c * c + c * c * c + c * c * c * c + \cdots \tag{7.64b}$$

is readily derived from Eqs. (7.63). We notice that if $B' = 0$, we recover the HNC approximation; then the graphical expansion of $t(1\ 2)$ in terms of f bonds on two white circles labeled 1 and 2 and one or more black ρ circles may be obtained from

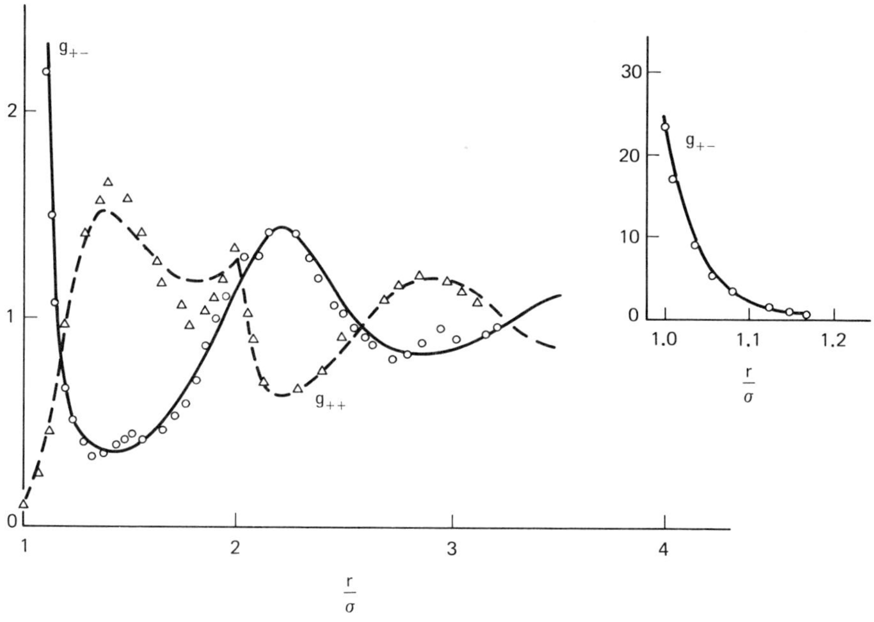

Figure 7.5 Monte Carlo and HNC approximations for restricted primitive model for a fused salt.

Pair correlation functions:

○○○, ——: $g_{+-}(r)$

△△△, ------: $g_{++}(r) = g_{--}(r)$

State: $T_r = 0.0148$ [see Eq. (7.61a)], $\eta = 0.4$

The solid and dashed curves are HNC results. The circles and triangles are MC results. (From Larsen, 1979.)

the graphical expansion of the exact $t(1\ 2)$ by suppressing any graph with one or more residuals which are bridge graphs. We also notice that if $B' = \ln(1 + t) - t$, we recover the PY approximation. Finally, if $B'(1\ 2)$ is a specified function of r_{12}, we have the situation encountered in the RHNC approximation, which is discussed in Chapter 9.

Now we assume that some approximation t_j to the t function is given and we discuss the use of Eqs. (7.64) to refine this function, the algorithm of Rossky and Dale (1980).

$$c_j = (1 + f)e^{t_j + B'_j} - t_j - 1 \tag{7.65a}$$

$$t_{j+1} = c_j * c_j + c_j * c_j * c_j + c_j * c_j * c_j * c_j + \cdots \tag{7.65b}$$

The numerical computation of $c_j(r)$ for given $f(r)$, $t_j(r)$, and $B'_j(r)$ is quite straightforward. One represents each function as a series of discrete values, say $f(n\hat{r})$, where n is an integer ranging from 1 to about 1000 and $\hat{r}$ is small enough, say 0.05 Å, so that $|f((n + 1)\hat{r}) - f(n\hat{r})|$ is always small compared to unity. The computation in

Eq. (7.65b) is also easy if we use the convolution theorem (Appendix 1) with a fast Fourier transform computational routine to evaluate the sum of the infinite series in k space.

At first sight we have no reason to expect that finding the iterative solution of the integral equation with the help of the algorithm in Eqs. (7.65) should be advantageous compared with other iterative procedures that might be proposed. Indeed, one cannot even tell merely by inspection that iterating in this way will give sequences of functions c_j and t_j that converge to the solution of the integral equation approximation in Eqs. (7.63). However, much more is known in the particular case that this iterative procedure is initiated with the function $t_0(r) = 0$ for all r. Then the Rossky–Dale algorithm is essentially the recursion relation for a systematic way of summing the infinite series in the graphical expansion of $h(1\ 2)$.

The Rossky–Dale algorithm has been generalized to apply to cases involving the renormalization of short-range as well as long-range forces and also to mixtures of species.

DO PROBLEMS 7.15–7.17

THE LITERATURE

General references for correlation function theory are Hansen and McDonald (1976), Percus (1964), Andersen (1975), Watts (1973), and Stell (1964, 1967). For the history of Eq. (7.15), its derivation, the next terms in the large r expansion of $c(r)$, and the critical region behavior, see Stell (1977).

The moment conditions for ionic solutions are due to Stillinger and Lovett (1968). The present derivation is similar to that of Mitchell et al. (1979). For the Fourier transform of the Coulomb potential, see Friedman (1962).

For analytic solutions of integral equations, see Thiele (1963), Wertheim (1964, 1980), Baxter (1971), Waisman and Lebowitz (1972), and Blum (1980).

For comparison of various approximation methods applied to the primitive model of ionic systems, see Outhwaite (1975), Rasaiah et al. (1972), and Valleau et al. (1980).

PROBLEMS

7.1. Derive Eq. (7.5) from (7.4) with the help of Lemma 1 of Section 6.6.

7.2. Using the results in Appendix 1, show that if

$$f = \circ\!\!-\!\!\circ = \frac{e^{-sr}}{r}$$

then

$$f * f = \circ\!\!-\!\!\bullet\!\!-\!\!\circ = \frac{2\pi\rho}{s} e^{-sr}$$

and

$$\frac{f * f}{f} \longrightarrow \infty \qquad \text{as } r \longrightarrow \infty$$

so $f * f$ is longer range than f.

7.3. Using the results in Appendix 1, find

$$\frac{f * f}{f} \qquad \text{as } r \longrightarrow \infty$$

in the case in which f is Gaussian.

7.4. For a multicomponent system the equations in Section 7.1 are generalized in the way described in connection with Eq. (6.75). Show that in this case the OZ equation becomes

$$h_{ss'}(1\ 2) = c_{ss'}(1\ 2) + \sum_{s''} \int c_{ss''}(1\ 3) d(3) \rho_{s''} h_{s''s'}(3\ 2) \tag{7.66}$$

where the sum is over species.

7.5. **(a)** Show that $f * g = g * f$ using Eq. (A1.4).

(b) Show that $f_1 * f_2 * \cdots * f_n$ is unchanged if the order of the functions in the convolution is changed.

7.6. **(a)** Show that, in the notation of Appendix 1, one can write the Ornstein–Zernicke equation for a one-component system as

$$(1 + \rho\tilde{h})(1 - \rho\tilde{c}) = 1 \tag{7.67}$$

or

$$\tilde{h} = \tilde{c} + \tilde{c}\rho\tilde{c} + \tilde{c}(\rho\tilde{c})^2 + \cdots \tag{7.68}$$

(b) In the multicomponent case it is convenient to introduce

$$H_{ss'}(1\ 2) \equiv (\rho_s \rho_{s'})^{1/2} h_{ss'}(1\ 2) \tag{7.69}$$

$$C_{ss'}(1\ 2) \equiv (\rho_s \rho_{s'})^{1/2} c_{ss'}(1\ 2) \tag{7.70}$$

Show that in this case the Ornstein–Zernicke equation may be written

$$(\mathbf{1} + \tilde{\mathbf{H}}) \cdot (\mathbf{1} - \tilde{\mathbf{C}}) = \mathbf{1} \tag{7.71}$$

where $\mathbf{1}$ is the unit matrix, $\tilde{\mathbf{H}}$ is the matrix whose elements are $\tilde{\mathrm{H}}_{ss'}(k)$ and $\tilde{\mathbf{C}}$ is the matrix whose elements are $\tilde{\mathrm{C}}_{\ s'}(k)$.

7.7. Verify Eq. (7.12) by differentiating with respect to h to get the more familiar expansion

$$\frac{1}{1+h} = 1 - h + h^2 - h^3 + \cdots \tag{7.72}$$

which in turn may be verified by multiplying both sides by $1 + h$. In a similar way show that Eq. (7.71) implies

$$\tilde{\mathbf{H}} = \tilde{\mathbf{C}} + \tilde{\mathbf{C}} \cdot \tilde{\mathbf{C}} + \tilde{\mathbf{C}} \cdot \tilde{\mathbf{C}} \cdot \tilde{\mathbf{C}} + \cdots \tag{7.73}$$

7.8. **(a)** Noting that Eq. (4.41) may be written

$$\frac{\kappa_T \rho}{\beta} = 1 + \rho\tilde{h}(0) \tag{7.74}$$

then from Eq. (7.67) derive

$$\frac{\beta}{\rho\kappa_T} = 1 - \rho\tilde{c}(0) \tag{7.75}$$

(b) Similarly, combine Eqs. (4.39) and (7.71) with (1.65) to derive

$$\beta\frac{\partial\mu_s}{\partial\rho_{s'}} = \frac{\delta_{ss'}}{\rho_s} - \tilde{c}_{ss'}(0) \tag{7.76}$$

7.9. Take the Fourier transform [as in Eq. (A1.1)] of both sides of Eq. (7.22) to get (7.23). Notice the integration by parts, for example,

$$\int_V d^3r\, e^{i\mathbf{k}\cdot\mathbf{r}}\frac{\partial^2\psi(\mathbf{r})}{\partial x^2} = \int_S d^2r\, e^{i\mathbf{k}\cdot\mathbf{r}}\frac{\partial\psi(\mathbf{r})}{\partial x} - ik_x\int_V d^3r\, e^{i\mathbf{k}\cdot\mathbf{r}}\frac{\partial\psi(\mathbf{r})}{\partial x} = \text{etc.}$$

7.10. Show that $\hat{M}$ does give a power series in k, where the moments are coefficients as indicated in Eq. (7.44).

7.11. Evaluate M_0 and M_2 using Eq. (7.43) and the Debye–Hückel function, Eq. (7.30). Are Eqs. (7.46) also satisfied by the EXP approximation, Eq. (7.37)? Or by the approximation $h_{ss'}(r) = [-\beta e_s e_{s'} e^{-\kappa(r-d)}/\epsilon r(1+\kappa d)]\theta(r-d)$, where d is called the "distance of closest approach"?

7.12. Often, the electroneutrality condition (7.46c) is written in terms of $\int g(r)\, d^3r$ rather than $\tilde{h}(0)$. Which is correct?

7.13. Assuming two species labeled + and −, write out Eqs. (7.24) to (7.27) in detail.

7.14. For an ionic system the Gibbs–Duhem equation reads, for $\rho = \Sigma_s\, \rho_s$,

$$d[(1-\phi)\rho] + \rho d\ln\gamma_\pm = 0 \tag{7.77}$$

where $\phi = \beta p/\rho$ is the osmotic coefficient and $\gamma_\pm$ the mean ionic activity coefficient. Show that Eq. (7.31) leads to the Debye–Hückel limiting law, $\ln\gamma_\pm = -A\sqrt{\rho}$, that is familiar in physical chemistry, then find the theoretical expression for A. Finally, can you get the same result more directly from Eq. (7.76)? Consider the last question in relation to the case of a mixture of two kinds of ions with charges e and $-e$.

7.15. **(a)** For an ionic system, is the PY equation consistent with the moment conditions?
(b) How about the HNC equation? [*Hint:* In each case make a cluster expansion of $c_{ss'}(r)$ and see if it goes to $-\beta u_{ss'}(r)$ at large r.]

7.16. Show that for a nonionic system the pressure calculated either by PY or HNC is exact only through the term $\rho^3 B_3$ [see Eq. (6.1)].

7.17. **(a)** Show that Eq. (7.47b) generates a set of graphs that is a subset of the graphs that represent the direct correlation function.
(b) Show that in the case of a fluid of hard spheres of diameter σ Eq. (7.47b) gives $c(r) = 0$ for $r > \sigma$.
(c) Draw three of the graphs that are in the exact $c(r)$ but not in $c_{\text{PY}}(r)$.

8

Molecular Correlation Functions

8.1 ATOMS OR MOLECULES

In those theories in which the coordinates of the particles play a role, we mostly study systems of structureless particles, so only three coordinates per particle suffice to fix the configuration of a system. Fortunately, most such theories are equally applicable to more general systems, an aspect that is mentioned in Section 3.3 and in connection with Eq. (6.78).

Now we introduce some of the mathematical tools that are valuable in the work of applying the statistical mechanical theories to systems of molecules with rigid internal structure such that the configuration of the system depends on two or three orientational coordinates as well as three center-of-mass coordinates for each molecule. We also illustrate the use of these techniques in several important problems relating to molecular systems.

First, however, we discuss the choices one has in constructing models to represent systems of molecules. Thus we consider the potential function U_N for several different kinds of model representations of a system of molecules. For clarity here it is advantageous to consider a particular fluid made up of polyatomic molecules; water is our example.

In one case U_N is the Born–Oppenheimer potential surface for N oxygen and $2N$ hydrogen atoms. Then the interpretation of the equations in Chapters 3 to 7 in terms of atomic coordinates (3 per O atom and 3 per H atom) is straightforward, but two serious problems can be anticipated. One is that in some regions of the BO surface, for example, those important in determining vibrations of chemical bonds, the curvature is so great that quantum "corrections" will be enormous [see Eq. (3.38)], thereby making the classical limit quite irrelevant. A particular example is the OH bond vibration with a frequency near 3000 cm^{-1} [see Eq. (3.39)]. A second serious problem concerns the approximation of U_N as pairwise additive, which is surely a very crude approximation. For example, the energy of H_3O with three O—H distances near 1.00 Å is nowhere near the sum of the energies of three O—H with 1.00 Å bond length, even after including the H—H pair potentials.

Nevertheless, a type of model Hamiltonian which may be considered for this problem has U_N as the sum of O—H, H—H, and O—O atom–atom pair potentials. Remarkably, it has been demonstrated that one can construct these potentials to get about the right structure for the isolated H_2O molecules as well as for the dense fluid, as shown for the central force model of water studied by Rahman et al. (1975). Of course, the theory of Chapters 4 to 7 is directly applicable to central force models.

In another case U_N is the potential of interaction of N rigid water molecules as in the ST2 and MCY models for water (Stillinger, 1975). In this case we neglect the changes in the bond vibrations due to the intermolecular forces. Doing so may be a useful approximation because water's molecular bond geometry and bond vibrations are nearly the same in condensed phases as in isolated molecules.

The further developments in this chapter are for models of fluids as assemblies of rigid molecules with orientation-dependent interactions.

8.2 INVARIANT EXPANSIONS

For two rigid molecules the pair potential has the form

$$u(1\ 2) = u(r_{1\ 2}, \Omega_1, \Omega_2) \tag{8.1}$$

where $r_{1\ 2}$ is the center-to-center distance and $\Omega_i = \alpha_i, \beta_i, \gamma_i$ is the set of Eulerian angles (Fig. 8.1) that specify the orientation of a coordinate system fixed in molecule i with respect to a reference coordinate system which is at first chosen with its z axis fixed on the center-to-center line (Fig. 8.2a). In isotropic systems the correlation functions such as $g(1\ 2)$ as well as the potential $u(1\ 2)$ are invariant pairwise functions of the variables $r_{1\ 2}$, Ω_1, and Ω_2, meaning that these functions are unchanged if the pair of molecules is translated or rotated as a whole. For such functions there is a generalized Fourier transformation (Appendix 2) which converts convolution integrals to matrix products. Thus the Ornstein–Zernicke equation for a one-component system is

$$h(1\ 2) = c(1\ 2) + \rho' \int c(1\ 3) d(3) h(3\ 1) \tag{8.2a}$$

where $\rho' = N/8\pi^2 V$. In more detail we have

$$h(r_{1\ 2}, \Omega_1, \Omega_2) = c(r_{1\ 2}, \Omega_1, \Omega_2) + \rho' \int c(r_{1\ 3}, \Omega_1, \Omega_3)\, d^3r_3\, d^3\Omega_3\, h(r_{3\ 2}, \Omega_3, \Omega_2) \tag{8.2b}$$

which is converted by generalized Fourier transformation to

$$\mathbf{H} = \mathbf{C} + \mathbf{C} \cdot \mathbf{H} \tag{8.3a}$$

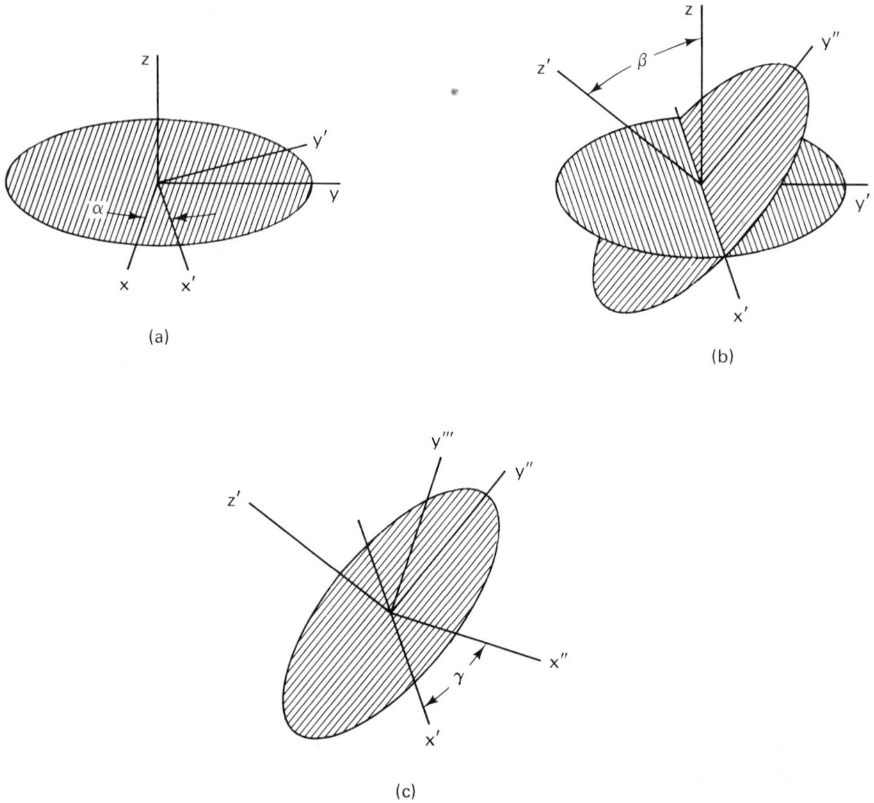

Figure 8.1 Eulerian angles for the rotation of coordinate system x, y, z to x'', y''', z'. (a) Rotation about z by angle α changing x to x' and y to y'. (b) Rotation about x' by angle β changing y' to y'' and z to z'. (c) Rotation about z' changing x' to x'' and y'' to y'''.

or

$$(\mathbf{1}+\mathbf{H})\cdot(\mathbf{1}-\mathbf{C})=\mathbf{1} \tag{8.3b}$$

where $\mathbf{H}$ and $\mathbf{C}$ are matrices whose elements are, respectively,

$$(l,\, m|\mathbf{H}|l',\, m') = \rho' \hat{h}(k)^{ll'}_{mm'n} \tag{8.3c}$$

$$(l,\, m|\mathbf{C}|l',\, m') = \rho' \hat{c}(k)^{ll'}_{mm'n} \tag{8.3d}$$

where the transforms $\hat{h}$ and $\hat{c}$ are defined in Appendix 2. The indices l and m basically derive from the spherical harmonic $Y^m_l(\alpha,\, \beta)$, where α and β are two of the Eulerian angles, just as in the quantum-mechanical theory of angular momentum. In applications to mixtures the indices of matrix elements must be extended from $|l,\, m)$ to $|s,\, l,\, m)$, where s is the species of molecule, and then ρ' in Eqs. (8.3c) and (8.3d) is replaced by $(\rho'_s \rho'_{s'})^{1/2}$.

While these generalized Fourier transforms enable us to exploit the Ornstein–Zernicke equation for molecular (i.e., orientation-dependent) correlation functions, there is another general way to proceed that seems easier to grasp intuitively and

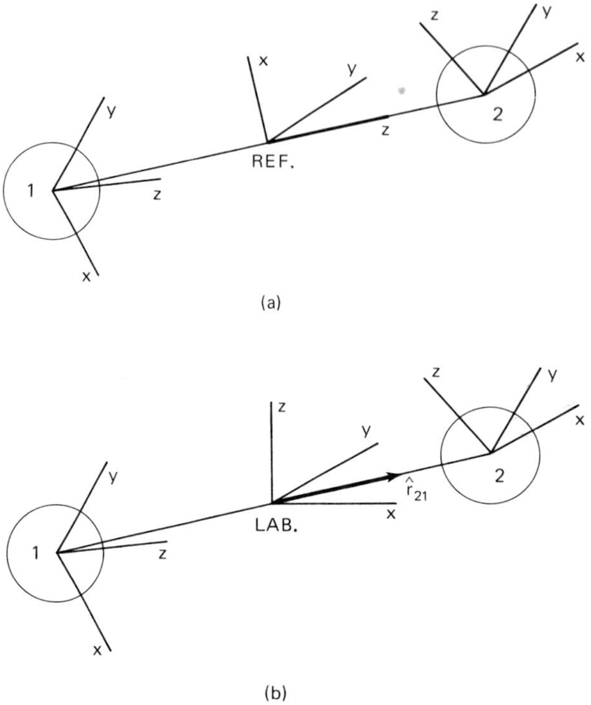

Figure 8.2 Coordinate systems for molecular correlation functions.

There is a Cartesian coordinate system fixed in molecule 1 and one fixed in molecule 2; they rotate with their respective molecules.

(a) The reference frame is chosen with its z axis on the $\mathbf{r}_2 - \mathbf{r}_1$ vector. Then Ω_n is the set of Eulerian angles for the rotation from the reference frame to the system fixed in molecule n. Obviously, any function of only Ω_1, Ω_2, and $|\mathbf{r}_2 - \mathbf{r}_1|$ is an invariant pairwise function.

(b) An alternative way to choose the coordinates. Here Ω_n is the set of Eulerian angles for the rotation from a laboratory-fixed frame to a frame fixed in molecule n. Besides Ω_1 and Ω_2 the configuration depends on $\hat{r}_{21} = (\mathbf{r}_2 - \mathbf{r}_1)/|\mathbf{r}_2 - \mathbf{r}_1|$, the direction of $\mathbf{r}_2 - \mathbf{r}_1$ in the laboratory-fixed frame.

easier to use in some cases. This is the invariant expansion theory of Blum and Torruella (1972).

Any invariant pairwise function $F(1\ 2)$, such as $u(1\ 2)$, $h(1\ 2)$, and $c(1\ 2)$ in isotropic systems, has an *invariant expansion* given in general form in Eq. (A2.6). The *rotational invariants* $\phi^{l_1 l_2 L}_{m_1 m_2}$ $(\Omega_1, \Omega_2, \hat{r}_{2\ 1})$ are generalized spherical harmonics that depend on the rotations Ω_i specified in Fig. 8.2b. Only the invariants with $m_1 = m_2 = 0$ need be used if the electric dipole lies on a molecular symmetry axis in each molecule, which is true for H_2O, but not for N_2O, for example. Some examples of $m_1 = 0 = m_2$ rotational invariants are given in Table 8.1. When the terms in Table 8.1 are sufficient, as for spherical molecules with charges or dipoles at their centers, the invariant expansion reduces to

$$F(1\ 2) = F^S(r_{1\ 2}) + F^{E_1}(r_{1\ 2})E_1(1\ 2) + F^{E_2}(r_{1\ 2})E_2(1\ 2) \\ + F^{\Delta}(r_{1\ 2})\Delta(1\ 2) + F^D(r_{1\ 2})D(1\ 2) \qquad (8.4)$$

In the case in which $F(1\ 2) = u(1\ 2)$, the pair potential, the expressions for the coefficients $u^{\alpha}(r_{1\ 2})$ associated with electrostatic interactions are given in Table 8.1.

The Fourier transform of a term in an invariant expansion has the form [see Eq. (A2.16)]

$$\int d^3 r_{1\ 2} \exp(i\mathbf{k}\cdot\mathbf{r}_{1\ 2}) F^{ll'L}_{mm'}(r_{1\ 2}) \Phi^{ll'L}_{mm'}(1\ 2) = \tilde{F}^{ll'L}_{mm'}(k) \tilde{\Phi}^{ll'L}_{mm'}(1\ 2) \qquad (8.5)$$

where $\tilde{F}^{ll'L}_{mm'}(k)$ is a Hankel transform [Eq. (A2.11)]. It is the same as the Appendix 1 Fourier transforms only in the $L = 0$ case but is the appropriate form here because

TABLE 8.1 LOWEST-ORDER FACTORS IN INVARIANT EXPANSIONS AND THEIR FOURIER TRANSFORMS

$l_1 l_2 L$	Space	Explicit $\phi_{00}^{l_1 l_2 L}(1\ 2)$[a]	Explicit $u_{00}^{l_1 l_2 L}$[b]
000	r	1	$u^s(r) + e_1 e_2/r = u^S(r)$
	k	1	$\tilde{u}^s(k) + 4\pi e_1 e_2/k^2 = \tilde{u}^S(k)$
101	r	$\hat{\mu}_1 \cdot \hat{r} = E_1(1\ 2)$	$\mu_1 e_2/r^2 = u^{E_1}(r)$
	k	$\hat{\mu}_1 \cdot \hat{k} = \tilde{E}_1(1\ 2)$	$4\pi i \mu_1 e_2/k = \tilde{u}^{E_1}(k)$
011	r	$\hat{\mu}_2 \cdot \hat{r} = E_2(1\ 2)$	$-e_1 \mu_2/r^2 = u^{E_2}(r)$
	k	$\hat{u}_2 \cdot \hat{k} = \tilde{E}_2(1\ 2)$	$-4\pi i e_1 \mu_2/k = \tilde{u}^{E_2}(k)$
110	r	$\hat{\mu}_1 \cdot \hat{\mu}_2 = \Delta(1\ 2)$	$0 = u^{\Delta}(r)$
	k	$\hat{\mu}_1 \cdot \hat{\mu}_2 = \tilde{\Delta}(1\ 2)$	$0 = \tilde{u}^{\Delta}(k)$
112	r	$3\hat{\mu}_1 \cdot \hat{r}_1 \hat{\mu}_2 \cdot \hat{r}_2 - \hat{\mu}_1 \cdot \hat{\mu}_2 = D(1\ 2)$	$-\mu_1 \mu_2/r^3 = u^D(r)$
	k	$3\hat{\mu}_1 \cdot \hat{k} \hat{\mu}_2 \cdot \hat{k} - \hat{\mu}_1 \cdot \hat{\mu}_2 = \tilde{D}(1\ 2)$	$4\pi \mu_1 \mu_2/3 = \tilde{u}^D(k)$

[a] $\hat{r}$ is the unit vector along $\mathbf{r}_2 - \mathbf{r}_1$ and $\hat{k}$ is the unit vector along $\mathbf{k}$. $\hat{\mu}_\alpha$ is a unit vector on the dipole at $\alpha = 1$ or 2.

[b] e_α is the charge at $\alpha = 1$ or 2. μ_α is the dipole moment at $\alpha = 1$ or 2 and $r \equiv |\mathbf{r}_2 - \mathbf{r}_1|$, all for the special case in which u is the potential for the interaction of a pair of spheres with embedded charges and dipoles. u^s is the short-range, angle-independent part of the potential, including the core repulsion.

the variation of the direction of $\mathbf{r}_{1\ 2}$ implicit in the integration modulates the rotational invariant. Also, while $\Phi_{mm'}^{ll'L}(1\ 2)$ is a given function of Ω_1, Ω_2, and the unit vector $\hat{r}_{2\ 1}$, its transform $\tilde{\Phi}_{mm'}^{ll'L}(1\ 2)$ is the same function of Ω_1, Ω_2, and the unit vector $\hat{k} = \mathbf{k}/|\mathbf{k}|$. See Table 8.1 for examples. Finally, the Fourier transform of the simple invariant expansion in Eq. (8.4) is

$$\tilde{F}(1\ 2) \equiv \int d^3 r_{1\ 2} \exp\,[i\mathbf{k} \cdot (\mathbf{r}_2 - \mathbf{r}_1)] F(1\ 2) \tag{8.6a}$$

$$\begin{aligned} &= \tilde{F}^S(k) + \tilde{F}^{E_1}(k)\tilde{E}_1(1\ 2) + \tilde{F}^{E_2}(k)\tilde{E}_2(1\ 2) \\ &\quad + \tilde{F}^{\Delta}(k)\tilde{\Delta}(1\ 2) + \tilde{F}^D(k)\tilde{D}(1\ 2) \end{aligned} \tag{8.6b}$$

The Ornstein–Zernicke equation for molecular correlation functions [i.e., Eq. (8.2b)] has the Fourier transform [defined in Eq. (8.5)]

$$\tilde{h}(1\ 2) = \tilde{c}(1\ 2) + \rho' \int \tilde{h}(1\ 3)\, d\Omega_3\, \tilde{c}(3\ 2) \tag{8.7}$$

with the orientational convolution still unresolved; only the $\mathbf{r}$ variable is transformed in Eq. (8.5). However, it is easy to construct the "multiplication table" for convolutions given in Table 8.2. It is important to notice that the rotational invariants 1, E_1, E_2, Δ_1, and D form a closed group under commutation,† so there is a certain consistency in the truncated expansion in Eq. (8.4) if the pair potentials have no nonzero terms besides the ones listed in Table 8.1. In the next section we investigate an even simpler model in which the E terms are absent from the potential and we exploit the fact that 1, Δ, and D form a closed group under convolution.

† That is, the convolution of any two members of the group is also a member of the group.

TABLE 8.2 MULTIPLICATION TABLE GIVING THE CONVOLUTION AS IN EQ. (8.7)[a]

$$\tilde{F}(1\ 2)=\frac{1}{8\pi^2}\int \tilde{A}(1\ 3)\, d\Omega_3\, \tilde{B}(3\ 2)$$

	$B(3\ 2)$				
$A(1\ 3)$	1	$E_3(3\ 2)$	$E_2(3\ 2)$	$\Delta(3\ 2)$	$D(3\ 2)$
1	1	0	$E_2(1\ 2)$	0	0
$E_1(1\ 3)$	$E_1(1\ 2)$	0	$\frac{D(1\ 2)+\Delta(1\ 2)}{2}$	0	0
$E_3(1\ 3)$	0	$\frac{1}{3}$	0	$\frac{1}{3}\,E_2(1\ 2)$	$\frac{2}{3}\,E_2(1\ 2)$
$\Delta(1\ 3)$	0	$\frac{1}{3}\,E_1(1\ 2)$	0	$\frac{1}{3}\,\Delta(1\ 2)$	$\frac{1}{3}\,D(1\ 2)$
$D(1\ 3)$	0	$\frac{2}{3}\,E_1(1\ 2)$	0	$\frac{1}{3}\,D(1\ 2)$	$\frac{1}{3}\,[D(1\ 2)+2\Delta(1\ 2)]$

[a] $\tilde{F}(1\ 2)$ is in the body of the table.

DO PROBLEMS 8.1 TO 8.3.

8.3 MSA FOR DIPOLAR HARD SPHERES

In the mean spherical approximation (Section 7.4) applied to dipolar hard spheres, we assume that

$$\begin{aligned} h(1\ 2) &= -1 && \text{if } r_{1\ 2}<\sigma \\ c(1\ 2) &= \frac{\beta\mu_1\mu_2}{r_{1\ 2}^3}\,D(1\ 2) && \text{if } r_{1\ 2}>\sigma \end{aligned} \tag{8.8}$$

It is interesting to examine these closure conditions in terms of the invariant expansion and see that they lead to the data in Table 8.3. The second column merely recapitulates Eq. (8.8), while the next three show how the closure is reflected in the S, Δ, and D projections. The entries for $\bar{h}^D$ and $\bar{c}^D$ will be discussed later.

TABLE 8.3 CLOSURE DATA FOR THE MSA FOR DIPOLAR HARD SPHERES

$r<\sigma$	$h(r)=-1$	$h^S(r)=-1$	$h^\Delta(r)=0$	$h^D(r)=0$	$\bar{h}^D(r)=-3K$
$r>\sigma$	$c(r)=-\beta u^D(r)D(1\ 2)$	$c^S(r)=0$	$c^\Delta(r)=0$	$c^D(r)=-\beta u^D(r)$	$\bar{c}^D(r)=0$

Wertheim (1971) showed how to obtain the exact $h(1\ 2)$ for the Ornstein–Zernicke equation with this closure by proceeding nearly along the following lines. It is the one example of the "exact results" of modern statistical mechanics that we shall follow in detail.

First we substitute the invariant expansions for $\tilde{h}(1\ 2)$ and $\tilde{c}(1\ 2)$ in Eq. (8.7) and do the convolutions according to Table 8.2. In the resulting equation we have many terms, each with one rotational invariant factor, either $\tilde{D}(1\ 2)$, $\tilde{\Delta}(1\ 2)$, or 1. Because of the mutual orthogonality of rotational invariants with respect to the inner product [Eq. (A2.8)], the coefficients of 1, D, and Δ give, respectively, these equations in k space,

$$\tilde{h}^S = \tilde{c}^S + \rho \tilde{h}^S \tilde{c}^S \tag{8.9a}$$

$$\tilde{h}^D = \tilde{c}^D + \frac{\rho}{3}(\tilde{h}^D \tilde{c}^D + \tilde{h}^D \tilde{c}^\Delta + \tilde{h}^\Delta \tilde{c}^D) \tag{8.9b}$$

$$\tilde{h}^\Delta = \tilde{c}^\Delta + \frac{\rho}{3}(\tilde{h}^\Delta \tilde{c}^\Delta + 2\tilde{h}^D \tilde{c}^D) \tag{8.9c}$$

We notice that the spherical projections $\tilde{h}^S(k)$ and $\tilde{c}^S(k)$ are only found in Eq. (8.9a), so the spherical integral equation is decoupled from the others. Moreover, an inverse Fourier transformation converts (8.9a) to an Ornstein–Zernike equation governing the spherical projections only,

$$h^S(r) = c^S(r) + h^S * c^S \tag{8.10}$$

where we use the shorthand for convolutions introduced in Eq. (7.7). With the closure conditions for h^S and c^S specified in Table 8.3, this equation is the PY approximation for a fluid of simple hard spheres of diameter σ at concentration ρ, the same ρ as in Eq. (8.9). The analytical solution for the direct correlation function for this case is given in Eq. (7.50).

Remarkably enough, the remaining problem in calculating the structure and thermodynamics of dipolar hard spheres under MSA can also be reduced to two PY equations for simple hard spheres if we transform Eqs. (8.9b) and (8.9c) back to r space in order to exploit the remaining closure conditions in Table 8.3. However, this transformation requires some care. Thus by consulting Table 8.1 and Eq. (8.5), we see that $\tilde{F}^S(k)$ and $\tilde{F}^\Delta(k)$ are obtained from the r-space functions by integrating with $j_0(kr)$, the normal kernel for the Fourier transform of a spherical function of r in three-dimensional space. [See Eqs. (A1.5) and (A1.6), recalling that $j_0(x) = (\sin x)/x$.] In these cases the k-space to r-space transforms yield $F^S(r)$ and $F^\Delta(r)$, respectively, as shown by Eqs. (A1.5) and (A1.6). However, the same transformations applied to $\tilde{F}^D(k)$ leads to the integral

$$\frac{1}{2\pi^2}\int_0^\infty k^2\,dk\,j_0(kr)(-4\pi)\int_0^\infty R^2\,dR\,j_2(kR)F^D(R) \equiv \bar{F}^D(r) \tag{8.11}$$

where $\bar{F}^D(r)$ may not be the same as $F^D(r)$. The integral over k is a standard Bessel function type giving

$$\int_0^\infty k^2\,dk\,j_0(kr)j_2(kR) = \frac{\pi}{2}\left[\frac{3\theta(R-r)}{R^3} - \frac{\delta(R-r)}{R^2}\right] \tag{8.12}$$

and then Eq. (8.11) gives

$$\bar{F}^D(r) = F^D(r) - 3\int_r^\infty \frac{dR\,F^D(R)}{R} \tag{8.13}$$

From these results we conclude that the r-space forms of Eqs. (8.9b) and (8.9c) are

$$\bar{h}^D(r) = \bar{c}^D(r) + \tfrac{1}{3}[\bar{h}^D * \bar{c}^D + \bar{h}^D * c^\Delta + h^\Delta * \bar{c}^D] \tag{8.14a}$$

$$h^\Delta(r) = c^\Delta(r) + \tfrac{1}{3}[c^\Delta * h^\Delta + 2\bar{c}^D * \bar{h}^D] \tag{8.14b}$$

where again we use the shorthand for convolutions introduced in Eq. (7.7). We decouple this pair of integral equations by introducing the linear combinations

$$H_+(r) \equiv a[\bar{h}^D(r) + \tfrac{1}{2}h^\Delta(r)], \qquad H_-(r) = a[\bar{h}^D(r) - h^\Delta(r)] \tag{8.15}$$

and similarly for $C_\pm$ in terms of $\bar{c}^D$ and c^Δ. The coefficient a, the same in all four equations, is to be determined later. So we obtain the integral equations

$$H_+(r) = C_+(r) + \frac{2}{3a} H_+ * C_+ \tag{8.16a}$$

$$H_-(r) = C_-(r) - \frac{1}{3a} H_- * C_- \tag{8.16b}$$

Finally, we need to take note of the behavior of the functions $H_\pm$ at $r < \sigma$ and the functions $C_\pm$ at $r > \sigma$. The basic data are summarized in Table 8.3. Besides we have, in view of Eq. (8.13),

$$\begin{aligned} \bar{c}^D(r) &= c^D(r) - 3\int_r^\infty \frac{dR\, c^D(R)}{R} \\ &= 0 \qquad\qquad r > \sigma \end{aligned} \tag{8.17a}$$

$$\begin{aligned} \bar{h}^D(r) &= h^D(r) - 3\int_r^\infty \frac{dR\, h^D(R)}{R} \\ &= 0 - 3\int_\sigma^\infty \frac{dR\, h^D(R)}{R} \qquad r < \sigma \end{aligned} \tag{8.17b}$$

$$\equiv -3K \qquad\qquad r < \sigma \tag{8.17c}$$

where the last equation defines K, a function of state but not of r. These results are entered in Table 8.3. From Eq. (8.17) and the data in Table 8.3, we find that

$$H_\pm(r) = -3Ka \qquad \text{if } r < \sigma \tag{8.18a}$$

so we now choose $a = 1/3K$ to have the MSA closure conditions for these composite correlation functions.

$$H_\pm(r) = -1 \qquad \text{if } r < \sigma \tag{8.18b}$$

$$C_\pm(r) = 0 \qquad \text{if } r > \sigma \tag{8.18c}$$

These conditions on H_+ and C_+ together with the equation [see Eq. (8.16a)]

$$H_+(r) = C_+(r) + 2KH_+ * C_+ \tag{8.19a}$$

are satisfied by a function $H_+(r)$ that is the same as $h^{PY}(r, 2K\rho)$, which is $h(r)$ in the PY approximation for hard spheres of diameter σ at concentration $2K\rho$ (i.e., an exactly known result).

Conditions (8.18) on H_- and C_-, together with the equation [see Eq. (8.16b)],

$$H_-(r) = C_-(r) - KH_- * C_- \tag{8.19b}$$

are satisfied by a function $H_-(r)$ that is the same as $h^{\text{PY}}(r, -K\rho)$, the h function in the PY approximation for hard spheres of diameter σ and concentration $-K\rho$, again an exactly known result; even though a negative concentration such as $-K\rho$ does not have an obvious physical meaning, the PY approximation for a hard-sphere fluid is known to be valid over some range of negative as well as positive densities (Wertheim, 1971).

Finally, from Eqs. (8.15) we find

$$h^{\Delta}(r) = 2K[h^{\text{PY}}(r, 2K\rho) - h^{\text{PY}}(r, -K\rho)] \tag{8.20a}$$

$$\bar{h}^{D}(r) = K[2h^{\text{PY}}(r, 2K\rho) + h^{\text{PY}}(r, -K\rho)] \tag{8.20b}$$

and the inverse of Eq. (8.13),

$$h^{D}(r) = \bar{h}^{D}(r) - \frac{3}{r^3}\int_0^r R^2\, dR\, \bar{h}^{D}(R) \tag{8.20c}$$

A further result (Wertheim, 1971)

$$3y = \frac{(1+4\alpha)^2}{(1-2\alpha)^4} - \frac{(1-2\alpha)^2}{(1+\alpha)^4} \tag{8.21}$$

enables K to be calculated in terms of ρ, T, and the model parameters σ and μ since y is the reduced concentration of dipoles

$$y \equiv \frac{4\pi\beta\mu^2\rho}{9} \tag{8.22}$$

and

$$\alpha \equiv \eta K \tag{8.23}$$

where

$$\eta \equiv \frac{\pi}{6}\rho\sigma^3 \tag{8.24}$$

is the packing fraction. The excess free energy, calculated from the work of increasing the dipole moments from zero to the final value (Nienhuis and Deutch, 1972), is given by

$$\frac{\beta A^{\text{ex}}}{N} = \left(\frac{\beta A^{\text{ex}}}{N}\right)^0 - \frac{1}{\eta}\left[8\frac{(1+\alpha)^2}{(1-2\alpha)^4} + \frac{(2-\alpha)^2}{(1+\alpha)^4}\right] \tag{8.25a}$$

where the first term on the right pertains to the PY result for the hard spheres without the dipoles. In Fig. 8.3 the MSA result for the excess free energy is compared with certain other approximations (Rushbrooke et al., 1973).

From h^D or h^{Δ} one can also obtain the MSA approximation to the dielectric

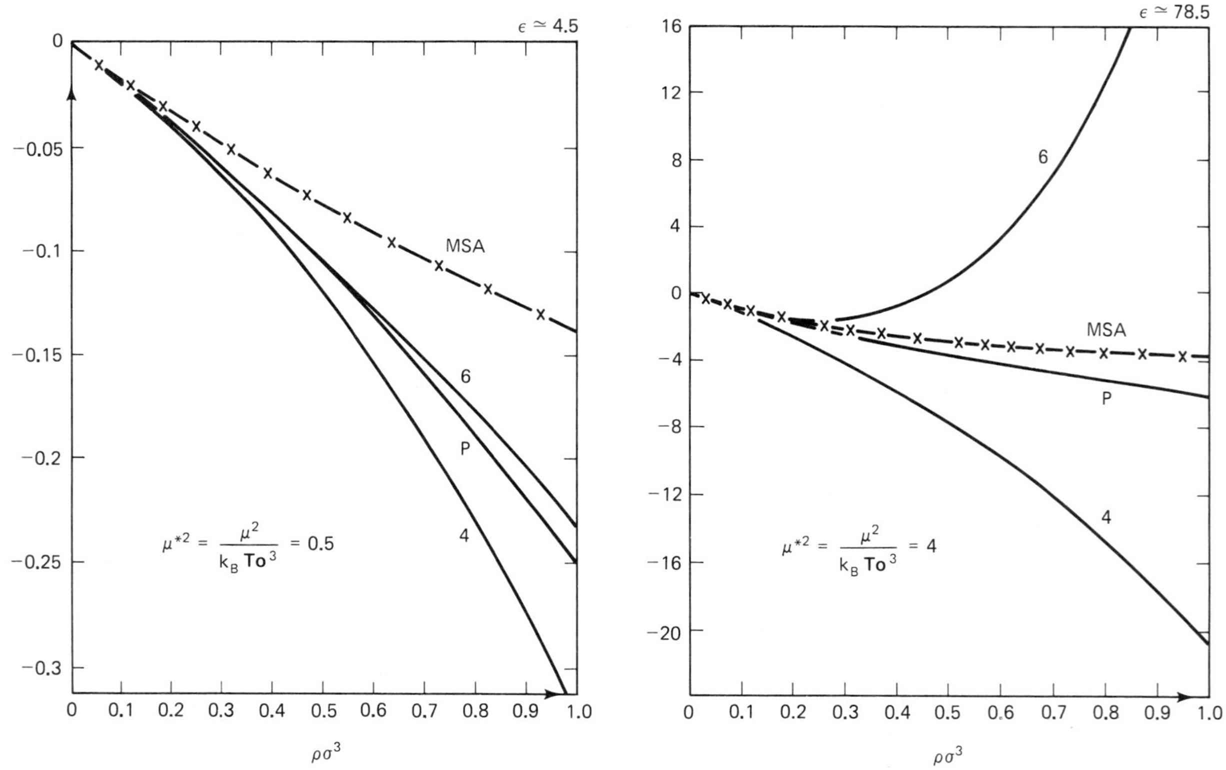

Figure 8.3 Reduced excess free energy $\beta[A - A_0]/N$ of dipolar hard spheres in several approximations: MSA, 4 = perturbation to fourth order in μ, 6 = perturbation to sixth order in μ, P = Padé approximant based on 4 and 6. A_0 pertains to the same system in the $\mu = 0$ limit. ϵ is the dielectric constant at $\rho\sigma^3 = 1$ under MSA. (After Rushbrooke et al., 1973.)

constant by using the theory in the next section. The result is

$$\epsilon = \frac{(1 + 4\alpha)^2(1 + \alpha)^4}{(1 - 2\alpha)^6} \tag{8.25b}$$

It is remarkable that α is determined solely by y [Eq. (8.21)] and y does not depend on σ, so under MSA the dielectric constant of the dipolar hard-sphere fluid does not depend on the molecular diameters. In Fig. 8.4 this approximation is compared with the LHNC approximation described in Chapter 9 and with Onsager's approximation, which is based on a model in which a dipolar molecule in a spherical cavity in a dielectric medium interacts with the mean field that it elicits from the medium.

8.4 THE DIELECTRIC CONSTANT

Here we identify the dielectric constant in terms of its role in shielding ions dissolved in the dielectric medium rather than its role in the polarization response of the dielectric medium to external fields. For this purpose we consider a system of polar molecules, species w and dipole moment μ, mixed with very few ions, of species i and charge e_i. While some simplifications are made here, the method gives essentially the same relation of ϵ to c^Δ for all systems of unpolarizable molecules (Friedman, 1982).

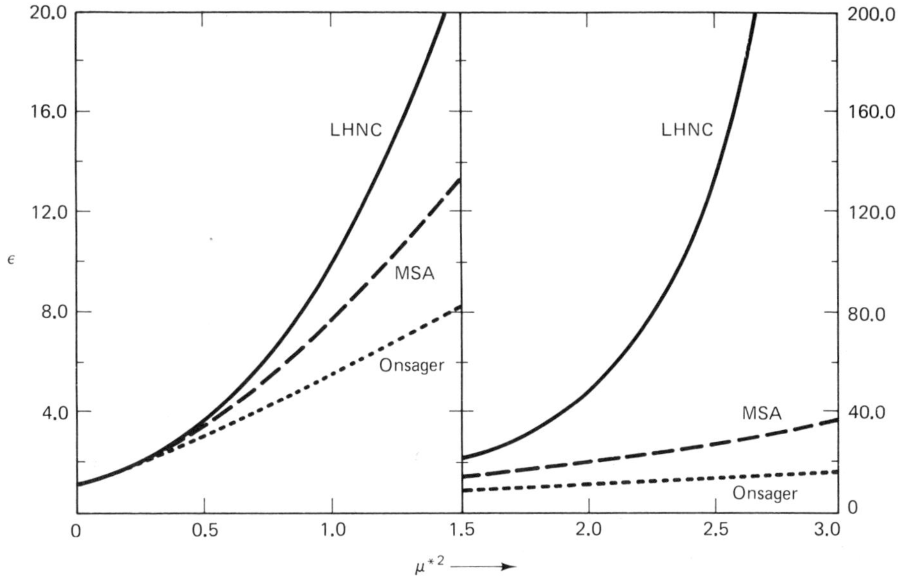

Figure 8.4 Dielectric constant of a fluid of dipolar hard spheres at $\rho\sigma^3 = 0.8$. The reduced dipole strength is given by $\mu^2/k_BT\sigma^3 \equiv \mu^{*2}$. The results of three approximation methods are indicated. (From Patey, 1977.) The MC results for dipolar hard-sphere models are not free of controversy but seem to agree with the MSA at least up to $\mu^{*2} = 2$ (Stell et al., 1981). For water, $\mu^{*2} \sim 4$.

For the ion–dipole mixture the Ornstein-Zernicke equations are two-component versions of Eq. (8.2):

$$h_{ii} = c_{ii} + c_{iw} * h_{wi} + c_{ii} * h_{ii} \tag{8.26a}$$

$$h_{iw} = c_{iw} + c_{iw} * h_{ww} + c_{ii} * h_{iw} \tag{8.26b}$$

$$h_{ww} = c_{ww} + c_{ww} * h_{ww} + c_{wi} * h_{iw} \tag{8.26c}$$

where $*$ represents the convolution spelled out in Eq. (8.2b). These equations describe the system at the Born–Oppenheimer level (Section 4.3). Suppose that we could "see" only the ion species i and not the polar species w. Then we might associate h_{ii}, which is the same in either case, with a solvent-averaged direct correlation function $\bar{c}_{ii}$ given by the equation†

$$h_{ii} = \bar{c}_{ii} + \bar{c}_{ii} * h_{ii} \tag{8.27}$$

In view of the development from Eq. (7.8) to (7.15), it follows that there is a solvent-averaged pair potential function $\bar{u}_{ii}(1\ 2)$ that stands in the same relation to $h_{ii}(1\ 2)$ and $\bar{c}_{ii}(1\ 2)$ in a given state ρ_w, ρ_i, T as $u(1\ 2)$ stands to $h(1\ 2)$ and $c(1\ 2)$ in a given ρ, T state of a one-component system (see Section 10.3). It is sufficient for our purpose to note that this implies

$$\bar{c}_{ii}(r) \longrightarrow -\beta \bar{u}_{ii}(r) \qquad \text{as } r \longrightarrow \infty \tag{8.28a}$$

and

$$\bar{u}_{ii}(r) \longrightarrow \frac{e_i^2}{\epsilon_\rho r} \qquad \text{as } r \longrightarrow \infty \tag{8.28b}$$

where ϵ_ρ is the dielectric constant of the ion–dipole mixture. Clearly, in the $\rho_i = 0$ limit ϵ_ρ is the dielectric constant ϵ of the fluid comprised of polar molecules alone.

In the $\rho_i \longrightarrow 0$ limit we find, from Eqs. (8.26) and (8.27),

$$\begin{aligned}\bar{c}_{ii} &= h_{ii} \\ &= c_{ii} + c_{iw} * h_{wi} \\ &= c_{ii} + c_{iw} * (c_{wi} + h_{ww} * c_{wi}) \\ &= c_{ii} + c_{iw} * (1 + h_{ww}) * c_{wi} \\ &= c_{ii} + c_{iw} * (1 - c_{ww})^{-1} * c_{wi}\end{aligned} \tag{8.29}$$

where we use $(1 + h_{ww}) * (1 - c_{ww}) = 1$, the Ornstein–Zernicke equations for the pure solvent. One may interpret $(1 - c_{ww})^{-1}$ as 1 plus the sum of all convolution chains of c_{ww}-bonds since $1 + h$ has this meaning in the context of the Ornstein–Zernicke equation.

We also have, in view of Eq. (7.15) and Table 8.1,

† The barred notation $\bar{c}$ introduced here has no relation to $\bar{F}^D$ introduced in Eq. (8.11).

$$\left.\begin{aligned} c_{iw}(1\ 2) &= \frac{\beta e_i \mu}{r_{1\ 2}^2} E_2(1\ 2) && (8.30a)\\ c_{wi}(1\ 2) &= -\frac{\beta e_i \mu}{r_{1\ 2}^2} E_1(1\ 2) && (8.30b)\\ c_{ii}(1\ 2) &= -\frac{\beta e_i^2}{r_{1\ 2}} && (8.30c)\\ c_{ww}(1\ 2) &= \frac{\beta \mu^2}{r_{1\ 2}^3} D(1\ 2) && (8.30d) \end{aligned}\right\} \text{ as } r_{1\ 2} \longrightarrow \infty$$

with the Fourier transforms in the $k \longrightarrow 0$ regime given by

$$\tilde{c}_{iw}(1\ 2) = \frac{4\pi i \beta e_i \mu}{k} \tilde{E}_2(1\ 2) + \cdots \tag{8.31a}$$

$$\tilde{c}_{wi}(1\ 2) = -\frac{4\pi i \beta e_i \mu}{k} \tilde{E}_1(1\ 2) + \cdots \tag{8.31b}$$

$$\tilde{c}_{ii}(1\ 2) = -\frac{4\pi \beta e_i^2}{k^2} + \cdots \tag{8.31c}$$

$$\tilde{\tilde{c}}_{ii}(1\ 2) = -\frac{4\pi \beta e_i^2}{k^2 \epsilon_\rho} + \cdots \tag{8.31d}$$

$$\tilde{c}_{ww}(1\ 2) = \tilde{c}^{\,S}_{ww}(k=0) + \tilde{c}^{\,\Delta}_{ww}(k=0)\tilde{\Delta}(1\ 2) - \frac{4\pi\beta\mu^2}{3} D(1\ 2) + \cdots \tag{8.31e}$$

where the neglected terms are arbitrarily small compared with those exhibited, provided that k is small enough. These equations provide the means to use Eq. (8.29) to relate ϵ_ρ to the system's Hamiltonian.

Equations (8.31a), (8.31b), and (8.31c) follow from the corresponding Eqs. (8.30), while Eq. (8.31d) follows from Eq. (8.28). Equation (8.31e) requires some detailed discussion because the Hankel transform (last column of Table 8.1) of r^{-3} is independent of k. So we begin with the invariant expansion for a system of dipolar molecules with the dipole along a molecular symmetry axis;

$$\begin{aligned} c_{ww}(1\ 2) = {} & c^S_{ww}(r_{1\ 2}) + c^\Delta_{ww}(r_{1\ 2})\Delta(1\ 2) \\ & + \left[sc^D_{ww}(r_{1\ 2}) + \frac{\beta\mu^2}{r_{1\ 2}^3} \right] D(1\ 2) + \cdots \end{aligned} \tag{8.32}$$

Still higher terms do not affect the result for ϵ and will be neglected here for simplicity. The term $sc^D_{ww}(r_{1\ 2})D(1\ 2)$ is a presumed contribution of many-body origin, with the same symmetry as Eq. (8.30d) but shorter in range. Thus we must have, in view of Eq. (8.30d),

$$\lim_{r=\infty} r^3 sc^D_{ww}(r) = 0 \tag{8.33}$$

It follows that

$$\int_0^\infty r^2\, dr\, s c^D_{ww}(r)$$

is finite and that

$$\int_0^\infty r^2\, dr\, s c^D_{ww}(r) j_L(kr) \tag{8.34}$$

vanishes at $k = 0$ unless $L = 0$. Similar short-range behavior applies to the spherical part $c^S_{ww}(r)$ and to $c^\Delta_{ww}(r)$. These considerations lead to Eq. (8.31e).

Now we combine Eqs. (8.31) with Eq. (8.29) to get the following equation in the $k \to 0$ regime, corresponding to a dominant contribution from a very widely separated pair of ions:

$$\tilde{\tilde{c}}_{ii}(1\ 2) = -\frac{4\pi\beta e_i^2}{k^2}(1 - 4\pi\beta\mu^2 Q) \tag{8.35a}$$

where

$$Q \equiv \tilde{E}_2 * \tilde{E}_1 + \tilde{E}_2 * \tilde{c}_{ww} * \tilde{E}_1 + \tilde{E}_2 * \tilde{c}_{ww} * \tilde{c}_{ww} * \tilde{E}_1 + \cdots \tag{8.35b}$$

This result follows from the interpretation of $(1 - c_{ww})^{-1}$ given in the discussion of Eq. (8.29).

In view of Table 8.2 and Eqs. (8.31) we find, all for small k,

$$\tilde{E}_2 * \tilde{E}_1 = \frac{\rho}{3} \tag{8.36a}$$

$$\tilde{E}_2 * \tilde{c}_{ww} = (\tfrac{1}{3}\rho\tilde{c}^\Delta_{ww} - 2y)\,\tilde{E}_2 \tag{8.36b}$$

where y is the reduced dipole strength defined in Eq. (8.22). So we find that

$$\begin{aligned} Q &= \tfrac{1}{3}\rho \sum_{n=0}^{\infty} (\tfrac{1}{3}\rho\tilde{c}^\Delta_{ww} - 2y)^n \\ &= \frac{1}{3}\rho \frac{1}{1 + 2y - \frac{1}{3}\rho\tilde{c}^\Delta_{ww}} \end{aligned} \tag{8.37}$$

which, with Eqs. (8.35a) and (8.31d), leads to the alternative forms

$$\frac{1}{\epsilon} = 1 - \frac{3y}{1 + 2y - \frac{1}{3}\rho\tilde{c}^\Delta_{ww}(0)} \tag{8.38a}$$

$$\frac{1}{\epsilon} = \frac{1 - y - \frac{1}{3}\rho\tilde{c}^\Delta_{ww}(0)}{1 + 2y - \frac{1}{3}\rho\tilde{c}^\Delta_{ww}(0)} \tag{8.38b}$$

We write ϵ in place of ϵ_ρ because the limit of zero ion concentration was taken earlier. Thus $\tilde{c}^\Delta_{ww}(0) \equiv [\tilde{c}^\Delta_{ww}(k)]_{k=0}$ is characteristic of the structure of the pure solvent. This same result can be written in many more quite different-looking ways. A theory of Kirkwood (1939), if restricted to systems of unpolarizable molecules, gives

$$\frac{(\epsilon-1)(2\epsilon+1)}{9\epsilon}=y(1+\langle\cos\theta\rangle) \tag{8.38c}$$

where $\langle\cos\theta\rangle = \frac{1}{3}\rho\tilde{h}^{\Delta}(k=0)$ is the average of the cosine of the angle between the dipole of one solvent molecule and all the others; $1+\langle\cos\theta\rangle = g_K$, the *Kirkwood g factor*. As shown in the problems, Eq. (8.38c) can be derived from Eq. (8.38b), as can

$$\frac{(\epsilon-1)^2}{3\epsilon}=-y\rho\tilde{h}^{D}_{ww}(k=0) \tag{8.38d}$$

While all of these equations are exact for the dielectric constant of a pure fluid of polar, unpolarizable molecules, those in terms of $\tilde{c}^{\Delta}$ retain their validity if the medium is an electrical conductor, owing to the presence of ions, while those in terms of h^{Δ} or h^{D} are strictly limited to perfect electric insulators (Levesque et al., 1980).

DO PROBLEMS 8.4 TO 8.6.

8.5 SITE–SITE AND ATOM–ATOM CORRELATION FUNCTIONS

As we have found above, it is quite possible to obtain useful results in the theory of molecular correlation functions in spite of the formidable aspect of the required mathematical machinery, namely the generalized Fourier transforms. An additional problem is that it is difficult to visualize the information carried in the molecular correlation functions $h(1\ 2)$ and their dependence on six coordinates: the center-to-center distance and five angles to fix the orientations in the most general case. The coefficients h^S, h^{Δ}, h^D, and so on, of the invariant expansion are projections of $h(1\ 2)$ that are merely functions of r but even so they do not, except for h^S, easily lead to a picture of the structure. Fortunately, there is another class of projections of $h(1\ 2)$ that are much easier to use to visualize the structure, namely the site–site correlation functions. Here we define these functions as projections of the molecular $h(1\ 2)$ and show their relation to the dielectric constant and the partial structure factors that determine the diffraction of neutrons. In the following section we consider a different route to the theory of the site–site correlation functions beginning with interaction site models for molecular pair potentials.

The correlation function $h_{\alpha\beta}(R)$ connecting a site (usually an atom) α in molecule a with a site β in molecule b may be written as a convolution integral (see Fig. 8.5)

$$h_{\alpha\beta}(r)=\int f_{\alpha a}(\alpha\ \ 1)d(1)h_{ab}(1\ \ 2)d(2)f_{b\beta}(2\ \ \beta) \tag{8.39}$$

where we define

$$f_{\alpha a}(\alpha\ \ 1)\equiv\delta(\mathbf{r}_\alpha-\mathbf{d}_\alpha-\mathbf{r}_a),\qquad f_{b\beta}(2\ \ \beta)\equiv\delta(\mathbf{r}_\beta-\mathbf{d}_\beta-\mathbf{r}_b) \tag{8.40}$$

where $\mathbf{r}_a$ is the location of the center of molecule a, $\mathbf{r}_\alpha$ is the location of site α, and where, in the integration, $\mathbf{r}_\beta$ and $\mathbf{r}_\alpha$ are fixed with $r=|\mathbf{r}_\beta-\mathbf{r}_\alpha|$. We remark that

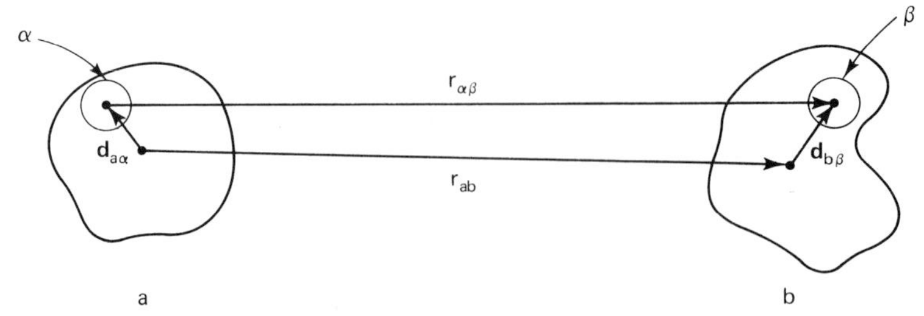

Figure 8.5 Diagram for the passage from molecular correlation functions to site–site correlation functions.

the notation s is often used in place of f for the intramolecular bonds. Applying generalized Fourier transforms to evaluate the convolution integral (Appendix 2) gives the one nonzero matrix element of the transforms of the site–site correlation function

$$\hat{h}_{\alpha\beta}(k)^{00}_{000} = \int \tilde{h}_{\alpha\beta}(R) e^{i\mathbf{k}\cdot\mathbf{R}}\, d^3R \\ \equiv \tilde{h}_{\alpha\beta}(k) \tag{8.41}$$

in terms of the molecular correlation function

$$\tilde{h}_{\alpha\beta}(k) = \sum_{ll'mm'} i^{l-l'} j_l(kd_\alpha) j_{l'}(kd_\beta)[2l+1)(2l'+1)]^{1/2} D^l_{0m}(d_\alpha \leftarrow a) \\ \times D^{*l'}_{0m'}(d_\beta \leftarrow b) \hat{h}_{ab}(k)^{ll'}_{mm'0} \tag{8.42}$$

where the vector $\mathbf{d}_\alpha$ to site α from the center of molecule a has length d_α and orientation in the molecular coordinate system a corresponding to the rotation $d_\alpha \leftarrow a$, and similarly for $\mathbf{d}_\beta$. Thus from the definitions of the internal "bonds" $f_{\alpha a}$ and $f_{b\beta}$ in Eq. (8.40), we obtain the following transforms:

$$\hat{f}_{\alpha a}(k)^{0l}_{0m0} = i(2l+1)^{1/2} j_l(kd_\alpha) D^l_{0m}(d_\alpha \leftarrow a) \tag{8.43}$$

$$\hat{f}_{b\beta}(k)^{l0}_{m00} = i^{-l}(2l+1)^{1/2} j_l(kd_\beta) D^{*l}_{0m}(d_\beta \leftarrow b) \tag{8.44}$$

the transforms with other indices being identically zero. Applying the convolution theorem, Eq. (A2.2) to Eq. (8.39), then gives Eq. (8.42).

The site–site correlation functions $\tilde{h}_{\alpha\beta}(k)$ are clearly related to the partial structure factor $S_{\alpha\beta}(k)$ that is measurable by a diffraction experiment if α and β are sites where nuclei are located [see Eq. (5.27)]. However, it must be noticed that the site–site correlation function $\tilde{h}_{\alpha\beta}(k)$ is not, in general, the same as the atom–atom correlation function seen in a diffraction experiment. To explain this point we temporarily adopt an unconventional notation, using $\tilde{h}^{\text{out}}_{\alpha\beta}(\mathbf{k})$ to represent the site–site correlation function established by Eq. (8.42), $\tilde{h}^{\text{in}}_{\alpha\beta}(k)$ to represent the correlation function for two sites within a given molecule, and $\tilde{h}^{\text{all}}_{\alpha\beta}(k)$ to represent the atom–atom correlation function "seen" by radiation in a diffraction experiment, that is,

$$\rho S_{\alpha\beta}(k) = \rho_\alpha \delta_{\alpha\beta} + \rho_\alpha \rho_\beta \tilde{h}^{\text{all}}_{\alpha\beta}(k) \tag{8.45}$$

The relation between the two correlation functions is most simply expressed in r space:

$$h_{\alpha\beta}^{\text{all}}(r) = h_{\alpha\beta}^{\text{in}}(r) + h_{\alpha\beta}^{\text{out}}(r) \tag{8.46}$$

Thus Eq. (8.42) yields only contributions $h_{\alpha\beta}^{\text{out}}$ that come from correlations between α and β in different molecules. In the general case there are intramolecular correlations counted in $h_{\alpha\beta}^{\text{in}}$ that must be included to account for the diffraction data. For a rigid molecule one has

$$h_{\alpha\beta}^{\text{in}}(r) = \sum_{j=1}^{n_{\alpha\beta}} \frac{1}{4\pi\rho_m d_{\alpha\beta_j}^2}\,\delta(r - d_{\alpha\beta_j}) \tag{8.47}$$

where the sum is over the distinct vectors from an atom of species α to one of species β within the molecule, $d_{\alpha\beta_j}$ is the length of the jth such vector, and ρ_m is the concentration of the molecules.

We can also express the dielectric constant in terms of site–site correlations. Thus expanding $\tilde{h}_{\alpha\beta}(k) = \tilde{h}_{\alpha\beta}^{\text{out}}(k)$ in powers of k, we find to order k^2, from Eq. (8.42),

$$\begin{aligned}\tilde{h}_{\alpha\beta}(k) = \Bigg[& 1 - \frac{k^2 d_\alpha^2}{6}\left(1 - \frac{k^2 d_\beta^2}{6}\right)\left(1 + k\frac{\partial}{\partial k} + \frac{1}{2}k^2\frac{\partial^2}{\partial k^2}\right)\hat{h}_{ab}(k)_{000}^{00} \\ & + ik\frac{d_\alpha}{3^{1/2}}D_{0m}^1(d_\alpha \leftarrow a)\left(1 + k\frac{\partial}{\partial k}\right)\hat{h}_{ab}(k)_{m\,00}^{10} \\ & - ik\frac{d_\beta}{3^{1/2}}D_{0m}^{*1}(d_\beta \leftarrow b)\left(1 + k\frac{\partial}{\partial k}\right)\hat{h}_{ab}(k)_{0m0}^{01} \\ & + \frac{k^2}{3}d_\alpha d_\beta D_{0m}^1(d_\alpha \leftarrow a)D_{0m'}^{*1}(d_\beta \leftarrow b)\hat{h}_{ab}(k)_{mm'0}^{11}\Bigg]_{k\to 0}\end{aligned} \tag{8.48}$$

where summation over m and m' is implied. If both a and b are electrically neutral, terms like $\Sigma_\alpha\, e_\alpha$ vanish if we sum over all of the charge sites in molecule a, so we have

$$\begin{aligned}&\sum_{\alpha\beta} e_\alpha e_\beta \tilde{h}_{\alpha\beta}(k) \\ &\qquad = \tfrac{1}{3}k^2 \sum_{\alpha,\beta}\sum_{m,m'} e_\alpha d_\alpha e_\beta d_\beta D_{0m}^1(d_\alpha \leftarrow a)D_{0m'}^{*1}(d_\beta \leftarrow b)h_{ab}(k)_{mm'0}^{11}\end{aligned} \tag{8.49}$$

All of the rest of the terms either are higher order in k or vanish in the sum over all of the charge sites in a and in b.

If both a and b are sufficiently symmetrical, all but the $m = 0 = m'$ terms in (8.35) vanish. More generally, the other terms mutually cancel because we have

$$\sum_\alpha e_\alpha \mathbf{d}_\alpha \cdot \hat{x} = 0 \qquad \text{and} \qquad \sum_\alpha e_\alpha \mathbf{d}_\alpha \cdot \hat{y} = 0 \tag{8.50}$$

by virtue of the choice of the vector dipole in molecule a as the z axis of the coordinate system fixed in molecule a, $\hat{x}$ and $\hat{y}$ being unit vectors on the other axes of this coordinate system. Then we notice that we have

$$\sum_{\alpha} e_\alpha d_\alpha D^1_{0\pm 1}(d_\alpha \leftarrow a) = \frac{-1}{2^{1/2}} \sum_{\alpha} e_\alpha \mathbf{d}_\alpha \cdot (\hat{x} \pm i\hat{y}) = 0 \tag{8.51}$$

where also

$$\sum_{\alpha} e_\alpha d_\alpha D^1_{00}(d_\alpha \leftarrow a) = \sum_{\alpha} e_\alpha \mathbf{d}_\alpha \cdot \hat{z} = \mu_a$$

is the magnitude of the dipole moment of molecule a. So Eq. (8.49) gives

$$[\hat{h}_{ab}(k)^{11}_{000}]_{k=0} = \frac{3}{\mu_a \mu_b} \left[\sum_{\alpha,\beta} \frac{e_\alpha e_\beta \tilde{h}_{\alpha\beta}(k)}{k^2} \right]_{k=0} \tag{8.52a}$$

$$= -\frac{1}{2\mu_a \mu_b} \sum e_\alpha e_\beta \int h_{\alpha\beta}(r) r^2 \, d^3 r \tag{8.52b}$$

In Eq. (8.52a) the $k = 0$ limit is finite even though $h_{\alpha\beta}(k) \neq 0$ at $k = 0$; indeed, detailed examination of the limit leads to Eq. (8.52b). Now to get the dielectric constant we turn to one of the forms derivable from Eq. (8.38a), namely Eq. (8.57) (see Problem 8.6).

$$\frac{\epsilon - 1}{\epsilon} = 3y + 3y\rho_w \hat{h}_{ww}(0)^{11}_{000} \tag{8.53}$$

$$= 3y - \frac{2\pi}{3} \beta \rho_w^2 \sum_{\alpha\beta} e_\alpha e_\beta \int h_{\alpha\beta}(r) r^2 \, d^3 r \tag{8.54}$$

This remarkable result, first obtained by Høye and Stell (1976) and by Chandler (1977), shows how the dielectric constant of a fluid could be determined by diffraction if the molecular polarity were due to discrete charges localized at atomic centers, and these charges were known.

DO PROBLEMS 8.7 AND 8.8.

8.6 INTERACTION SITE MODELS AND THE SITE–SITE ORNSTEIN–ZERNIKE EQUATION

An interaction site model or ISM is one in which the molecule–molecule pair potential $u(1\ 2)$ is a sum of central force interactions between sites fixed in each of the molecules. Thus we have

$$u(1\ 2) = \sum_{\alpha=1}^{m} \sum_{\beta=1}^{m} u_{\alpha\beta}(r_{\alpha\beta}) \tag{8.55}$$

where there are m sites per molecule and the sums are over sites in molecules 1 and 2, respectively. For simplicity we assume here that the molecules are rigid, all of one species, and with labeled sites, so the m sites in a molecule are distinguishable even if they would not be if the labels were removed.

The ST2 model for water developed by Stillinger and coworkers (see Stillinger and Rahman, 1974) is a much studied ISM in which there are five sites in each

H_2O. They are the oxygen nucleus (o), the two protons (p), and two other sites (e) representing the centers of each of the two lone pairs of electrons in H_2O. The site–site potential u_{oo} is a 6–12 potential with the parameters appropriate to fluid Ne. The other $u_{o,a}$ potentials are zero. Each u_{pp}, u_{ee}, and u_{pe} is a Coulombic term protected by a "switching function" which turns the interaction off when the two charge sites get very close to each other, so there are no configurations in which the site–site potential u_{pe} goes to $-\infty$.

As illustrated by this example, the sites in an ISM need not be atom locations. A whole class of models of the ISM type called TIPS models,. in which each site is either an atom or a group of atoms such as CH_3, has been developed by Jorgenson (1981) for studies of solution phenomena in organic chemistry.

The ST2 and TIPS models have been studied mostly by MD or MC simulation methods. We now turn to the study of interaction site models by other methods in which some measure of analytical theory is used to minimize the computational requirements.

With an ISM it is quite natural to use the site–site correlation functions defined in Eq. (8.39); they may also be defined as follows in view of Fig. 8.5:

$$1 + h_{\alpha\beta}(|\mathbf{R} - \mathbf{R}'|) = V^2\langle\delta(\mathbf{r}_{1a} - \mathbf{R})\ \delta(\mathbf{r}_{2b} - \mathbf{R}')\rangle \tag{8.56}$$

For h^{out} as defined in Eq. (8.46), $1a$ and $2b$ refer to sites in two distinct molecules 1 and 2, while for h^{in} one should replace $2b$ by $1b$ in Eq. (8.56) because the two sites are then in a given molecule.

The form of the Ornstein–Zernicke equation that is appropriate for an ISM may be derived as follows. We first treat the N-molecule system as a mixture of N particles of each of the species $1, 2, \ldots, \alpha, \ldots, m$ by introducing matrices $\mathbf{H}$ and $\mathbf{C}$ with the elements [see Eqs. (7.69)–(7.71)]

$$H_{\alpha\beta} = \rho_m h_{\alpha\beta}(r) \tag{8.57a}$$

$$C_{\alpha\beta} = \rho_m c_{\alpha\beta}(r) \tag{8.57b}$$

where ρ_m is the particle number density N/V of the molecules. The Ornstein–Zernicke equation for the mixture is

$$\mathbf{H} = \mathbf{C} + \mathbf{C} * \mathbf{H} \tag{8.57c}$$

$$= \mathbf{C} + \mathbf{C} * \mathbf{C} + \mathbf{C} * \mathbf{C} * \mathbf{C} + \cdots \tag{8.57d}$$

where the star indicates a matrix product and a convolution as in

$$(\mathbf{C} * \mathbf{C})_{\alpha\alpha'} = \sum_{\beta} \int C_{\alpha\beta}(1\ \ 3) d(3) C_{\beta\alpha'}(3\ \ 2) \tag{8.58}$$

To adapt this general OZ equation for the mixture to our purpose, we need to make the "out" and "in" distinctions for $\mathbf{H}$ and $\mathbf{C}$ as for $h_{\alpha\beta}$ in Eq. (8.46). Thus we write

$$c^{\text{all}}(r) = c^{\text{in}}(r) + c^{\text{out}}(r) \tag{8.59}$$

where $c^{\text{in}}(r) = 0$ for $r > \sigma_{\alpha\beta}$ and $c^{\text{out}}(r) = 0$ for $r < \sigma_{\alpha\beta}$, and similarly for the decomposition of h^{all}. Then we classify the graphical contributions according to the "in" or "out" characteristics of the correlation functions. Representing **H** as a double line ═══ and **C** as a single line ───, we write Eq. (8.57d) as

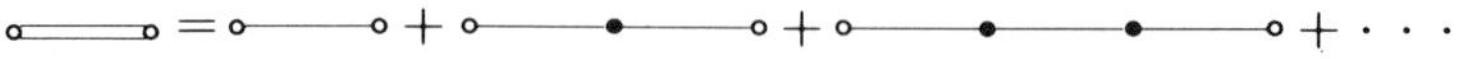

(8.60)

where we have, on the right, **C** bonds and white and black circles of unit weight. We then introduce the separation of "in" parts of the correlation functions, indicated by an "in" over the corresponding bond, from the "out" parts, indicated by an "out" over the bond, to get the following expansions of $\mathbf{H}^{\text{out}}$ and $\mathbf{H}^{\text{in}}$ in terms of graphs with $\mathbf{C}^{\text{out}}$ and $\mathbf{C}^{\text{in}}$ bonds on white and black circles of unit weight.

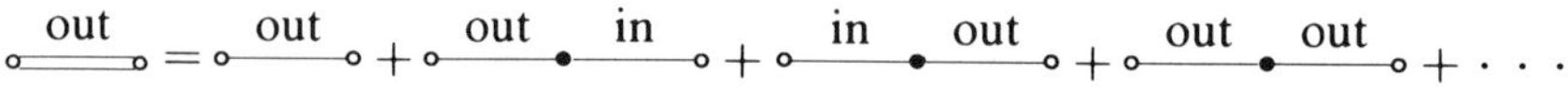

(8.61a)

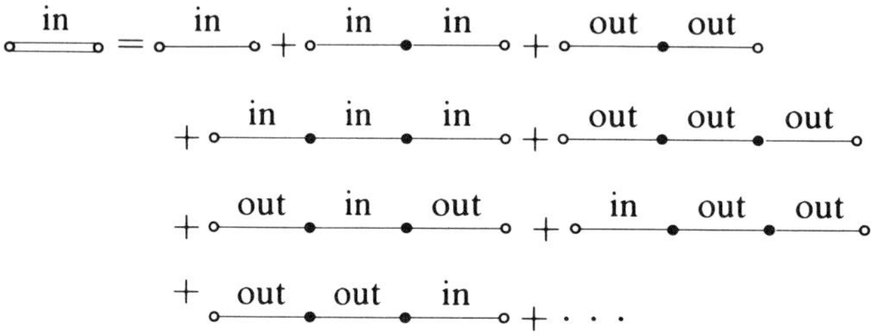

(8.61b)

In each graph contributing to $\mathbf{H}^{\text{out}}$ the two white circles are in distinct molecules, while in each graph contributing to $\mathbf{H}^{\text{in}}$ the two white circles are in the same molecule. On this basis the graph series for $\mathbf{H}^{\text{out}}$ can be topologically reduced. Thus consider a term in the graphical expansion of $\mathbf{H}^{\text{out}}$ in which there is a single $\mathbf{C}^{\text{in}}$ bond and any number of $\mathbf{C}^{\text{out}}$ bonds. There will be other graphs which are the same as this graph in all respects except in the connections between the two circles connected by the $\mathbf{C}^{\text{in}}$ bond in the original graph. The collection of all of these graphs is given by the original graph modified just by replacing the $\mathbf{C}^{\text{in}}$ bond by an $\mathbf{H}^{\text{in}}$ bond, in view of Eq. (8.61b). By applying the same topological reduction to graphs of Eq. (8.61a) in which there are two or more $\mathbf{C}^{\text{in}}$ bonds, each of which connects sites on a different molecule than the others, we can obtain the graphical series

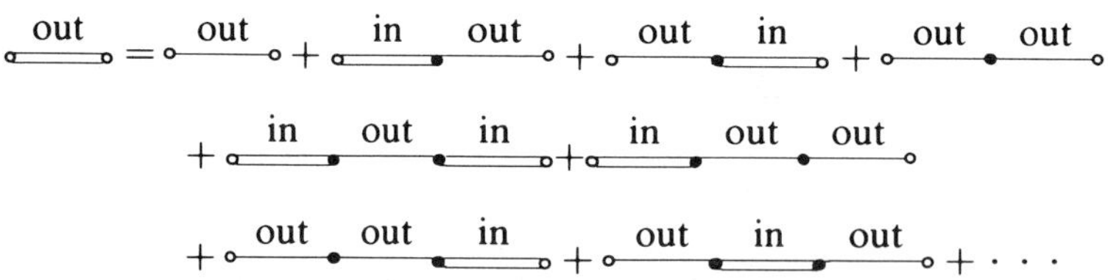

(8.62)

in which $\mathbf{H}^{\text{out}}$ is expressed as a sum of chains of $\mathbf{C}^{\text{out}}$ and $\mathbf{H}^{\text{in}}$ bonds with no circles intersected by two of the latter.

Equation (8.62) is a graphical form of the site–site Ornstein–Zernike equation (SSOZ equation). A more compact way to express it is to introduce omega bonds defined as follows:

$$\boldsymbol{\omega}(r) = \mathbf{1}\delta(\mathbf{r}) + \mathbf{H}^{\text{in}}(r) \tag{8.63}$$

where the diagonal matrix elements of $\mathbf{H}^{\text{in}}$ are all zero, so that $\boldsymbol{\omega}$ has matrix elements

$$\omega_{\alpha\beta}(r) = \delta_{\alpha\beta}\delta(\mathbf{r}) + h^{\text{in}}_{\alpha\beta}(r) \tag{8.64}$$

Then it is easy to verify that Eq. (8.62) represents the iterative solution of the integral equation form of the SSOZ equation:

$$\mathbf{H}^{\text{out}} = \boldsymbol{\omega} * \mathbf{C}^{\text{out}} * \boldsymbol{\omega} + \boldsymbol{\omega} * \mathbf{C}^{\text{out}} * \boldsymbol{\omega} * \mathbf{H}^{\text{out}} \tag{8.65}$$

where the $*$ operation signifies both a matrix product and a convolution as in Eq. (8.58).

The SSOZ equation is still more compact when written in terms of matrices whose matrix elements are Fourier transforms of $\omega_{\alpha\beta}(r)$ and the "out" parts of the correlation functions $h_{\alpha\beta}(r)$ and $c_{\alpha\beta}(r)$ [see Eq. (7.71)]. Thus in k space the SSOZ equation is

$$\tilde{\mathbf{H}}^{\text{out}}(k) = \tilde{\boldsymbol{\omega}}(k)\cdot\tilde{\mathbf{C}}^{\text{out}}(k)\cdot\tilde{\boldsymbol{\omega}}(k) + \tilde{\boldsymbol{\omega}}(k)\cdot\tilde{\mathbf{C}}^{\text{out}}(k)\cdot\tilde{\boldsymbol{\omega}}(k)\cdot\tilde{\mathbf{H}}^{\text{out}}(k) \tag{8.66}$$

or, in still more compact form,

$$(\tilde{\boldsymbol{\omega}} + \tilde{\mathbf{H}}^{\text{out}})\cdot(\tilde{\boldsymbol{\omega}}^{-1} - \tilde{\mathbf{C}}^{\text{out}}) = \mathbf{1} \tag{8.67a}$$

which is striking in comparison with the corresponding OZ equation for mixtures, Eq. (7.71), which here is modified by affixing the label "all" to the correlation functions to indicate that they comprise both "in" and "out" parts, in distinction to Eq. (8.67a).

$$(\mathbf{1} + \tilde{\mathbf{H}}^{\text{all}})\cdot(\mathbf{1} - \tilde{\mathbf{C}}^{\text{all}}) = \mathbf{1} \tag{8.67b}$$

A remarkable conclusion about the large-r form of $c_{\text{ab}}(r)$ follows from the detailed study of the application of both equations (8.67) to a given system. Thus in view of Eq. (8.63) we may write

$$\begin{aligned}\mathbf{1} - \tilde{\mathbf{C}}^{\text{all}} &= (\mathbf{1} + \tilde{\mathbf{H}}^{\text{all}})^{-1} \\ &= (\tilde{\boldsymbol{\omega}} + \tilde{\mathbf{H}}^{\text{out}})^{-1} \\ &= \tilde{\boldsymbol{\omega}}^{-1} + (\mathbf{1} + \tilde{\mathbf{H}}^{\text{out}}\cdot\tilde{\boldsymbol{\omega}}^{-1})^{-1}\end{aligned} \tag{8.68a}$$

$$\mathbf{1} - \tilde{\mathbf{C}}^{\text{all}} \simeq \tilde{\boldsymbol{\omega}}^{-1} \tag{8.68b}$$

where the last step is actually exact in the "ideal limit" in which $u^{\text{out}}_{\alpha\beta}(r)$ vanishes for all α and β. We return below to a discussion of Eq. (8.68b).

Next we notice that $\boldsymbol{\omega}$ is known, in view of its definition in terms of $\mathbf{H}^{\text{in}}$ and the fact that the matrix elements of the latter are given in Eq. (8.47). We only have to set $n_{\text{ab}} = 1$ in the latter because in the present analysis the sites in a given molecule are labeled, so they surely are all distinct. The explicit results for the matrix elements of $\boldsymbol{\omega}$ and $\tilde{\boldsymbol{\omega}}$ are given here.

$$\omega_{\alpha\beta}(r) = \delta_{\alpha\beta}\delta(\mathbf{r}) + \frac{(1-\delta_{\alpha\beta})\delta(r - d_{\alpha\beta})}{4\pi\rho_m d_{\alpha\beta}{}^2} \tag{8.69a}$$

$$\tilde{\omega}_{\alpha\beta}(k) = \delta_{\alpha\beta} + (1-\delta_{\alpha\beta})j_0(kd_{\alpha\beta}) \tag{8.69b}$$

$$= \delta_{\alpha\beta} + (1-\delta_{\alpha\beta})[1 - 3(kd_{\alpha\beta})^2 + \cdot\cdot\cdot] \tag{8.69c}$$

where j_0 is the zero-order spherical Bessel function, $j_0(x) = \sin x/x$.

Now we specialize our discussion to the class of ISM with two sites in each molecule. Then we have, from Eq. (8.69c),

$$\tilde{\omega}^{-1} = \left[\frac{3}{(kd_{\alpha\beta})^2}\right] \begin{vmatrix} 1 & -1 \\ -1 & 1 \end{vmatrix} + O(k^0) \tag{8.70}$$

and Eq. (8.68b) becomes

$$C^{\text{out}}_{\alpha\beta} = \frac{1}{3(kd_{\alpha\beta})^2} + O(k^0) \tag{8.71a}$$

or

$$c^{\text{out}}_{\alpha\beta}(r) \longrightarrow \frac{1}{4\pi\rho_m d^2_{\alpha\beta} r} \qquad \text{as } r \longrightarrow \infty \tag{8.71b}$$

So we see that the site–site direct correlation function departs violently from the important rule $c(r) \longrightarrow -\beta u(r)$ as $r \longrightarrow \infty$ given in Eq. (7.15). Moreover, for charge–charge potentials, hard-sphere potentials, and many others, $h_{\alpha\beta}(r)$ is short range in the sense that $h_{ab}(k)$ is finite at $k = 0$. In such cases the term in $\mathbf{H}^{\text{out}}$ in Eq. (8.68a) vanishes at $k = 0$, so Eq. (8.68b) is exact in this limit. A more complete analysis by Cummings and Stell (1982) leads to the conclusion that Eq. (8.71b) is correct up to a factor that is independent of r but depends on the state ρ_m, T as well as on $u^{\text{out}}_{\alpha\beta}(r)$. They also extend the analysis to other classes of interaction site models and find that the site–site direct correlation function does *not* conform to Eq. (7.15) in most cases.

In the next section we discuss the application to interaction site models of a number of integral equation approximation methods which are not consistent with the correct large-r asymptotic form of $c_{\alpha\beta}(r)$ as established above. In many cases these methods are impressively successful in generating the correct pair correlation functions in the small-r range, so it seems likely that the remaining inconsistencies are really important only for long-range correlations and those global properties such as the dielectric constant that depend critically upon them.

Another topic that we might well have considered in this discussion of general aspects site–site correlation functions is the modification of the theory of the cluster expansion of $h(r)$ in terms of white circles, black ρ circles, and Mayer f bonds [Eq. (7.1)] that is needed for a graphical analysis of site–site correlation functions (Ladanyi and Chandler, 1975).

8.7 INTERACTION SITE APPROXIMATIONS

A particularly important class of interaction site models are the reference interaction site models (RISMs), namely those in which the site–site potentials have the hard-sphere form

$$u_{\alpha\beta}(r) = \begin{cases} \infty & \text{if } r < \sigma_{\alpha\beta} \\ 0 & \text{if } r > \sigma_{\alpha\beta} \end{cases} \tag{8.72}$$

We follow the nomenclature of Cummings and Stell (1982), who reserve the term RISM for the particular models that conform to this equation and introduce other names for the approximation methods used to treat ISM- and RISM-type models. However, it must be remarked that in the literature one finds the term RISM applied both to various aspects of interaction site models and to the approximation methods used to calculate their properties.

The reference interaction site approximation or RISA is the closure of the SSOZ equation for the RISM with the auxilliary conditions

$$h_{\alpha\beta}(r) = -1 \qquad \text{if } r < \sigma_{\alpha\beta} \tag{8.73a}$$

$$c_{\alpha\beta}(r) = 0 \qquad \text{if } r > \sigma_{\alpha\beta} \tag{8.73b}$$

which are similar to the PY approximation for simple hard spheres. The RISA integral equation can conveniently be solved numerically with results that are very successful in terms of the short-range structure established for the same models by simulation methods.

The PY-like approximation for more general interaction site models, which

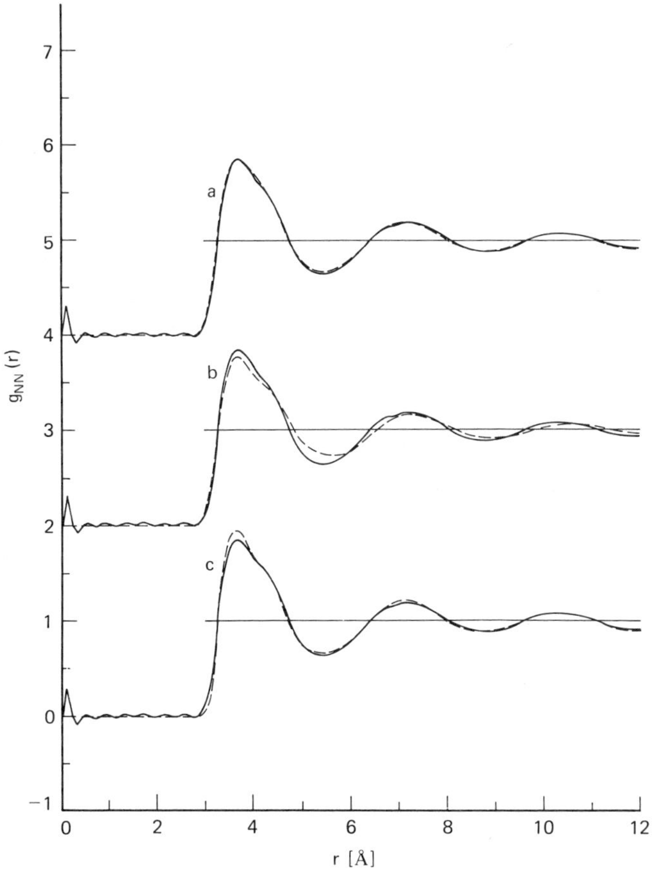

Figure 8.6 Intermolecular atom–atom pair correlation function of liquid nitrogen near the triple point (65K) determined by x-ray diffraction (solid line) compared with calculations by PY-ISA (dashed line) using various models for the interaction between sites in different molecules, with a site at each atomic center. (a) An exp-6 potential. (b) An exp-6 repulsive reference potential. (c) A 6–12 potential. (From Narten et al., 1980.)

we may call the PY-ISA, although it is most often called the RISM equation, is the SSOZ equation combined with

$$c_{\alpha\beta}(r) = f_{\alpha\beta}(r) y_{\alpha\beta}(r) \tag{8.74}$$

where f and y are defined as in Eq. (7.47b). The numerical solution for this approximation (Johnson and Hazoume, 1979) gives the functions $h_{\alpha\beta}(r)$ in quite good agreement with the results from simulation methods applied to the same models in the range of states corresponding to the liquid. A comparison of the PY-ISA results for several models with $h^S(r)$ determined experimentally by x-ray diffraction is shown in Fig. 8.6.

For systems in which long-range forces contribute to u^{out}, it is reasonable to expect that the SSOZ equation with an HNC-like closure will be superior to the PY-ISA. In fact, Pettit and Rossky (1982) and Hirata et al. (1983) find that the HNC-ISA applied to water-like models gives $h_{\alpha\beta}(r)$ in good agreement with simulation results and also gives promising results for a model for ions in a polar solvent.

Finally, it is important to remark that interaction site models and the related approximation methods can be extended to allow for nonrigid molecules, especially those with internal rotations, as shown by Chandler and Pratt (1976) and as applied by them to study the effect of solvents upon intramolecular rotational conformation.

DO PROBLEMS 8.9 AND 8.10.

THE LITERATURE

Various models for liquid H_2O, including the ST2 model mentioned in this chapter, together with the results from their study are reviewed by Stillinger (1975). References for generalized Fourier transforms are given in Appendix 2. The reviews by Adelman and Deutch (1975), Gray (1975), Wertheim (1979), and Stell et al. (1981), as well as the book by Gray and Gubbins (1982) and Chandler's (1977) article on dielectrics, provide background for the material in Sections 8.2 to 8.5. The modern theory of interaction site models begins with a paper by Chandler and Andersen (1972). The present account owes much to a report by Cummings and Stell (1982).

PROBLEMS

8.1. Express each of the rotational invariants $E_1(1\ 2)$, $\Delta(1\ 2)$, and $D(1\ 2)$ in terms of Wigner D^l_{mn} generalized spherical harmonics of Ω_1 and Ω_2, and the coordinates fixing the direction of $\mathbf{r}_2 - \mathbf{r}_1$ in the laboratory coordinate system (see Appendix 2 and notice $\hat{\mathrm{x}}\cdot\hat{\mathrm{y}} = \mathrm{D}^1_{00}(\mathrm{x} \leftarrow \mathrm{y})$).

8.2. Verify the $\Delta(1\ 3)$ row of Table 8.2.

8.3. Verify the orthogonality of $\Delta(1\ 2)$ and $\mathrm{D}(1\ 2)$.

8.4. Show that $\tilde{h}^{\Delta}$ can be eliminated from Eqs. (8.9) to get, in terms of $z = \frac{1}{3}\rho\tilde{c}^{\Delta}(0)$ and $y = -\frac{1}{3}\rho\tilde{c}^{D}(0)$,

$$\rho\tilde{h}^{D} = -\frac{3y}{(1-y-z)(1+2y-z)} \tag{8.75}$$

8.5. Using Eq. (8.75), convert Eq. (8.9) to read

$$\frac{1}{3}\rho\tilde{h}^{D}(0) = -\frac{y}{\epsilon}\left(\frac{\epsilon-1}{3y}\right)^{2} \tag{8.76}$$

8.6. From Eq. (8.38), in view of Eq. (A2.12), derive

$$1 + \frac{1}{3}\rho\tilde{h}^{\Delta}(0) + \frac{2}{3}\rho\tilde{h}^{D}(0) = \frac{\epsilon-1}{3y\epsilon} \tag{8.77}$$

and then, by comparison with Eq. (8.76),

$$1 + \frac{1}{3}\rho\tilde{h}^{\Delta}(0) = \frac{(\epsilon-1)(2\epsilon+1)}{9y\epsilon} \tag{8.78}$$

Notice that we have

$$\begin{aligned} \rho\tilde{h}^{\Delta}(0) &= \frac{\rho}{\int d(1)}\int d(1)d(2)h(1\,2)\hat{\mu}_1\cdot\hat{\mu}_2 \\ &= \frac{\rho}{\int d(1)}\int d(1)d(2)g(1\,2)\hat{\mu}_1\cdot\hat{\mu}_2 \\ &= \langle\cos\theta_{1\,2}\rangle \end{aligned} \tag{8.79}$$

and compare this result with Eq. (8.38c).

8.7. For a rigid model for HF, how is $h_{\alpha\beta}$ in Eq. (8.54) related to the partial structure factor $S_{\alpha\beta}(k)$ in the cases of $\alpha\beta$ = HH, HF, and FF?

8.8. Consider an equimolar mixture of particles of species a and b with pairwise interaction potentials

$$u_{ab} = (1-\delta_{ab})c\delta(r-d_{ab}) + \left(\frac{\sigma_{ab}}{r}\right)^{t} \tag{8.80}$$

(Cummings and Stell, 1982).

(a) Show that if c is a negative coefficient and $d_{ab} < \frac{1}{2}\sigma_{ab}$, this model in the $t = \infty$ limit will give diatomic molecules with sites $\alpha\beta$ at separation $d_{\alpha\beta}$, where α and β are locations of a and b atoms, respectively, and $d_{\alpha\beta} = d_{ab}$.

(b) If the model fluid generated above consists of diatomic molecules with no other species of particles, Eq. (8.47) will be satisfied with $n_{\alpha\beta} = 1$. Show that this implies that the coefficient c in Eq. (8.80) will depend on the state T, ρ_m.

(c) Show that if $d_{ab} > \sigma_{ab}$, the model in Eq. (8.80) will not give simply diatomic molecules.

8.9. Construct an alternative form of Eq. (8.61a) in which $h^{\text{out}}_{\alpha\beta}(1\,2)$ is expressed in terms of graphs with $c^{\text{in}}_{\alpha\beta}$ and $c^{\text{out}}_{\alpha\beta}$ bonds on two white circles of unit weight, one labeled α at location 1 and the other labeled β at location 2, together with any number of black ρ circles.

8.10. In a fluid of RISM diatomics of one kind the site–site correlation function has cusps (i.e., discontinuous derivatives $\partial/\partial r$ . . .) at $r = \sigma_{\alpha\beta} + d_{\alpha\beta}$.

(a) Explain how this comes about.
(b) Show that more generally $h_{\alpha\beta}(r)$ has cusps at

$$r = \sigma_{\alpha\gamma} + d_{\gamma\beta},\ \gamma = 1, \ldots, m \qquad \text{but } \gamma \neq \beta$$

(c) Show that the site–site y function $[y_{\alpha\gamma} = \exp(u_{\alpha\gamma}/k_B T)(1 + h_{\alpha\gamma})]$ has even more cusps (Ladanyi and Chandler, 1975).

9

Perturbation Theory

9.1 WHAT AND WHY?

Very often in theoretical studies it is inviting to begin with a reference system, with Hamiltonian H_0 and with known properties. Next one asks how these properties are affected by changing the Hamiltonian to $H_0 + H_1$. Indeed, the material in Chapters 4 to 8 can be discussed in this way; we identify H_0 with the kinetic energy and H_1 with the potential $U_N(1, \ldots, N)$, a perturbation we "turn on" to generate a description of a more or less realistic model fluid. So, while Chapters 4 to 8 were not presented as perturbation theory, they could well be characterized as equilibrium perturbation theory with respect to ideal gas reference systems.

An important further development is stimulated by examining Fig. 5.5 and the corresponding graphs of $g(r)$ (Figs. 4.3 and 5.3), where we see that the structure of the hard-sphere fluid may be close to the structure of a fluid derived from a more realistic model. So it is especially inviting to consider the class of perturbation theories in which H_0 is the Hamiltonian of a hard-sphere fluid, while

$$H_1 = U_N(1 \cdots N) - \sum_{\text{pairs}} u^{\text{HS}}(1\ 2) \tag{9.1}$$

is the difference between the potential function for some model of interest and the hard-sphere potential function for some chosen hard-sphere diameter (or "diameters" in the case of mixtures). This study is even more inviting in view of the fact that the structure and the excess thermodynamic coefficients of a hard-sphere fluid depend only on density, not on temperature. Consequently, the properties of a hard-sphere reference system can be represented very conveniently compared with more general reference systems for which temperature dependence also needs to be exhibited.

Finally, we are motivated to study the perturbation theory because it turns out that highly accurate numerical results for interesting models can be obtained in this way.

9.2 THE λ EXPANSION

In place of Eq. (9.1) we write

$$H_1 = \lambda W_N(1, \ldots, N) \tag{9.2}$$

where W_N is the perturbing term in the potential, and we look at the total potential function

$$U_{N,\lambda}(1, \ldots, N) = U_N(1, \ldots, N) + \lambda W_N(1, \ldots, N) \tag{9.3}$$

with the convention that when $\lambda = 0$, we have the potential function for the reference system (i.e., U_N), while when $\lambda = 1$ we have the potential function for the system of interest. Systems with λ in the range between zero and unity have the perturbing potential partly "turned on." The corresponding excess free energy is A_λ^{ex} given by the canonical ensemble equation

$$\exp(-\beta A_\lambda^{ex}) = V^{-N} \int \exp(-\beta U_{N,\lambda})\, d(1) \cdots d(N) \tag{9.4}$$

and it is natural to look at the derivative

$$\begin{aligned} \frac{\partial(-\beta A_\lambda^{ex})}{\partial \lambda} &= -\beta \frac{\int W_N e^{-\beta U_{N,\lambda}} d(1) \cdots d(N)}{\int e^{-\beta U_{N,\lambda}} d(1) \cdots d(N)} \\ &= -\beta \langle W_N \rangle_\lambda \end{aligned} \tag{9.5}$$

and higher derivatives and then construct a Taylor's series expansion

$$-\beta A_\lambda^{ex} = -\beta A_0^{ex} + \lambda\beta\langle W_N\rangle_0 + \tfrac{1}{2}\lambda^2\beta^2[\langle W_N^2\rangle_0 - \langle W_N\rangle_0^2] + \cdots \tag{9.6}$$

where $\langle f\rangle_0$ means the average of f over the ensemble of $\lambda = 0$ systems

$$\langle f\rangle_0 \equiv \frac{\int e^{-\beta U} f\, d(1) \cdots d(N)}{\int e^{-\beta U} d(1) \cdots d(N)} \tag{9.7}$$

We have seen an example of Eq. (9.6) in a case in which $\lambda = 1$ and $\langle W_N\rangle_0 = 0$ in Eq. (3.28) for the first quantum correction! The λ^2 term in Eq. (9.6) evidently is always positive (see Problems 9.1 to 9.3).

In the important case in which the perturbation W_N is pairwise additive,

$$W_N(1, \ldots, N) = \sum_{\text{pairs}} w(1\ 2) \tag{9.8}$$

[see Eq. (4.27)] we find that

$$-\left(\frac{\beta}{V}\right) A_\lambda^{ex} = -\left(\frac{\beta}{V}\right) A_0^{ex} + \frac{\lambda}{2}\beta\rho^2 \int d^3r\, g_0(r) w(r) + O(\lambda^2) \tag{9.9}$$

where the higher-order terms involve three-point and higher correlations, even for pairwise additive potentials. In this equation reference system properties ($\lambda = 0$) are denoted by subscript zero. With $\lambda = 1$ this equation is often called *the* perturbation theory. It is very useful for a qualitative estimate of the importance of a perturbation even when higher terms in λ are not negligible in Eq. (9.6).

9.3 THE GIBBS–BOGOLIUBOV INEQUALITY

Bounds on the perturbed free energy A_λ^{ex} can be derived as follows. First we show that for any two spatial distribution functions F_N and G_N, both normalized to unity on a given space,

$$\int F_N(1, \ldots, N) d(1) \cdots d(N) = \int G_N(1, \ldots, N) d(1) \cdots d(N) = 1 \tag{9.10}$$

we have

$$\begin{aligned} &\int F_N(\ln F_N - \ln G_N) d(1) \cdots d(N) \\ &= \int G_N \left\{ \frac{F_N}{G_N} \ln \frac{F_N}{G_N} - \frac{F_N}{G_N} + 1 \right\} d(1) \cdots d(N) \\ &\geq 0 \end{aligned} \tag{9.11}$$

because the quantity $\{\cdots\}$ cannot be negative (i.e., for any real nonnegative x the function $x \ln x - x + 1$ is nonnegative).

Now we may identify **F** with the $\lambda = 0$ system,

$$F_N = \frac{e^{-\beta(U_N - A_0^{ex})}}{V^N} \tag{9.12a}$$

and G_N with the distribution function for the $\lambda > 0$ case,

$$G_N = \frac{e^{-\beta(U_{N,\lambda} - A_\lambda^{ex})}}{V^N} \tag{9.12b}$$

and apply Eq. (9.11) to find

$$-\beta[\langle U_N \rangle_0 - A_0^{ex}] \geq -\beta[\langle U_{N,\lambda} \rangle_0 - A_\lambda^{ex}] \tag{9.13}$$

or, after rearrangement,

$$A_\lambda^{ex} \leq A_0^{ex} + \lambda \langle W_N \rangle_0 \tag{9.14}$$

which is known as the Gibbs–Bogoliubov inequality. By identifying F_N and G_N the other way around we find that

$$A^{ex} \geq A_0^{ex} + \lambda \langle W_N \rangle_\lambda \tag{9.15}$$

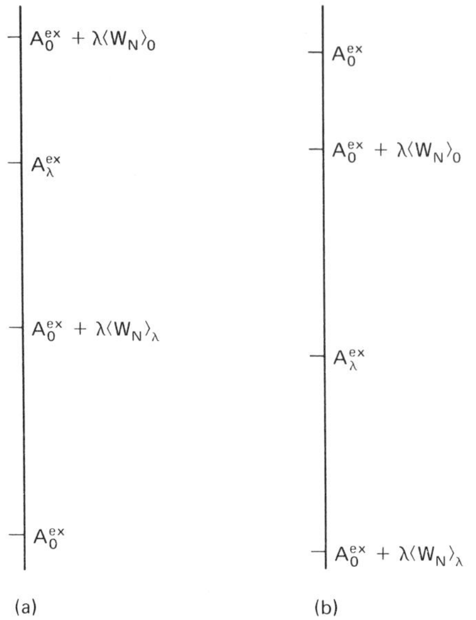

Figure 9.1 Gibbs–Bogoliubov inequalities illustrated.
(a) $\langle W_N\rangle$ positive
(b) $\langle W_N\rangle$ negative

The significance of these inequalities is elucidated by making simple graphs as in Fig. 9.1, from which it is apparent that knowledge of $\langle W_N\rangle_0$ and $\langle W_N\rangle_\lambda$ provides bounds on $A_\lambda^{ex} - A_0^{ex}$. Of course, it is expected that $\langle W_N\rangle_0$ is much easier to estimate than $\langle W_N\rangle_\lambda$.

9.4 BLIP FUNCTION THEORY

Here it is assumed that the potential $U_N(1, \ldots, N)$ is pairwise additive [Eq. (4.27)] and the pair potential $u(1\ 2)$ is repulsive, soft, and relatively short range. The r^{-12} term in the 6–12 potential is an example. In such a case the structure of the fluid is surely close to that of a suitably chosen hard-sphere fluid, but nevertheless the perturbation method described above does not seem very promising. Thus if we write

$$u(r) = u_d(r) + w(r) \tag{9.16}$$

where u_d is the pair potential for hard spheres of diameter d, then $w(r)$ has an infinite range (see Fig. 9.2) and Eq. (9.9) is not a sensible beginning for a perturbation expansion.

On the other hand, the Mayer f functions [Eq. (6.40)] $f(r)$ for the soft-sphere system and $f_d(r)$ for the hard-sphere reference system are much more similar if d is well chosen, as shown in Fig. 9.2. Their difference

$$\delta f(r) \equiv f(r) - f_d(r) \tag{9.17}$$

is zero everywhere except near $r = d$, where there is a "blip" of finite amplitude. These observations suggest that we should construct a perturbation theory in terms of a cluster expansion, so we begin with Eq. (6.75b), the cluster expansion for $-\beta A^{ex}$, and there substitute every f bond by $f_d + \delta f$ and find that

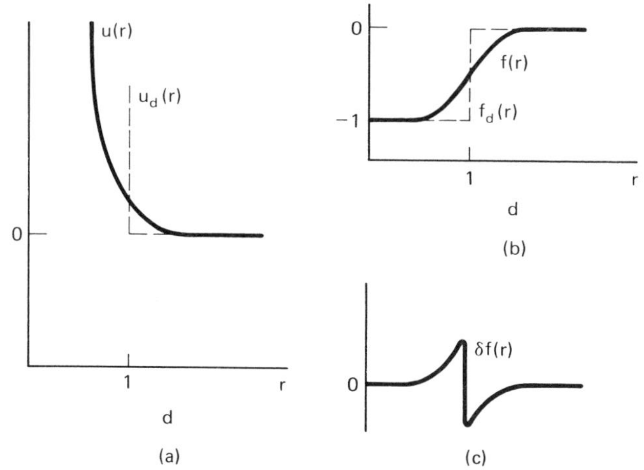

Figure 9.2
(a) Illustrating Eq. (9.16)
(b) Corresponding Mayer f functions
(c) Blip function, Eq. (9.17)

$$-\beta A^{\text{ex}} = -\beta A_d^{\text{ex}} + \text{sum over all simple, distinct, irreducible graphs of } f_d \text{ and } \delta f \text{ bonds on two or more black } \rho \text{ circles with one or more } \delta f \text{ bonds} \tag{9.18}$$

because $-\beta A_d^{\text{ex}}$ comprises all of the graphs with only f_d bonds. Here A_d^{ex} is the free energy of the hard-sphere reference system.

Next we notice that all of the graphs in Eq. (9.18) with one δf bond can be constructed in another way. Begin with the cluster expansion of $-\beta A_d^{\text{ex}}$, eliminate one f_d bond in each graph in all possible ways, and replace it by a δf bond. In view of the derivation of Eq. (6.39) and the definition of $y(r)$ in Eq. (7.4), we obtain the *blip function expansion*

$$-\beta A^{\text{ex}} = -\beta A_d^{\text{ex}} + \tfrac{1}{2}\rho^2 V \int y_d(r)\delta f(r)\, d^3r + \cdots \tag{9.19a}$$

It is rather easy to formulate the higher-order terms as well as to generalize to other reference systems and to make a perturbation theory of $g(r)$ in terms of the reference system correlation function $g_d(r)$,

$$g(r) = e^{-\beta u(r)} y_d(r) + \cdots \tag{9.19b}$$

but we next consider the cluster theory of perturbation expansions more generally, after introducing the application to perturbation theory of the process called *renormalization* or *topological reduction.* This is the process of collecting an infinite number of graphs with bonds of one type to give a small number of graphs with bonds of a less elementary type, as in the formulation of Eq. (9.19b) or in the derivation of the Debye-Hückel equation in Section 7.2.

Just one additional aspect of blip function theory should be mentioned here, namely that the choice of the reference hard-sphere diameter is a matter of convenience. It may be chosen so that the integral in Eq. (9.19a) vanishes, so the first correction to the reference-system free energy is second order in $\delta f(r)$. With this choice for d, which is used in the WCA (Weeks, Chandler, and Andersen) theory to be described below, the blip function theory is remarkably successful, as shown in Table 9.1.

The self-consistency test illustrated there proves to be a rigorous test of the approximation method. It is also important to notice the extent to which the d of the reference hard-sphere fluid is dependent on the state.

TABLE 9.1 REDUCED PRESSURE OF FLUID WITH $\beta u(r) = (\sigma/r)^{12}$

	$\beta p/\rho$			
$\sigma^3\rho$	MC[a]	WCA/A[b]	WCA/v[c]	d/σ[d]
0.1414	1.45	1.45	1.45	1.0663
0.2828	2.12	2.12	2.16	1.0614
0.4243	3.12	3.12	3.23	1.0552
0.5657	4.58	4.57	4.85	1.0477
0.7071	6.66	6.71	7.23	1.0390
0.8485	9.56	9.89	10.60	1.0295

[a] Monte Carlo results of Hansen (1970).

[b] p from thermodynamic differentiation of A^{ex}. A^{ex} from blip function theory [Eq. (9.19)] with WCA choice of d (Andersen et al., 1971).

[c] From virial theorem using blip function formula for $g(r)$ (Andersen et al., 1971).

[d] Relative equivalent hard-sphere diameter as determined by the WCA criterion.

Source: After Table IX of Hansen and McDonald (1976).

DO PROBLEMS 9.1 TO 9.4.

9.5 TOPOLOGICAL REDUCTION

Having employed topological reduction or renormalization in Chapters 7 and 8 and Section 9.4, we give here a detailed definition of this technique. Here we closely follow Andersen (1977).

Suppose that we have a graphical series, which we call M, whose sum represents some quantity of interest, which we call F. All members of M have the same number of white circles which have the same labels and all are free of articulation circles and topologically different from one another. The bonds in M are $b_1, b_2, \ldots, b_q$. Suppose that we have another graphical series, which we call N. All members of N have two white circles, labeled 1 and 2, and the bonds in N are some or all of the set $b_1, \ldots, b_q$. All graphs in N are free of articulation circles and topologically different from one another. The sum of all the graphs in N is some function $G(x_1, x_2)$. A topological reduction is a process that allows us to conclude that F, which is defined as a sum of a certain set M of topologically different graphs containing $b_1, b_2, \ldots, b_q$ bonds, is also equal to the sum of another set of topologically different graphs P which contain $b_1, \ldots, b_q$ and G bonds. The new series P contains, in general, "fewer" graphs than does M, and the two series P and M have different topological specifications. Also, in general, one graph in P is equal to the sum of a number of diagrams in M.

For example, with $F = -\beta A^{ex}$, $b_1 = f_d$, $b_2 = \delta f$, and $G(1\ 2) = y_d(1\ 2)$, this definition characterizes the blip function expansion in Section 9.4. For another example, with F equal to the sum of all chains of c bonds and $G(1\ 2) = h_{ss'}(1\ 2)$, this definition characterizes the derivation of Debye screening. [Eq. (7.30)].

Topological reduction plays a role in many developments in statistical mechanics following the ionic solution theory of Mayer (1950), where the technique was used to convert a certain series M, in which every term is divergent in the thermodynamic limit, to another series P with much more convenient properties.

Next we describe an explicit but rather widely useful framework for exploiting the possibilities of various techniques of topological reduction.

DO PROBLEM 9.5.

9.6 CLUSTER PERTURBATION THEORY

We again divide the pair potential into a reference part and a perturbation part, here with the notation

$$u(1\ 2) = u_0(1\ 2) + u_1(1\ 2) \tag{9.20}$$

and proceed as though the free energy A_0 and pair correlation function $g_0(1\ 2)$ of the reference system are known. We introduce the reference system cluster function

$$f_0(1\ 2) \equiv \exp\left[-\beta u_0(1\ 2)\right] - 1 \tag{9.21}$$

and the perturbation "bond"

$$\phi(1\ 2) = -\beta u_1(1\ 2) \tag{9.22}$$

in terms of which the full cluster function $f(1\ 2) = \exp\left[-\beta u(1\ 2)\right] - 1$ is given by

$$f(1\ 2) = f_0(1\ 2) + [1 + f_0(1\ 2)]\left[\phi(1\ 2) + \frac{1}{2!}\phi(1\ 2)^2 + \frac{1}{3!}\phi(1\ 2)^3 + \cdots\right] \tag{9.23}$$

It is easy to verify that the cluster expansion of the free energy may be written [see Eq. (6.75b)]

$$-\beta A^{ex} = \text{sum of all distinct irreducible graphs on two or more black } \rho \text{ circles with each pair of circles connected by zero or one } f_0 \text{ bonds and by zero or more } \phi \text{ bonds} \tag{9.24}$$

Thus the graphs in Eq. (9.24) are not "simple" and the $\nu!$ factor in Eq. (6.43) is needed for the first time in this text.

The sum of all of the graphs in Eq. (9.24) with no ϕ bonds is $-\beta A_0^{\text{ex}}$. Collectively, the graphs in Eq. (9.24) with one ϕ bond comprise the "high-temperature approximation" term of $-\beta A^{\text{ex}}$.

$$Va_{\text{HTA}} = \text{sum of all distinct irreducible graphs on two or more black } \rho \text{ circles with one } \phi \text{ bond (each pair of circles, including the pair connected by the } \phi \text{ bond, is connected by zero or one } f_0 \text{ bonds)} \tag{9.25}$$

$$= \frac{1}{2}\rho^2 \int d(1)d(2)\phi(1\ 2)g_0(1\ 2)$$

where the last step is a topological reduction based on Eq. (7.4); all distinct irreducible graphs of f_0 bonds on two white circles 1 and 2 with one or more black ρ circles comprise $y_0(1\ 2)$. Since circles 1 and 2 are also connected by $\phi(1\ 2)[1 + f_0(1\ 2)]$, we thus reduce the sum of all the specified graphs to $\phi(1\ 2)g_0(1\ 2)$.

Continuing the same topological reduction procedure we obtain the cluster perturbation theory.

$$-\beta A^{\text{ex}} = -\beta A_0^{\text{ex}} + Va_{\text{HTA}} + \text{sum of all distinct irreducible graphs on two or more black } \rho \text{ circles with zero or one } h_0 \text{ bonds and zero or more } \phi \text{ bonds between each pair of circles, with at least two } \phi \text{ bonds altogether and with no } h_0 \text{ bond redundancies} \tag{9.26}$$

We define an *F-bond redundancy* as a residual (a piece between an articulation pair of circles) that has two or more F bonds and no bonds of any other type (Fig. 9.3). Any h_0 bond redundancies are excluded in Eq. (9.26) because after collecting all possible patterns of f_0 bonds and black ρ circles between white circles 1 and 2, one has exactly $h_0(1\ 2)$; there is no remainder consisting of $h_0 * h_0$ and more complicated combinations of h_0 bonds connecting the white circles.

By substituting Eq. (9.23) in the graphical expansion for h [Eq. (7.3)] and making the same topological reductions, we find

$$h(1\ 2) = h_0 + \text{the sum of all distinct irreducible graphs on two white 1 circles labeled 1 and 2 and one or more black } \rho \text{ circles, with zero or one } h_0 \text{ bond and zero or more } \phi \text{ bonds connecting each pair of circles and with no } h_0 \text{ bond redundancies} \tag{9.27}$$

Now we ask about the strategy for a better approximation for $A^{\text{ex}} - A_0^{\text{ex}}$ than is provided by the high-temperature approximation. Is it best to calculate the contributions of successively higher powers of ϕ, in the spirit of the λ expansion, or to renormalize the series in Eq. (9.26) to look like the blip function expansion and then calculate the contributions of successively higher powers of δf? Or should we renormalize the series in Eq. (9.26) or (9.27) in still another way? This problem

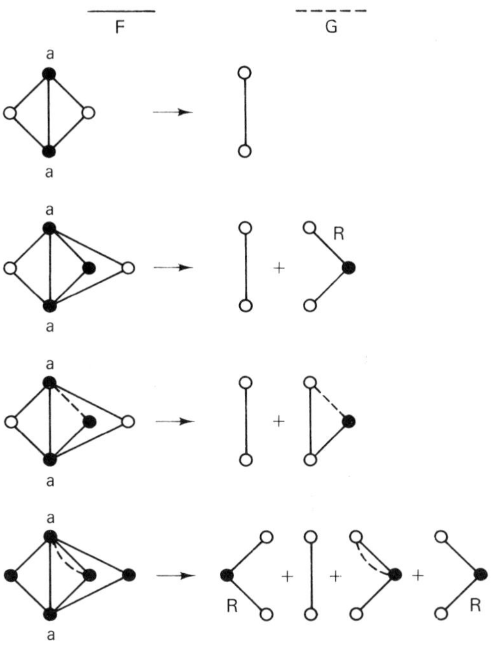

Figure 9.3 Illustrations of F-bond redundancies. In each graph on the left the circles designated a a comprise an articulation pair and the arrow leads to the set of residuals of this pair (Section 6.6). The bonds are either F or G bonds, as indicated. The residuals which are F-bond redundancies are labeled R.

has been extensively explored. It seems fair to say that definite conclusions can be obtained only for certain special choices for $u_0(1\ \ 2)$ and $u_1(1\ \ 2)$, but that one further renormalization step appears to be useful for many models. Thus we write

$$-\beta A^{\text{ex}} = -\beta A_0^{\text{ex}} + Va_{\text{HTA}} + Va_{\text{RING}} + \cdots \tag{9.28}$$

where the "ring" terms derive from a certain subset of the sum of graphs in Eq. (9.25).

$$Va_{\text{ring}} \equiv \text{sum of all rings of } h_0 \text{ and } \phi \text{ bonds on two or more black } \rho \text{ circles, with at least one } \phi \text{ bond and with no } h_0 \text{ redundancies} \tag{9.29}$$

In this case the exclusion of h_0 redundancies simply results in the exclusion of any graph in which any circle is intersected by two h_0 bonds. To represent a_{ring} neatly, we introduce a new type of bond called a *hypervertex function.* In this case it is

$$H_0(1\ \ 2) = h_0(1\ \ 2) + \frac{\delta(1\ \ 2)}{\rho} \tag{9.30}$$

If we define [see Eq. (7.7) notation]

$$\begin{aligned} p(1\ \ 2) &= \int H_0(1\ \ 3)\rho d(3)\phi(3\ \ 2) \\ &= H_0 * \phi \end{aligned} \tag{9.31}$$

then

$$\begin{aligned} a_{\text{ring}} &= \text{sum of all rings of } p \text{ bonds on two or more black circles} \\ &= -\frac{1}{2}\frac{1}{(2\pi)^3}\int d^3k\,\{\rho\tilde{p}(k) + \ln[1-\rho\tilde{p}(k)]\} \end{aligned} \tag{9.32}$$

where we use a result derived in Problem 9.5. We also notice that $\tilde{p}(k)$ is simply

$$\tilde{p}(k) = \tilde{\phi}(k)[1 + \rho\tilde{h}_0(k)] \tag{9.33}$$

The corresponding partial sum of the graphs of $h(1\ 2)$ is

$$\begin{aligned} \mathscr{C}(1\ 2) &= \text{sum of all distinct graphs on two white 1 circles labeled 1 and 2 connected by chains of zero or more black } \rho \text{ circles and } h_0 \text{ and } \phi \text{ bonds, with at least one } \phi \text{ bond and with no } h_0 \text{ redundancies} \\ &= H_0 * \phi * H_0 + H_0 * \phi * H_0 * \phi * H_0 + \cdots \end{aligned} \tag{9.34}$$

which has the Fourier transform

$$\begin{aligned} \mathscr{C}(k) &= \frac{\tilde{H}_0\rho^2\tilde{\phi}\tilde{H}_0}{1-\rho^2\tilde{\phi}\tilde{H}_0} \\ &= \frac{(1+\rho\tilde{h}_0)\tilde{p}}{1-\rho\tilde{p}} \end{aligned} \tag{9.35}$$

Then we have

$$\begin{aligned} h(1\ 2) = h_0(1\ 2) + {} & \text{sum of all distinct irreducible graphs between two white circles labeled 1 and 2 and zero or more black } \rho \text{ circles, with zero or one } h_0 \text{ bonds and zero or more } \mathscr{C} \text{ bonds between each pair of circles, with at least one } \mathscr{C} \text{ bond and with no } h_0 \text{ bond redundancies and no black circles connected only to two } \mathscr{C} \text{ bonds} \end{aligned} \tag{9.36}$$

We close this section by noticing that most approximations to the sum in Eq. (9.36),

$$h(1\ 2) \cong h_0(1\ 2) + \mathscr{C}(1\ 2) \tag{9.37}$$

for example, are inconsistent with the core condition [Eq. (6.44)]. This observation motivates the developments in the next section.

DO PROBLEMS 9.6 to 9.11.

9.7 OPTIMIZED CLUSTER PERTURBATION THEORY

We notice that Eq. (9.37), which apparently does not satisfy the core condition on $h(1\ 2)$, could be brought into agreement if we could specify

$$\mathscr{C}(r) = 0 \qquad \text{for } r < \sigma \tag{9.38}$$

where σ is the core diameter such that $h_0(r) = -1$ for $r \leq \sigma$. This may be possible because $\mathscr{C}(r)$ is a functional of $u_1(r)$ while $u_1(r)$ is arbitrary in the range $r < \sigma$ because configurations in which $r < \sigma$ make no contribution to the properties of the system because $u(r) = \infty$ for $r < \sigma$.

It can be shown that adjusting $u_1(r)$ in $r < \sigma$ to satisfy Eq. (9.38) is indeed possible and that this operation also minimizes the positive quantity a_{RING}.

In practice the optimization may be carried out by minimizing a_{ring} with respect to variations of $u_1(r)$ inside the core, the latter being represented by a series on suitable basis functions (e.g., Legendre polynomials) within the core.

The resulting approximation to A^{ex}, that is,

$$-\beta A^{\text{ex}}/V = -\beta A_0^{\text{ex}}/V + a_{\text{HTA}} + a_{\text{RING}} \tag{9.39}$$

with the minimized a_{ring} is called the optimized random phase approximation (ORPA), while "ORPA + B_2"

$$-\beta A^{\text{ex}}/V = -\beta A_0^{\text{ex}}/V + a_{\text{HTA}} + a_{\text{RING}} + \rho^2 B_2' \tag{9.40}$$

includes all of the terms missing in Eq. (9.39) having $\mathscr{C}$ and h_0 bonds on two black ρ circles, giving the modified second virial coefficient

$$B_2' = \tfrac{1}{2}\int d^3r\,[g_0(r)\sum_{n\geq 3}\frac{\mathscr{C}(r)^n}{n!} + \tfrac{1}{2}h_0(r)\mathscr{C}(r)^2] \tag{9.41}$$

With the optimized $\mathscr{C}(r)$ used in Eq. (9.64) we have the EXP approximation to $h(r)$, which is roughly equivalent to ORPA + B_2. The B_2 term is important when ρ is relatively small and the perturbation $u_1(r)$ large.

DO PROBLEMS 9.12 AND 9.13.

9.8 WCA THEORY

The perturbation theory of Weeks et al. (1971), which is remarkably successful for model monatomic liquids, typically with the 6-12 pair potential of Eq. (4.25), is described here.

In WCA theory the pair potential is separated into reference and perturbation parts in the particular way shown in Fig. 9.4. The reference system potential $u_0(r)$ is purely repulsive and sharply r-dependent: the potential of the "harsh repulsive force" in the language of WCA. By comparison $u_1(r)$ is a weak and slowly varying "tail."

With the WCA separation of the potential, even the λ expansion carried to first order gives an accurate result for the free energy of the full ($u_0 + u_1$) system relative to the reference (u_0) system at liquid densities. For example (Verlet and Weis, 1972), at $T^* = 1/e_0\beta = 0.72$ and $\rho^* \equiv \rho\sigma^3 = 0.85$, simulation gives $\beta A_0^{\text{ex}}/N = 4.49$. With the first-order λ expansion this gives $\beta A^{\text{ex}}/N = -4.84$, which agrees with the $u_0 + u_1$ simulation result within 1%.

Of course, to use the λ expansion to estimate $A^{\text{ex}} - A_0^{\text{ex}}$ we need accurate

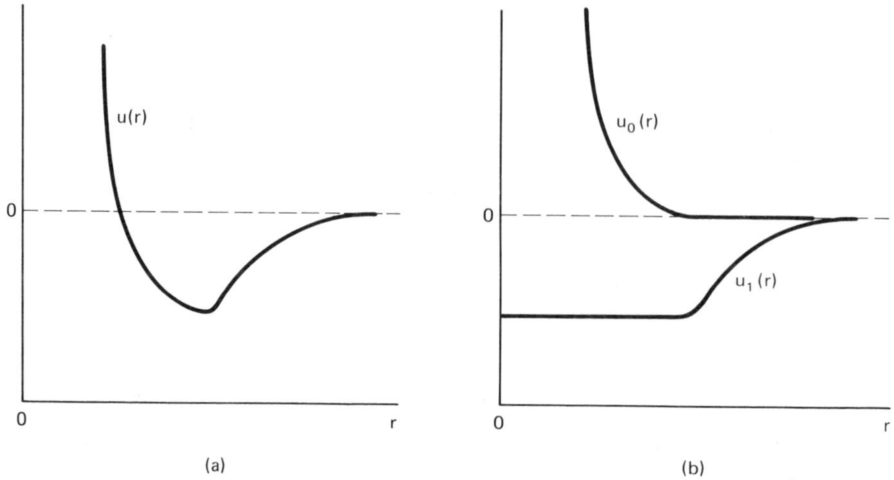

Figure 9.4 Separation of the pair potential in WCA theory.
(a) Typical pair potential function.
(b) Separation according to the sign of the force.

$g_0(r)$ and A_0^{ex}, the correlation function and free energy of the reference system with pair potential u_0. Now we write

$$u_0(r) = u_d(r) + u_b(r) \tag{9.42}$$

where $u_d(r)$ is the hard-sphere potential for spheres of diameter d and $u_b(r)$ is the remainder defined by this equation. To get the $g_0(r)$ and A_0^{ex} from the known values $g_d(r)$ and A_d^{ex} for the hard-sphere reference system, we use the blip function theory, now written as

$$-\beta A_0^{ex} \simeq -\beta A_d^{ex} \tag{9.43}$$

where d is chosen so that the integral over $\delta f(r)$ [now $f_0(r) - f_d(r)$] in Eq. (9.19) vanishes. The corresponding equation for $g_0(r)$ is [see Eq. (9.19b)]

$$g_0(r) = e^{-\beta u_0(r)} y_d(r) \tag{9.44}$$

With this combination of decompositions into reference potential and perturbation and use of topological reduction for efficient calculation of the perturbation effects, the WCA theory enables one to estimate the equilibrium properties of the 6–12 fluid over the range of states that characterize the liquid phase using, as the irreducible input, the structure and free energy of the hard-sphere fluid.

At less than liquid densities and unless the temperature is very high, the effect of the u_1 perturbation on the u_0 system becomes too large to be adequately given by the first-order λ approximation; here in the WCA theory the λ approximation is replaced by optimized cluster theory. Unfortunately, this replacement is far from straightforward because "optimization" of the chain sum $\mathscr{C}$ is prescribed only for perturbation from a hard-sphere fluid, so to use OCT we have to choose a new hard-sphere fluid to represent the u_0 reference system. Fortunately, there is an alternative procedure involving adaptation of the MSA to systems with soft cores, which is described in the next section.

9.9 INTEGRAL EQUATION PERTURBATION THEORY

The mean spherical approximation comprises the exact Ornstein–Zernicke equation (7.48a):

$$h = c + c * h \tag{9.45a}$$

the exact core condition (7.58a):

$$h(r) = -1 \qquad \text{if } r < \sigma \tag{9.45b}$$

and the approximation (7.58b):

$$c(r) = -\beta u(r) \qquad \text{if } r > \sigma \tag{9.45c}$$

With $u = u_1 + u_0$ as in Eq. (9.20) and in the case that u_0 is the pair potential for hard spheres of diameter σ, then in Eq. (9.45c) the potential is just the perturbation u_1. Then, too, Eq. (9.45) has the following interpretation: with $c = 0$ for $r > \sigma$ one has the PY approximation to the reference hard-sphere fluid, while inclusion of $c(r) = -\beta u_1(r)$ for $r > \sigma$ gives an approximation to the effect of u_1 as a perturbation on the hard-sphere potential. Indeed, in this case the MSA is equivalent to ORPA, the optimized random phase approximation (Problem 9.12).

The HNC approximation [Eq. (7.63)] is obtained from the Ornstein–Zernicke equation and the exact expression [Eq. (7.11)]

$$c = h - \ln(1 + h) - \beta u + B \tag{9.46a}$$

by omitting the sum of bridge graphs B. When $u = u_0 + u_1$, we have

$$c_0 = h_0 - \ln(1 + h_0) - \beta u_0 + B_0 \tag{9.46b}$$

for the reference system and Eq. (9.46a), with $u = u_0 + u_1$, governs the perturbed system. The RHNC (reference hypernetted chain) approximation of Lado (1964) is to replace (9.46a) by

$$c = h - \ln(1 + h) - \beta(u_0 + u_1) + B_0 \tag{9.46c}$$

Because the bridge graphs are relatively short-range functions, the approximation $B = B_0$ seems promising when the reference system has nearly the same short-range structure as the full system.

In one application of the RHNC approximation, the system of interest is the one-component plasma, point ions in a neutralizing background. In the reference system the Coulomb pair potential is replaced by a screened (damped) Coulomb potential and B_0 is derived from Monte Carlo simulation. The simulation is more reliable in the reference system than it is for the full system when the latter has long-range forces. In this case the RHNC structure and energies are in good agreement with the results of MC simulation for the full system (Table 9.1).

Madden and Rice (1980) have proposed an extension of the MSA to systems with continuous potentials which may be described as an application of the RHNC approximation. If the PY equation is assumed to apply to the reference system, one can combine Eqs. (7.48b) and (7.11) to obtain the PY approximation to B_0 which is substituted in Eq. (9.46c) to get

$$c = -\beta u_1 + (1 - e^{\beta u_0})g_0 + \left[\delta h - \ln\left(1 + \frac{\delta h}{g_0}\right)\right] \tag{9.47}$$

where $\delta h = h - h_0$.

If the separation $u = u_0 + u_1$ is made on the WCA prescription, then in suitable cases $g \simeq g_0$. On this basis the $[\cdot\ \cdot\ \cdot]$ terms are neglected to get a form of MSA extended to apply to systems with no hard core.

$$\begin{aligned} c &= -\beta u_1 + (1 - e^{\beta u_0})g \\ h &= c + h * c \end{aligned} \tag{9.48}$$

This extended MSA proves to be a very satisfactory solution for the problem discussed at the end of Section 9.8.

Another application of the HNC equation in a perturbation mode is to the fluid comprising dipolar hard spheres, which we describe in the invariant expansion form [see Eq. (8.4)] as repeated here for h:

$$h(1\ 2) = h^S(r) + D(1\ 2)h^D(r) + \Delta(1\ 2)h^\Delta(r) \tag{9.49}$$

where $r = |\mathbf{r}_1 - \mathbf{r}_2|$, where D and Δ are defined in Table 8.1, and where higher terms in the invariant expansion are neglected. We consider the dipolar term in the pair potential as a perturbation on the hard-sphere fluid

$$u(1\ 2) = u_d(r) - \frac{\mu^2 D(1\ 2)}{r^3} \tag{9.50}$$

where u_d is the hard-sphere pair potential for spheres of diameter d. So in terms of the invariant expansion we have

$$u^S(r) = u_d(r), \qquad u^\Delta(r) = 0, \qquad u^D(r) = \frac{-\mu^2}{r^3} \tag{9.51}$$

In the HNC approximation for this model we have

$$c(1\ 2) = h(1\ 2) - \ln[1 + h(1\ 2)] - \beta u(1\ 2) \tag{9.52}$$

together with the Ornstein–Zernicke equation. The LHNC approximation of Patey (1977) is obtained by substituting Eq. (9.49) in Eq. (9.52) and treating h^D and h^Δ as small perturbations in expanding $\ln(1 + h)$ to get the closure for each invariant component:

$$c^S = h^S - \ln(1 + h^S) - \beta u^S \tag{9.53a}$$

$$c^D = h^D\left(1 - \frac{1}{1 + h^0}\right) + \beta u^D \tag{9.53b}$$

$$c^{\Delta} = h^{\Delta}\left(1 - \frac{1}{1 + h^{0}}\right) \tag{9.53c}$$

These results can be used together with the corresponding decomposition of the OZ equation, already given in Eq. (8.9), to get a closed set of integral equations that can be solved by the usual Fourier transform techniques as described in the original report. Of course, one does not have to solve for $h^{S}(r)$, the hard-sphere correlation in the HNC approximation, but can replace this function by the accurate correlation function for the hard-sphere fluid. The linearized hypernetted chain (LHNC) approximation described here was independently derived at about the same time but in two entirely different ways by Wertheim (1973) and by Henderson and Gray (1978). This coincidence illustrates the unity of the equilibrium perturbation theory. Although, as described in this chapter, the theory takes many different forms, several of these may lead to a given algorithm for approximating the structure and energetics.

THE LITERATURE

In this chapter we have drawn heavily on the discussion of perturbation theories by Hansen and McDonald (1976) and Andersen (1977), but much more may be learned from these accounts, from Smith (1974), from Andersen et al. (1976), and from Stell (1977), especially concerning the extensive theory of approximations to the cluster perturbation series. Improvement of the λ approximation methods by interpolation between small λ and large λ is reviewed by Stell (1977). For further information about the integral equation perturbation theories, see Ceperley and Chester (1977) for RHNC, Lado et al. (1983) for an optimized version of RHNC, Madden (1981) for the optimized cluster theory, and Patey (1977 and 1978) for LHNC.

PROBLEMS

9.1. Graph $x(\ln x - 1)$ as a function of x. Calculate its slope as a function of x. Can you prove that $x(\ln x - 1)$ is not less than -1 for real positive x?

9.2. Extend the blip function equation (9.19a) to give $-\beta(A^{\mathrm{ex}} - A_d^{\mathrm{ex}})$ in fully renormalized form (i.e., as a cluster expansion with h_d and y_d δf bonds on black ρ circles).

9.3. See what happens if you apply Eq. (9.19) to a model in which $w(r)$ is Coulombic. Will the multicomponent version of Eq. (9.19) help?

9.4. In Table 9.1 the temperature for which these results are derived does not appear except in the reduced pressure $\beta p/\rho$. Explain. Refer to the literature cited if necessary.

9.5. A common example of topological reduction is the chain sum

$S_c(1\ 1') =$ 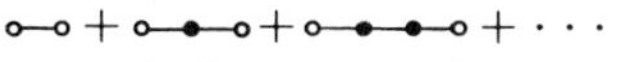

= sum of all chains of f bonds and black ρ circles between white ρ circles 1 and 1′ (9.54)

Closely related but more difficult to evaluate is the ring sum

$S_R =$  $+ \cdots$
$\quad = \text{sum of all rings of } f \text{ bonds and black } \rho \text{ circles}$ (9.55)

(a) Show that the chain sum has the Fourier transform

$$\tilde{S}_c(k) = \frac{\rho^2 \tilde{f}(k)}{1 - \rho \tilde{f}(k)} V \tag{9.56}$$

where

$$\tilde{f}(k) \equiv \int d^3r\, f(r) e^{i\mathbf{k}\cdot\mathbf{r}} \tag{9.57}$$

(b) Show that the definition of S_R implies that

$$S_R = \sum_{n \geq 2} \frac{\rho^n}{2n} \int f(1\,2) f(2\,3) \cdots f(n\,1) d(1) \cdots d(n) \tag{9.58}$$

To see the relation of S_R to S_c, it is convenient to introduce a modified ring sum defined as

$$\begin{aligned} S_{\text{MR}} &= \int d(1) d(1') f(1\,1') S_c(1\,1') \\ &= \sum_{n \geq 2} \rho^n \int f(1\,2) f(2\,3) \cdots f(n-1\,n) f(n\,1) d(1) \cdots d(n) \end{aligned} \tag{9.59}$$

so we have $S_{\text{MR}} = (2\rho)\partial S_R / \partial \rho$.

(c) Using the convolution theorem (Appendix 1), derive

$$\begin{aligned} \frac{S_R}{V} &= \frac{1}{(2\pi)^3} \int d^3k \sum_{n \geq 2} \frac{\rho^n}{2n} \tilde{f}(k)^n \\ &= -\frac{1}{2} \frac{1}{(2\pi)^3} \int d^3k\, [\rho \tilde{f}(k) + \ln(1 - \rho \tilde{f}(k)] \end{aligned} \tag{9.60}$$

9.6. Compare Eq. (9.25) with the corrections to $A^{\text{ex}} - A_0^{\text{ex}}$ formulated in Eqs. (9.9) and (9.19). Which of the three differs from the others?

9.7. What collection of terms in Eq. (9.26) gives the first-order blip function expansion in Eq. (9.19)?

9.8. An important tool in cluster perturbation theory is the model in which u_1 in Eq. (9.20) and (9.22) has the form (Stell 1977)

$$u_1(r) = -\gamma^3 \psi(\gamma r) \tag{9.61}$$

where γ is called the Kac inverse range parameter; for $\gamma \longrightarrow 0$ the perturbation becomes infinitely weak and infinitely long range. We also assume that $\psi(x) \longrightarrow 0$ as $x \longrightarrow \infty$ and $\int_0^\infty x^2\, dx\, \psi(x)$ is not infinite. Prove the following statements for this model.

(a) $\int d^3r\, \phi(r)$ is independent of γ.

(b) A ring of ϕ bonds on black ρ vertices is proportional to γ^3.

(c) a_{RING} in Eq. (9.28) is proportional to γ^3.

(d) a_{HTA} is independent of γ as $\gamma \longrightarrow 0$.

(e) $\mathscr{C}(1\,2)$ is proportional to γ^3.

9.9. For the model in Problem 9.8. prove that, neglecting only terms of order γ^6 and higher, we have

$$h(1\ 2) = h_0(1\ 2)[1 + \phi(1\ 2)] + \mathscr{C}(1\ 2) \tag{9.62}$$

9.10. The term *hypervertex* refers to the role of $H_0(1\ 2) \equiv$  [see Eq. (9.30)] in the following representation of $\mathscr{C}(1\ 2)$, where $\phi(1\ 2) =$ —— (Hansen and McDonald, 1976).

$\mathscr{C}(1\ 2) =$ 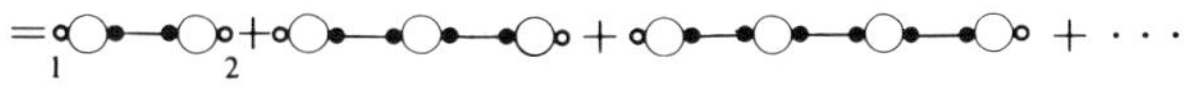 (9.63)

which we may think of as all chains of ϕ bonds on H_0 hypercircles. Formulate this sum of chains in words.

9.11. Work out the topological reduction, based on Lemma 1 in Section 6.6, that leads to

$$g(1\ 2) = g_0(1\ 2) \exp \mathscr{C}(1\ 2) + \cdots \tag{9.64}$$

Neglect of the missing terms is called the EXP approximation if we use the optimized $\mathscr{C}$ (Section 9.7). Notice that (9.64) satisfies the core condition in any case.

9.12. Within the approximation

$$-\beta A^{\text{ex}}/V = -\beta A_0^{\text{ex}}/V + a_{\text{HTA}} + a_{\text{RING}} \tag{9.65a}$$

show that

$$\frac{\delta(-\beta A^{\text{ex}}/V)}{\delta\phi(\mathbf{r})} = \tfrac{1}{2}\rho^2\mathscr{C}(\mathbf{r}) \qquad \text{if } r < \sigma \tag{9.65b}$$

which justifies the optimization condition, Section 9.7.

9.13. Show that ORPA is equivalent to the MSA if in ORPA the reference system is the hard-sphere fluid represented in the PY approximation (Hansen and McDonald, 1976).

9.14. Write exact expressions for u_0 and u_1 derived from the 6–12 potential in Eq. (4.25) according to the WCA separation.

9.15. An important model for ionic systems is charged soft spheres with the pair potential

$$u_{ab}(r) = K\left(\frac{r_a^* + r_b^*}{r}\right)^9 + \frac{e_a e_b}{\epsilon r} \tag{9.66}$$

where a and b are ionic species labels, r_a^* and r_b^* are ion size parameters, and K and ϵ are suitable constants. How would you divide up the potentials to apply ORPA to this model?

9.16. Equations (9.9) and (9.19a) are special cases of the first-order expansion of A^{ex} under a more general perturbation theory based on functional expansions (Smith, 1974). The simplest result of this theory is that for any functional $s[u(1\ 2)]$ of the pair potential, the first-order perturbation is

$$A^{\text{ex}} - A_0^{\text{ex}} = \tfrac{1}{2}\rho^2 \int g_0(1\ 2)\left[\frac{\delta u(1\ 2)}{\delta s(1\ 2)}\right]_0 \times [s(1\ 2) - s_0(1\ 2)]d(1)d(2) \tag{9.67}$$

where the subscript 0 specifies the reference system.

(a) What is the functional $s[u]$ for which (9.67) reduces to (9.9)? to (9.19)?

(b) Show that Eq. (9.67) results from the functional expansion of A^{ex} in terms of $s[u]$.

10

Equilibrium Theory of Solutions

10.1 KIRKWOOD–BUFF THEORY

We recall that the grand ensemble theory for many-component systems leads to Eq. (1.59)

$$\langle N_s N_{s'} \rangle - \langle N_s \rangle \langle N_{s'} \rangle = \left(\frac{V}{\beta}\right) \frac{\partial \rho_s}{\partial \mu_{s'}} = \left(\frac{V}{\beta}\right) M_{ss'}$$

for the relation between composition fluctuations in an open system and the partial derivative on the right, which is at fixed V and T and fixed chemical potentials of all of the component species except s'. This derivative is the matrix inverse, in the sense defined in Section 1.13, of the more convenient thermodynamic coefficient $\partial \mu_{s'}/\partial \rho_s$, which is a partial derivative at fixed values of the temperature and the densities (i.e., concentrations) of all species except s. Eliminating the composition fluctuations between this equation and Eq. (4.38),

$$\langle N_s N_{s'} \rangle - \langle N_s \rangle \langle N_{s'} \rangle = V \left[\rho_s \delta_{ss'} + \rho_s \rho_{s'} \int d^3 r \, h_{ss'}(r) \right]$$

we find that

$$M_{ss'} = \beta \left[\rho_{ss'} \delta_{ss'} + \rho_s \rho_{s'} \int h_{ss'}(r) \, d^3 r \right] \tag{10.1a}$$

for the matrix elements of $|M| = |L|^{-1}$, where the matrix elements $L_{ss'} = \partial \mu_s / \partial \rho_{s'}$ are the convenient thermodynamic coefficients mentioned above. Now by thermodynamic analysis, while giving great care to the partial derivative relations, we can express many experimentally accessible thermodynamic coefficients in terms of the

$\mathbf{M}_{ss'}$, thus providing a basis for the molecular interpretation of the experimental coefficients. One example may be found in Eq. (1.89); others taken from Kirkwood and Buff (1951) follow. They apply to a mixture of two species, typically with index 1 for the solvent and 2 for the solute,

$$\overline{V}_1 = \left(\frac{\partial V}{\partial N_1}\right)_{T,P,N_2}$$

$$= \frac{1 + (G_{22} - G_{12})\rho_2}{\rho_1 + \rho_2 + \rho_1\rho_2(G_{11} + G_{22} - 2G_{12})} \tag{10.1b}$$

$$\beta\rho_1\left(\frac{\partial\mu_1}{\partial\rho_1}\right)_{T,P} = \frac{1}{1 + \rho_1(G_{11} - G_{12})} \tag{10.1c}$$

$$-(1 - x_2)\left(\frac{\partial\mu_1}{\partial x_2}\right)_{T,P} = x_2\left(\frac{\partial\mu_2}{\partial x_2}\right)_{T,P}$$

$$= \frac{1}{1 + \rho_1 x_2(G_{11} + G_{22} - 2G_{12})} \tag{10.1d}$$

$$\beta\left(\frac{\partial\pi_1}{\partial\rho_2}\right)_{T,\mu_1} = \frac{1}{1 + \rho_2 G_{22}} \tag{10.1e}$$

where $G_{ss'}$ is the zeroth moment of $h_{ss'}(r)$ [i.e., the integral in Eq. (10.1a)], $\overline{V}_1$ is the partial molecular volume of component 1, x_2 is the mole fraction of component 2, and π_1 is the osmotic pressure measured with component 1 as the solvent. A complementary set of equations may be obtained by interchanging the indices 1 and 2 everywhere.

It may be noticed that $G_{ss'}$ is the $k = 0$ limit of the Appendix 1 type of Fourier transform $\tilde{h}_{ss'}(k)$, which in turn is very closely related to the partial structure factor at $k = 0$ [see Eq. (5.27)]. One application of Kirkwood–Buff theory is to supplement diffraction studies by providing the $k = 0$ limit of the partial structure factors in terms of thermodynamic coefficients, which often are quite accurately known. In the case of ionic solutions, the thermodynamic data can be related only to combinations of partial structure factors at $k = 0$ because the ρ_s are not all independently variable (Beeby, 1973).

If neither s nor s' is ionic, then in the infinite dilution limit $G_{ss'}$ approaches $[1 + \delta_{ss'}]$ times the second virial coefficient $B_{ss'}$ defined in Eq. (6.8). However, if both s and s' are ionic species, then $B_{ss'}$ does not exist (Problem 6.4). On the other hand, if s is ionic and s' not, then $B_{ss'}$ exists but is not the infinite dilution limit of $G_{ss'}$, a surprising manifestation of the long-range Coulombic forces in the solution (Friedman et al., 1973)!

Next we turn to the solution theories of McMillan–Mayer and of Adelman which deal with the pair correlation functions $h_{ab}(r)$ rather than merely their zeroth moments, but where a and b are limited to be solute species only. So these theories are applicable to systems for which there is a rationale for classifying the species as solute or solvent components. Within this limitation they enable us to apply all of the methods of pair correlation function theory to determine the structures of the solutions as well as their thermodynamic coefficients.

10.2 McMILLAN–MAYER THEORY OF SOLUTIONS

We first consider a mixture of two species of molecules, one designated as solute (activity = z, number of molecules = K) and the other designated as solvent (activity = y, number of molecules = L). For this mixture, at the Born–Oppenheimer level of description (Section 4.3) the grand partition function is

$$\Xi(z, y, V, T) = \sum_{K,L} \frac{z^K}{K!}\frac{y^L}{L!} \text{x} \int \exp\,[-\beta U_{K+L}(1, \ldots, K; 1, \ldots, L)]d(1) \cdots d(K)d(1) \cdots d(L) \tag{10.2a}$$

$$= \sum_K \frac{z^K}{K!}\int d(1) \cdots d(K)\left[\sum_L \frac{y^L}{L!}\int \exp\,(-\beta U_{K+L})d(1) \cdots d(L)\right] \tag{10.2b}$$

This equation may be compared to the K-particle correlation function

$$g_K(1, \ldots, K) = \frac{z^K}{\rho^K}\frac{1}{\Xi(z, y, V, T)}\sum_{K',L}\frac{z^{K'}}{K'!}\frac{y^L}{L!}\int \exp\,(-\beta U_{K+K'+L})d(1) \cdots d(K')d(1) \cdots d(L) \tag{10.2c}$$

for K solute particles at fixed locations in the solution characterized by activities y and z. Here ρ is the solute concentration $\langle K/V\rangle$ in the mixture. Equation (10.2c) is straightforward generalization of Eq. (4.23) to the correlation function of two or more solute particles in a system of two components. (See Problem 4.5.)

In the $z = 0$ limit we have $\gamma \equiv z/\rho \neq 1$ because $y \neq 0$. In this case

$$[g_K(1, \ldots, K)]_{z=0} = \frac{\gamma^K}{\Xi(0, y, V, T)}\sum_L \frac{y^L}{L!}\int \exp\,(-\beta U_{K+L})\, d(1) \cdots d(L) \tag{10.3}$$

Compare the $1 \cdots L$ integral in Eq. (10.3) with the one in Eq. (10.2b) with given y. Combine the equations to eliminate the $1 \cdots L$ integral:

$$\Xi(z, y, V, T) = \sum_K \frac{z^K}{K!}\int d(1) \cdots d(K)\, \Xi(0, y, V, T)\frac{1}{\gamma^K}[g_K(1, \ldots, K)]_{\substack{z=0\\ y>0}} \tag{10.4}$$

Now we introduce the notation for the solvent-averaged potential for K solute molecules at specified locations in the pure solvent ($z = 0$, $y > 0$)

$$\overline{U}_K(1, \ldots, K) \equiv -k_B T \ln\,[g_K(1, \ldots, K)]_{\substack{z=0\\ y>0}} \tag{10.5}$$

and rearrange Eq. (10.4) to read

$$\Xi(z, y, V, T) = \Xi(0, y, V, T)\sum_K \frac{(z/\gamma)^K}{K!}\int \exp\,(-\beta \overline{U}_K)d(1) \cdots (K) \tag{10.6}$$

Finally, we notice that if we redefine the solute activity

$$a \equiv \frac{z}{\gamma} = z \lim_{z=0} \frac{\rho}{z} \tag{10.7a}$$

then we have

$$\lim_{z=0} \frac{a}{\rho} = 1 \tag{10.7b}$$

We also define

$$\Xi_{\text{osm}}(a, V, T) \equiv e^{\beta p_{\text{osm}} V} \equiv \frac{\Xi(z, y, V, T)}{\Xi(0, y, V, T)} \tag{10.7c}$$

and so we arrive at the basic equation of McMillan–Mayer theory.

$$\Xi_{\text{osm}} = \sum_K \frac{a^K}{K!} \int \exp\,[-\beta \bar{U}_K(1, \ldots, K)]d(1) \cdots d(K) \tag{10.8}$$

It is remarkable that Eq. (10.8) has exactly the same form as Eq. (4.3). This simplicity is a bit deceptive because there is an important complication in $\bar{U}_{\mathbf{K}}$ compared with $U_{\mathbf{K+L}}$, as discussed at the end of this section. Until then we assume that both U and $\bar{U}$ are pairwise additive.

In Chapter 6 we derived the density expansions of p and A^{ex} from the basic equation

$$\Xi = \sum \frac{z^N}{N!} \int e^{-\beta U_N} d(1) \cdots d(N) \tag{10.9}$$

It seems clear that if instead we began with Eq. (10.8), we would get (recall that $\rho = \rho_{\text{solute}}$ here)

$$\beta p_{\text{osm}} = \rho + \rho^2 \bar{B}_2 + \rho^3 \bar{B}_3 + \cdots \tag{10.10}$$

where $\bar{B}_n$ is just like B_n, except that the f bonds are replaced by

$$\bar{f}(1\,2) = e^{-\beta \bar{u}(1\,2)} - 1 \tag{10.11}$$

bonds, where $\bar{u} = \bar{U}_2$ is the solvent-averaged pair potential. Also we would get, in place of Eq. (6.75b),

$-\beta \bar{A}^{\text{ex}} =$ sum of all simple, distinct, ALDC graphs of $\bar{f}$ bonds on black ρ circles

=  + ⋯ (10.12)

where

$$\bar{A}^{\text{ex}} \equiv A_{\text{total}} - A^{\text{id}} \tag{10.13}$$

is the total free energy in the given ρ, T, V state minus A^{id}, the free energy of a hypothetical state of the same ρ, T, V but where there are no solute–solute interactions and where the solute–solvent interactions are the same as at $\rho = 0$. The formulation in Eq. (10.13) corresponds exactly to the usual practice of dividing A into ideal and excess parts in the thermodynamical theory of unsymmetrical mixtures. To the degree that $\bar{U}_N$ is not pairwise additive, additional types of bonds will appear in Eq. (10.12), as discussed later.

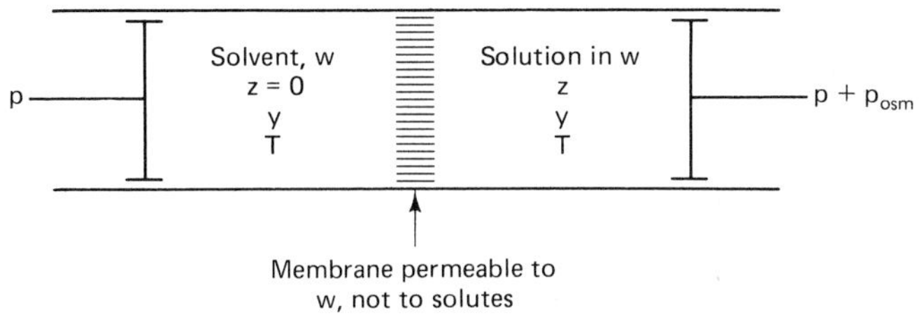

Figure 10.1 Osmotic pressure experiment related to the McMillan–Mayer theory.

We consider the osmotic equilibrium of the solution with the pure solvent (Fig. 10.1). If we change the solvent activity y by adding solute to the solution, we can maintain equilibrium by changing p_{osm} at fixed p. Notice, however, that in this process the concentration of solvent in the solution compartment is free to change, and presumably decreases. But the McMillan–Mayer theory solvent-averaged potentials $\bar{U}_N$ are determined in the $z = 0$ limit, where the solvent density in the solution is the same as on the left if p is fixed.

Now we notice that the theory from Eqs. (10.2a) to (10.13) is easily generalized to the case of many solute and solvent species as long as each distinct species is designated either as solute or as solvent. In place of Eq. (10.8) we now have, using the notation in Eq. (1.45),

$$\Xi_{\text{osm}} = \sum_{\mathbf{K}} \frac{\mathbf{a}^{\mathbf{K}}}{\mathbf{K}!} \int e^{-\beta \bar{U}_{\mathbf{K}}} \prod_{s} d(1_s) \cdots d(K_s) \tag{10.14}$$

where we integrate over the coordinates of all solute particles. Equation (10.12) is unchanged except that now the black circles imply summation over solute species $a, b, \ldots$ as well as integration over the volume. Some relevant details are given in the discussion of Eq. (6.76). Also in Fig. 10.1 the solvent activity y becomes the set of solvent activities $\mathbf{y} = y_1, y_2, \ldots$ and the solute activity z becomes the set of solute activities $\mathbf{z} = z_a, z_b, \ldots$. In the discussion of the McMillan–Mayer theory in the rest of this chapter, the mixed solute case will be explicitly considered; the mixed solvent generalization affects only the definition of $\bar{U}_K$, the potential energy for a solute configuration at given $\mathbf{y}$, so it does not change the equations explicitly.

From Eq. (10.8) following the derivation of Eq. (6.79) we get

$$g_{ab}(1\;2) = \exp[-\beta \bar{u}_{ab}(1\;2)]\,[1 +$$ 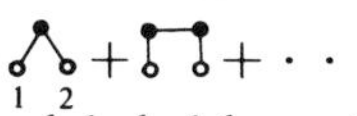 $$+ \cdots], \tag{10.15}$$
$\bar{f}$ bonds, white 1 circles labeled 1a and 2b, black ρ circles

where again summation over solute species is implied at the black ρ circles and where a and b are solute species. We do not write $\bar{g}_{ab}$ because the pair correlation function in Eq. (10.15) is the same as $g_{ss'}$ in Eq. (4.20) if s is solute species a and s' is solute species b. Thus the same correlation function is also given by

$$g_{ab}(1\;2) = e^{-\beta u_{ab}(1\;2)}[1 + \text{sum of all } h\text{-allowed graphs of } f \text{ bonds on white 1 circles labeled 1a and 2b and on one or more black } \rho \text{ circles, and with no direct bond between the white circles}] \tag{10.16}$$

where summation over *both* solute and solvent species is implied at each black ρ circle. Notice that $\bar{u}$, $\bar{f}$, and $\rho = \rho_{\text{solute}}$ in Eq. (10.15) become, respectively, u, f, and ρ_{all} in Eq. (10.16). Of course, we can go on in this way and quite directly transcribe most of the theory in Chapters 6 to 9, which we may characterize as at the Born–Oppenheimer level (i.e., at the level of a simple classical fluid; see Section 4.3), to obtain the corresponding theory at the McMillan–Mayer level. All of the elements of these transcriptions (i.e., changing ρ to ρ_{solute}, changing U_N to $\overline{U}_N$, etc.) are illustrated by the examples we just considered. In the case of MM-level molecular dynamics simulations, with the force on solute particle a expressed as $-\nabla_a \overline{U}_N$, rather than $-\nabla_a U_N$ as in Chapter 5, the diffusion coefficients and other dynamical coefficients calculated from the resulting trajectory in the phase of the solute particles are meaningless; to do better we must introduce features which appropriately represent the dynamical contributions of the solvent (Section 13.3). Fortunately, the MM equilibrium thermodynamic and structural averages, calculated from the MM-level molecular dynamics trajectories, will be as accurate as the corresponding B-O level results. The reason is that the equilibrium averages in the McMillan–Mayer theory depend only upon $\overline{U}_N$ and not on the kinetic part of the Hamiltonian.

Thus with respect to equilibrium properties, the McMillan–Mayer theory offers a powerful way to extend the applicability of the Born–Oppenheimer level results derived in earlier chapters by simply reinterpreting the notation.

As usual in the case of apparent bargains there is a price to pay, but before exploring that aspect, we should notice that many model studies at the McMillan–Mayer level, using approximation methods described in earlier chapters, have been quite successful in interpreting the thermodynamic excess functions of aqueous electrolytes in terms of quite simple models for the solvent-averaged ion–ion interactions (Friedman, 1981). Indeed, the results in Fig. 7.4 were obtained in the course of studies of this kind, and still others are referred to elsewhere in Chapter 7; thus in Chapter 7 we implicitly assumed the validity of the McMillan–Mayer theory in a way that is quite customary, so that we could use electrolyte solution studies to illustrate applications of several approximation methods.

A more stringent test of the same models will be to see whether, when adjusted to fit the thermodynamic data, they give ion–ion pair correlation functions in agreement with the experimental data which are becoming available (Fig. 10.2). In their report of the determination of the three ion–ion correlation functions in a $NiCl_2$ solution, Neilson and Enderby (1983) remark that the Ni–Ni correlation function is in qualita-

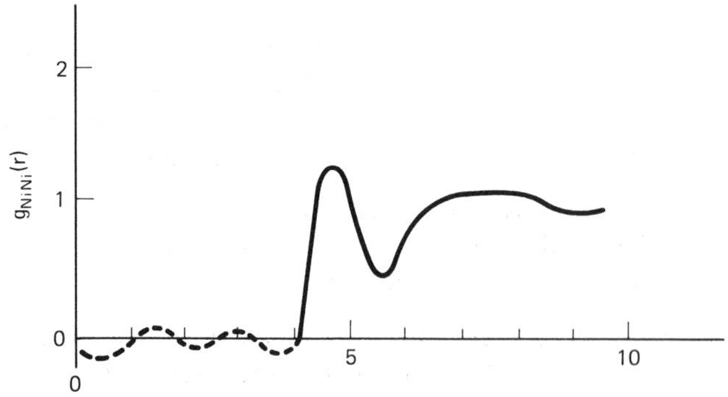

Figure 10.2 Cation–cation pair correlation function in a 4.35 molal solution of $NiCl_2$ in water determined by neutron diffraction using a second-order difference method (Neilson and Enderby, 1983). The dashed portion would be zero except for the limited range and limited accuracy of the partial structure factor, which is the quantity determined experimentally.

tive agreement with results for McMillan–Mayer level models of the kind mentioned above; obviously, much more can be done.

A subtle difference between Eqs. (10.15) and (10.16) is that the assumed pairwise additivity of $\bar{U}_N$, on which Eq. (10.15) is based, is a much less satisfactory approximation than the pairwise additivity of U_N, on which Eq. (10.16) is based. Thus, as explained following Eq. (4.27), a realistic Born–Oppenheimer N-point potential U_N can be expected to have deviations from pairwise additivity as a consequence of the prior average over electronic coordinates. We must expect that larger deviations from pairwise additivity will be exhibited by a realistic McMillan–Mayer N-point potential $\bar{U}_N$ as a consequence of averaging over solvent molecular coordinates as well as averaging over all electronic coordinates. As at the Born–Oppenheimer level, we can discuss *models* in which the N-point potentials, now the $\bar{U}_N$, are exactly pairwise additive. For such models Eqs. (10.12) and (10.15) are exactly valid as they are written. More generally, deviations from pairwise additivity can be taken into account, giving modified graphical expressions in which we have not only $\bar{f}(1\ 2)$ bonds but also bonds such as

$$\bar{f}_3(1\ 2\ 3) = e^{-\beta \bar{u}(1\ 2\ 3)} - 1 \tag{10.17}$$

where

$$\bar{u}(1\ 2\ 3) = \bar{U}_3(1\ 2\ 3) - \bar{U}_2(1\ 2) - \bar{U}_2(2\ 3) - \bar{U}_3(3\ 1) \tag{10.18}$$

is the deviation of the three-point potential from pairwise additivity. We also have $\bar{f}_4$, which carries the effects of deviations from 3-point additivity, and so on. Although incorporating $\bar{f}_3$, $\bar{f}_4$, . . . into the graph theory for $\bar{A}^{\text{ex}}$ and $g(1\ 2)$ would make the theory formally valid for quite general MM-level models, the complexity of that operation is discouraging to contemplate. It seems likely that MM-level theories will be useful mainly when applied to models in which $\bar{U}_N$ is pairwise additive. On the other hand, it is known that for ionic solutions realistic $\bar{f}_3(1\ 2\ 3)$, $\bar{f}_4(1\ 2\ 3\ 4)$, . . . have long-range contributions (Friedman, 1982), so at some stage the development of practical methods for treating nonpairwise-additive models will be required.

At the pair potential level, how does $\bar{U}_2(1\ 2)$ differ from $U_2(1\ 2)$ for the same fixed particles? The main effect, and the only one that is really well known, concerns the dielectric constant ϵ in the large-r form of the ion–ion pair potential [see Eq. (7.31)]. For the solvent-averaged potential $\bar{u}_{ss'}$ it is the dielectric constant of the solvent, rather than that of free space.

Another, more subtle effect, tends to make any two particles in any solvent, say even two green hard spheres in a solvent of red hard spheres, attract each other over a small range. This attraction is associated with the sharp and high peak in $g(r)$ at $r = \sigma$ in the hard-sphere fluid (Fig. 4.3).

10.3 ADELMAN'S THEORY OF SOLUTIONS

This theory already has been formulated in a simple way in Section 8.4 in terms of the correlation functions of an unsymmetrical mixture of a solute i with a solvent

w. The basic idea is that a solvent-averaged direct correlation function $\bar{c}_{ii}$ is determined by the solute–solute correlation function h_{ii} via the Ornstein–Zernicke equation,

$$h_{ii} = \bar{c}_{ii} + \bar{c}_{ii} * h_{ii} \tag{10.19}$$

which implies that there is an effective ion–ion pair potential

$$\beta u_{ii}^{\text{eff}} = h_{ii} - \bar{c}_{ii} + \ln (1 + h_{ii}) - \bar{B}_{ii} \tag{10.20}$$

which is written in analogy to Eq. (7.11). Here $\bar{B}_{ii} = \bar{B}_{ii}(1\ 2)$ is the sum of bridge graphs made of $f_{ii}^{\text{eff}} = (e^{-\beta u_{ii}^{\text{eff}}} - 1)$ bonds on white 1-circles of species i.

We recall that $B(1\ 2)$ can be expressed as a functional of $h(1\ 2)$, as discussed following Eq. (7.14). It follows that Eq. (7.11) and the Ornstein–Zernicke equation enable one to express $u(1\ 2)$ as a functional of $h(1\ 2)$, as one might wish to do to determine $u(1\ 2)$ from diffraction data for a one-component fluid. If carried out exactly, this procedure would give the exact pair potential for the fluid if and only if the real potentials $U_N(1, \ldots, N)$ were exactly pairwise additive. Otherwise, the procedure would give a state-dependent effective pair potential.

In the same way, Eqs. (10.19) and (10.20), together with the expression for $\bar{B}_{ii}$ as a functional of h_{ii}, generate an effective pair potential u_{ii}^{eff} that is dependent on the state of the system to which h_{ii} pertains (i.e., on the temperature, pressure, and solution composition). It follows that for a hypothetical one-component system with potentials given by

$$U_N^{\text{eff}}(1, \ldots, N) = \sum_{\text{pairs}} u_{ii}^{\text{eff}}(\text{r}) \tag{10.21}$$

the correlation function $h(1\ 2)$ derived from the Hamiltonian using the theories in Chapters 6 to 9 will be the same as $h_{ii}(1\ 2)$ as long as the state parameters also match. The compressibility and other functionals of h alone also will be the same, but other equations for thermodynamics, such as Eqs. (4.6) and (4.11), which are derived by varying the thermodynamic state, do not apply because of the state dependence of U_N^{eff} in Eq. (10.21).

More generally, $\{h_{ab}(1\ 2)\}$ determines $\{u_{ab}^{\text{eff}}(1\ 2)\}$, where $\{\ \cdots\ \}$ is the array formed from solute species pairs. Equivalently the structure of a solution can be thought of as determined by a set of solute–solute pair potential functions. This conclusion is at first startling because in Section 10.2 we found that solute–solute correlation functions are determined by solvent-averaged potential functions $\bar{U}_N$ that are not exactly pairwise additive. However, Adelman's solvent-averaged pair potential, which is determined from $h_{ab}(r)$ in a given thermodynamic state of the solution, depends on the composition of that state. On the other hand, the solvent-averaged potentials $\bar{U}_N(1, \ldots, N)$ of the McMillan–Mayer theory are all defined for states of infinite dilution; they do not depend on the concentration of the solution state in which they are used to calculate $h_{ab}(r)$. Thus we can choose either to work with the theory based on $\bar{U}_N$, which is not pairwise additive but is independent of solute concentration, or the theory relating to u_{ii}^{eff}, which is dependent on solute concentration, although it does not entail the use of higher n-point potentials. The dependence of u_{ii}^{eff} on solute concentration is the price of suppressing the dependence of higher n-point potentials.

10.4 THE LAWS OF DILUTE SOLUTIONS

The McMillan–Mayer theory gives a molecular formulation of the rules and coefficients which characterize the thermodynamic properties of dilute solutions. It is convenient in discussing these aspects to use the index w to specify the solvent and the indices $a, b, \ldots$ to specify solute species. Here we use mostly b for the solute index because a is the notation for activity as well [Eq. (10.7)]. So here a_b is the activity of solute species b.

First we look at Henry's law. We have, by construction, for a single solute,

$$\frac{a_b}{\rho_b} \longrightarrow 1 \qquad \text{as } \rho_b \longrightarrow 0 \tag{10.22}$$

where ρ_b is the solute concentration in the liquid solution. We also have

$$a_b = \frac{z_b}{\gamma_b} \longrightarrow \frac{\rho_{bG}}{\gamma_b} \qquad \text{as } \rho_b \longrightarrow 0 \tag{10.23}$$

where ρ_{bG} is the solute concentration in the gas phase at equilibrium and, provided that the vapor in equilibrium with the solution is also dilute (i.e., the solvent is not too volatile), we also have

$$\beta p_b = \rho_{bG} = \gamma_b a_b = \gamma_b \rho_b \qquad \text{as } \rho_b \longrightarrow 0 \tag{10.24}$$

where p_b is the partial pressure of the solute. Finally we have

$$p_b \longrightarrow \rho_b \left(\frac{\gamma_b}{\beta}\right) \qquad \text{as } \rho_b \longrightarrow 0 \tag{10.25}$$

which is *Henry's law.* Most often this is written with the concentration of solute expressed as mole fraction, but the change to get that is trivial.

In the *thermodynamic* formulation of Henry's law, the Henry's law coefficient

$$k_b = \frac{\gamma_b}{\beta} \tag{10.26}$$

is only required to be independent of ρ_b; its dependence on molecular properties is, of course, not given. To complete the *molecular* theory of Henry's law, we require γ_b expressed as a functional of the Hamiltonian of the system. From Eq. (10.3) in the $K = 1$ case we have

$$\begin{aligned} \gamma_b^{-1} &= \Xi(0, y, V, T)^{-1} \sum_L \frac{y^L}{L!} \int e^{-\beta U_{1_b+L}} d(1) \cdots d(L) \\ &= \Xi(0, y, V, T)^{-1} \sum_L \frac{y^L}{L!} \int_V e^{-\beta U_L - \beta \sum_i \phi(i)} d(1) \cdots d(L) \end{aligned} \tag{10.27}$$

where $\phi(i)$ is the potential acting on a solvent molecule at i due to the one solute molecule of species b fixed at some location in the interior of the volume V and where the sum $\sum_i = \sum_{i=1}^{L}$ extends over all solvent molecules. In writing the last step we introduce an approximation about the additivity of the solute–solvent interaction. Our conclusion is

$$\gamma_b^{-1} = \langle e^{-\beta \sum_i \phi(i)} \rangle_{y,V,T}$$

$$= \langle e^{-\beta \sum_i \phi(i)} \rangle_{N,V,T} \tag{10.28}$$

$$\gamma_b^{-1} = \frac{Z^*(N, V, T)}{Z(N, V, T)}$$

where we used the identity of ensemble averages (in the thermodynamic limit) to get the canonical ensemble form and where Z^* refers to the configuration integral for the system in the field of a fixed solute particle that is well inside. [See Eq. (4.17).]

It should be noticed that a theory of k_b is a theory of the thermodynamics of solvation of a particle. Equation (10.28) can be obtained more directly as well (Widom 1963).

It is a standard exercise in thermodynamics to show the equivalence of Henry's law, Raoult's law (as a limiting law at $\rho_b \longrightarrow 0$), and van't Hoff's equation:

$$\beta p_{\text{osm}} = \rho_b, \qquad \rho_b \longrightarrow 0 \tag{10.29}$$

Of course, we see that van't Hoff's equation as a limiting law follows directly from Eq. (10.10).

10.5 SECOND VIRIAL COEFFICIENTS IN SOLUTION

In solutions of a single solute dilute enough so that

$$\beta p_{\text{osm}} = \rho_b + \rho_b^2 \overline{B}_2 \tag{10.30}$$

is an adequate approximation, the functional dependence of $\overline{B}_2$ on the potentials

$$\overline{B}_2 = -\tfrac{1}{2} \int_V (e^{-\beta \overline{U}_2(r)} - 1)\, d^3r \tag{10.31}$$

$$e^{-\beta \overline{U}_2(r)} = v^2 z_{M+2}^{-1} \int_V e^{-\beta U_{M+2}}\, d(1) \cdots d(\mathrm{M}) \tag{10.32}$$

provides an explicit molecular theory of the first deviations from the dilute solution laws. In these equations the 2 refers to a set of two b particles, solute particles, while M is a set of solvent molecules such that $M/V = \rho_w$.[1]

As these equations show, the first deviations from dilute solution behavior are due to interactions of pairs of solute particles. As for the solute–solvent interaction, it determines the Henry's law constant but enters $\overline{B}_2$ only indirectly. The indirect

[1] Also, the thermodynamic limit is assumed.

effect comes through the interference of the solute–solvent interactions of two solute particles when they approach each other.

An interesting paradox derives from the foregoing considerations. Consider the solvent vapor pressure p_w of a particular solution in the composition range in which Eq. (10.30) is a good approximation. Now imagine a change in the solute species so that the solute–solvent interaction becomes more attractive. Intuitively, one expects p_w to be decreased, but in fact it may either increase or decrease, depending only on whether the change in solute–solvent interaction decreases $\bar{B}_2$ (more solute–solute attraction) or increases it (more solute–solute repulsion), and either of these things may indeed occur.

As an example, p_w is larger in a dilute aqueous solution of KCl than in one of argon provided that they are compared at

$$\rho_{\text{Ar}} = \rho_{\text{K}^+} + \rho_{\text{Cl}^-}$$

Certainly, the solute–solvent interaction is more attractive in the KCl solution than in the argon solution, but the solute–solute interaction is also more attractive in the former and this determines p_w.

All of these considerations are readily generalized to solutions having several solute species. For example, in a ternary solution, two solutes and a solvent, Eq. (10.10) becomes

$$\beta p_{\text{osm}} = \rho_a + \rho_b + \bar{B}_{aa}\rho_a^2 + \bar{B}_{ab}\rho_a\rho_b + \bar{B}_{bb}\rho_b^2 + \cdots \tag{10.33}$$

where

$$\bar{B}_{ab} = -\frac{1}{1+\delta_{ab}}\int \exp\left[-\beta\bar{U}_{ab}(r) - 1\right] d^3r \tag{10.34}$$

In such systems the experimenter often reports values for the coefficients in expansions such as

$$\mu_a(\rho_a, \rho_b) = \mu_a^\circ + kT\ln\rho_a + \mu_{aa}^{\text{ex}}\rho_a + \mu_{ab}^{\text{ex}}\rho_b + \mu_{aaa}^{\text{ex}}\rho_a^2 + \cdots \tag{10.35}$$

where μ_a is the chemical potential of species a, μ_a° is its value in the standard or reference state in which there are no solute–solute interactions, and μ_{aa}^{ex} and μ_{ab}^{ex} are the coefficients of interest. Beginning with Eq. (10.33) it is only an exercise in thermodynamics to express these in terms of $\bar{B}_{aa}$, $\bar{B}_{ab}$, and so on, so Eq. (10.34) provides the molecular theory of μ_{aa}^{ex} and so on.

The molecular theory of similar coefficients which occur in the concentration dependence of other thermodynamic properties (heats, volumes, thermal expansion, etc.) can be obtained from Eqs. (10.30) to (10.35) by strictly thermodynamic manipulations or by using equations like Eq. (4.42), but with identification of the various functions appropriate to solutions. For example, in the equation referred to, $-E^{\text{ex}}$ would be taken as the energy change in the experiment in which one begins with the solution of interest in the solution side of a long apparatus like the one in Fig. 10.1 and then pulls the solution piston out an infinite distance while keeping p fixed. It is thus very nearly the experimental heat of dilution.

In all these applications of the McMillan–Mayer theory, attention must be given to details of thermodynamics in making the comparison of properties calculated for systems in which the independent variables are $T, V, \mathbf{y}, \rho_a, \rho_b, \ldots$ with experimental data for systems in which the independent variables are $T, P, \rho_a, \rho_b, \ldots$ or $T, P, m_a, m_n, \ldots,$ where the m_a are mass concentrations (e.g., molalities).

THE LITERATURE

The fluctuation theory of solutions was developed by Kirkwood and Buff (1951) and, in a different way, by Hall (1971), while the theory of solvent-averaged functions is given by McMillan and Mayer (1945). The thermodynamic theory of the conversion from McMillan–Mayer states to the Lewis–Randall states that are better known in physical chemistry is developed by Friedman (1975). Adelman (1976b) gives a graphical prescription for the effective pair potential derived from the Ornstein–Zernicke equation for the solute, while more general aspects of his theory are found in Adelman (1976a).

PROBLEMS

10.1. $\bar{A}^{\text{ex}}$ depends on the solvent activity y. Where does y enter into Eq. (10.12)?

10.2. The equations above imply that $\exp(-\beta\bar{A}^{\text{ex}}) = \int \exp(-\beta\bar{U}_K)d(1) \cdots d(K)/V^K$, in analogy to Eq. (3.55). Derive the MM-level equations for the excess energy in analogy to Eq. (4.42) and the equation of state (for p_{osm}) in analogy to Eq. (4.35).

10.3. Show that the grand partition functions in Eq. (10.7c) pertain to two systems in osmotic equilibrium (i.e., with the same solvent activity y but different pressures).

10.4. **(a)** Write all of the graphs with two or three black ρ circles and two white circles in the h_{ii} bond prescription for $\bar{B}(1\ 2)$ in Eq. (10.20).

(b) What is the general equation for $\bar{B}(1\ 2)$ in terms of graphs with h_{ii} (1 2) bonds? (See "topological reduction," Sections 9.5 and 9.6.)

10.5. Derive the equation relating μ_{ab}^{ex} in Eq. (10.35) to $\bar{B}_{ab}$ in Eq. (10.33). Equation (7.77) may be useful in this exercise.

10.6. Consider dilute solutions of b and b' in a given solvent w and at the same temperature, pressure, and concentration (i.e., $\rho_b = \rho_{b'}$). Show that if p_{osm} is greater for the b solution, the solvent activity y and pressure p_w are greater for the b' solution.

10.7. **(a)** Show that in an electrically neutral ionic system the matrix $\mathbf{M}$ whose elements are $\partial\rho_s/\partial\mu_{s'}$ is singular (i.e., its determinant is zero).

(b) For electrically neutral $CaCl_2$ in water, derive the relations

$$4\mathrm{M}_{++} = 2\mathrm{M}_{+-} = \mathrm{M}_{--} \tag{10.36}$$

$$\left(\frac{\partial\bar{\mu}}{\partial\rho_+}\right)_{\rho_+/\rho_-} \equiv \left(\frac{\partial[\mu_+ + 2\mu_-]}{\partial\rho_+}\right)_{\rho_+/\rho_-} = \frac{1}{\mathrm{M}_{++}} \tag{10.37}$$

where the derivative of $\bar{\mu}$ is the measurable coefficient that is most closely related to $\mathbf{M}$ for a single electrolyte solution. [For mixtures, see Friedman and Ramanathan (1970).]

10.8. In solutions of two solutes (b and b' dissolved in w) in cases in which solute b is a slightly soluble liquid or solid, we can measure the solubility interaction coefficient

$$k_{bb'} = \left(\frac{\partial \ln \rho_b}{\partial \rho_{b'}}\right)_{\mu_b} \tag{10.38}$$

where μ_b is the chemical potential of the pure b phase in equilibrium with the solution, and ρ_b is the solubility of b in the b, b' solution in w. If b' is an electrolyte and b a nonelectrolyte, then $k_{bb'}$ is well known as the Setchenow coefficient. In any case it is often found that there is a regime of small $\rho_{b'}$ in which $k_{bb'}$ does not depend on b'. Here we assume that $k_{bb'}$ is measured in this regime and that neither b nor b' is an electrolyte.

(a) Show that

$$k_{bb'} = -\frac{(\partial \mu_b/\partial \rho_{b'})_{\rho_b}}{(\partial \mu_b/\partial \ln \rho_b)_{\rho_{b'}}} \tag{10.39}$$

(b) By applying Eqs. (10.35) and (10.33), show that

$$k_{bb'} = -\bar{B}_{bb'} + \cdots \tag{10.40}$$

where the remaining terms vanish if ρ_b is small enough.

(c) Show that

$$k_{bb'} = \frac{1}{\rho_b}\left(\frac{\partial \rho_b}{\partial \mu_{b'}}\right)_{\mu_b}\left(\frac{\partial \mu_{b'}}{\partial \rho_{b'}}\right)_{\mu_b} \tag{10.41}$$

(d) Combine Eq. (10.41) with the Kirkwood–Buff theory (Section 10.1) to derive

$$k_{bb'} = \int d^3r\, h_{bb'}(r) \tag{10.42}$$

(e) Compare Eqs. (10.40) and (10.42). Are they consistent?

(f) How much of the theory above is essentially the same if b or b' is an electrolyte? Indicate qualitatively what changes might be required.

11

Time Dependence

11.1 INTRODUCTION

In Chapters 4 to 10 we were concerned only with the theory of how the forces between the molecules of a system determine its thermodynamic properties and its equilibrium average structure as measured, at least in principle, in diffraction experiments. It is interesting to notice the point at which we gave up any possibility of learning about time-dependent processes. Apparently, this comes about through our use of the first postulate, Eq. (1.5), to avoid looking at the processes that take a system from one state to another. To develop a theory of time-dependent processes, we begin, in the following section, with the equations that do govern the change from one state to the next. For the rest we have chosen to emphasize the linear response theory. It reduces the calculation of suitable transport and relaxation coefficients, spectroscopic line shapes, and inelastic light scattering and neutron scattering cross sections to the study of fluctuations with time in equilibrium systems, and so brings us back to the study of equilibrium ensembles.

In the rest of this chapter we describe some fundamental aspects of time-dependent statistical mechanics. Because many of the important theoretical constructions are quite abstract, even without quantum mechanics, we begin with classical mechanics and only occasionally describe the corresponding quantal equations. However, we use a formulation of classical mechanics that makes the classical and quantal equations as similar as possible.

Our main objective in this chapter is to describe the basic theory of the time correlation functions that characterize the fluctuations in equilibrium systems. It should be noted that next, in Chapter 12, before going on to more technical matters, we exploit the Langevin equation as a mathematical model for fluctuating quantities in systems at or near equilibrium. We use it to illustrate how various equilibrium time correlation functions may be related to each other and to measurements, thus complementing the more abstract and formal development in the present chapter.

11.2 EQUATIONS OF MOTION IN CLASSICAL MECHANICS

The classical Hamiltonian function is [see Eqs. (3.9) and (3.40)]

$$H = \sum_{i}^{3N} \frac{p_i^2}{2m_i} + U_N(\mathbf{q}_1, \ldots, \mathbf{q}_{3N}) \tag{11.1}$$

where q_i is the ith location variable and $p_i = m_i \dot{q}_i$ is the conjugate momentum and m_i the mass. The dot means time derivative. [Here $\dot{q} = dq/dt$, but later, as in Eq. (11.11), the dot also denotes a partial derivative.] So we find Hamilton's equations of motion,

$$\frac{\partial H}{\partial p_i} = \dot{q}_i \text{ and } \frac{\partial H}{\partial q_i} = -\dot{p}_i, \text{ i} = 1, \ldots, 3\text{N} \tag{11.2}$$

The set of locations and momenta for an N-particle system is denoted as the phase Γ_N (see Problem 3.6)

$$\Gamma_N = q_1, p_1, \ldots, q_{3N}, p_{3N} \tag{11.3}$$

By a dynamical variable we mean one that depends only on the phase. Examples are H, its kinetic and potential parts, and the location or momentum of any particle or group of particles of the N-particle system. If A is any dynamical variable, its time development must be of the form

$$\frac{dA}{dt} = \left(\frac{\partial A}{\partial t}\right)_{\Gamma_N} + \sum_i \left(\frac{\partial A}{\partial q_i} \dot{\mathbf{q}}_i + \frac{\partial A}{\partial \mathbf{p}_i} \dot{\mathbf{p}}\right) \tag{11.4a}$$

with

$$\left(\frac{\partial A}{\partial t}\right)_{\Gamma_N} = 0 \tag{11.4b}$$

since A is fully determined by Γ_N. Introducing Kubo's notation for the Liouville operator $H^\times$ and using Eqs. (11.2), we have

$$\frac{dA}{dt} = H^\times A = -\sum_i \left(\frac{\partial H}{\partial q_i}\frac{\partial A}{\partial p_i} - \frac{\partial H}{\partial p_i}\frac{\partial A}{\partial q_i}\right) \equiv -[H, A]_{\text{PB}} \tag{11.5}$$

where $[\cdot\ \cdot\ \cdot]_{\text{PB}}$ denotes a *Poisson bracket,* the classical mechanics equivalent of the quantal commutator. In general, for any two dynamical variables X and Y, we define

$$[X, Y]_{\text{PB}} = \sum_i \left[\frac{\partial X}{\partial q_i}\left(\frac{\partial Y}{\partial p_i}\right) - \frac{\partial X}{\partial p_i}\left(\frac{\partial Y}{\partial q_i}\right)\right] \tag{11.6}$$

The explicit form for the Liouville operator is

$$H^{\times} = \sum_{i=1}^{N} \left(\frac{\mathbf{p}_i}{m_i} \cdot \frac{\partial}{\partial \mathbf{q}_i} + \mathbf{F}_i \cdot \frac{\partial}{\partial \mathbf{p}_i} \right) \tag{11.7}$$

where $F_i = \dot{p}_i$ is the force associated with coordinate i. So we can write the time derivative of A as

$$\dot{A} = \frac{dA}{dt} = H^{\times} A \tag{11.8}$$

which we may call *Heisenberg's equation of motion,* the usual name for the corresponding quantal equation $\dot{A} = (1/i\hbar)[H, A]$, where $[H, A]$ is the commutator. In these quantal expressions H is the Hamiltonian *operator,* Eq. (3.9).

Next we consider an ensemble of N-particle systems related to the one of interest, using "ensemble" in exactly the sense introduced in Section 1.2. We allow that the ensemble may be one of the equilibrium ensembles specified there. However, it may be a nonequilibrium ensemble. Just as an example of a nonequilibrium ensemble, we might consider a collection of N_r replica systems that comprised an equilibrium ensemble for times $t < t_0$. Then at t_0 a given perturbation (an external field, compression to a smaller volume, etc.) is applied to every system of the ensemble. Then for $t > t_0$ we have a nonequilibrium ensemble.

In any case we let $f_N(\Gamma_N, t)\, d\Gamma_N$ be the fraction of the systems of an ensemble in the region $d\Gamma_N$ around Γ_N. Its evolution is governed by

$$\begin{aligned} \frac{df_N}{dt} &= \left(\frac{\partial f_N}{\partial t}\right)_{\Gamma_N} + \sum_{i=1}^{3N} \left(\frac{\partial f^{(N)}}{\partial q_i} \dot{q}_i + \frac{\partial f^{(N)}}{\partial p_i} \dot{p_i} \right) \\ &= \left(\frac{\partial f_N}{\partial t}\right)_{\Gamma_N} + H^{\times} f_N \end{aligned} \tag{11.9}$$

Since $f_N(\Gamma_N, t)$ is not a dynamical variable, because it depends not only on Γ_N at time t but on the history of the ensemble at times earlier than t, we do not in general have $(\partial f_N/\partial t)_{\Gamma_N} = 0$. On the other hand, f_N is the locally varying concentration of replica systems in phase space.

Consider an ordinary concentration, say c_s, the concentration of sugar in a nonuniform solution of sugar in water. Since sugar molecules are *conserved* even as they diffuse from one place to another, we have

$$\dot{c}_s(\mathbf{R}) = -\nabla_{\mathrm{R}} \cdot [\mathbf{v}(\mathbf{R}) c_s(\mathbf{R})] \tag{11.10a}$$

where $c_s(\mathbf{R})$ is the concentration at location $\mathbf{R}$, $\mathbf{v}(R)$ is the average velocity of sugar molecules in the neighborhood of $\mathbf{R}$, and ∇_R is the gradient operator. Applying the same analysis to f_N, the local concentration of replica systems in phase space leads to the conservation equation

$$\left(\frac{\partial f_N}{\partial t}\right)_{\Gamma_N} = -\sum_i \left[\frac{\partial}{\partial q_i} (f_N \dot{q}_i) + \frac{\partial}{\partial p_i} (f_N \dot{p}_i) \right] \tag{11.10b}$$

where, compared to Eq. (11.10a), we have $3N$ location variables and $3N$ momentum variables instead of merely three location variables. As a consequence of Eq. (11.2), we have

$$\frac{\partial \dot{q}_i}{\partial q_i} = \frac{\partial^2 H}{\partial q_i \, \partial p_i} = -\frac{\partial \dot{p}_i}{\partial p_i}$$

so Eq. (11.10b) can be reduced to

$$\left(\frac{\partial f_N}{\partial t}\right)_{\Gamma_N} = -\sum_i \left[\left(\frac{\partial f_N}{\partial q_i}\right)\dot{q}_i + \left(\frac{\partial f_N}{\partial p_i}\right)\dot{p}_i\right]$$

By comparison with Eq. (11.9) we find $df_N/dt = 0$ and the Liouville equation

$$\dot{f}_N \equiv \left[\frac{\partial f_N}{\partial t}\right]_{\Gamma_N} = -H^\times f_N \tag{11.11a}$$

On the basis of Eq. (3.49) the *equilibrium* phase-space distribution function is given by

$$f_N^{\text{eq}}(\Gamma_N) = \frac{\exp\left[-\beta H(\Gamma_N)\right]}{\displaystyle\int d\Gamma_N \exp\left[-\beta H(\Gamma_N)\right]} \tag{11.11b}$$

Of course, an equilibrium distribution does not change with time:

$$\dot{f}_N^{\text{eq}}(\Gamma_N) = -H^\times f_N^{\text{eq}}(\Gamma_N) \simeq -H^\times e^{-\beta H} = \beta e^{-\beta H} H^\times H = 0 \tag{11.11c}$$

The comparison of the Liouville and Heisenberg equations is important: $\dot{f}_N$ and $\dot{A}$ are not defined in parallel ways and consequently there is a sign difference in the equations. Moreover, it is usually true in a given problem that the time-dependent statistical mechanics can be worked out either from Heisenberg's equation or from Liouville's equation. The choice corresponds closely to the choice of the "Heisenberg picture" or "Schrödinger picture" in time-dependent quantum mechanics.

The formal solutions to the Heisenberg and Liouville equations are

$$A(t) = e^{tH^\times} A(0) \qquad \text{and} \qquad f_N(\Gamma_N, t) = e^{-tH^\times} f_N(\Gamma_N, 0) \tag{11.12}$$

as the reader may verify. The exponential operators $\exp(\pm tH^\times)$ are called *streaming operators.* [Exponentiated operators are defined in Eq. (1.82).]

The equations

$$\dot{P}_N \equiv \left(\frac{\partial P_N}{\partial t}\right)_{\Gamma'_N, \Gamma_N} = -H^\times P_N(\Gamma'_N \overset{t}{\leftarrow} \Gamma_N) \tag{11.13a}$$

$$P_N(\Gamma'_N \overset{t}{\leftarrow} \Gamma_N) \underset{t=0}{\longrightarrow} \delta(\Gamma'_N - \Gamma_N) \tag{11.13b}$$

characterize the *Green's function* of the Liouville equation, the function that is governed by the equation and that has a delta function boundary condition. Often this Green's function P_N is called the *propagator.* In (11.13b) the delta function is the product of the Dirac delta functions of all the phase variables. Thus $P_N\left(\Gamma'_N \overset{t}{\leftarrow} \Gamma_N\right) d\Gamma'_N$ is the probability that a system at Γ_N at time zero will be in the small region $d\Gamma'_N$ at a time t later. In Eq. (11.13a) the operator $H^\times$ is assumed to operate on the late phase Γ'_N and not on the early phase Γ_N.

DO PROBLEMS 11.1 TO 11.4.

11.3 TIME CORRELATION FUNCTIONS. FORMAL THEORY

For any dynamical variable D we define the equilibrium ensemble average over D in an N-particle system as

$$\langle D \rangle \equiv \int d\Gamma_N f_N^{\text{eq}}(\Gamma_N) D(\Gamma_N) \tag{11.14}$$

In the particular case in which

$$D = A e^{tH^\times} B$$

where A and B are themselves dynamical variables, the equilibrium average is called a *time correlation function,*

$$\begin{aligned} \langle AB(t) \rangle &= \langle A(0) B(t) \rangle \\ &= \langle A e^{tH^\times} B \rangle \\ &= \int d\Gamma_N f_N^{\text{eq}}(\Gamma_N) A(\Gamma_N) e^{tH^\times} B(\Gamma_N) \end{aligned} \tag{11.15}$$

It is stationary, meaning that for any time t' we have

$$\langle B(t+t') A(t') \rangle = \langle B(t) A \rangle = \langle B(0) A(-t) \rangle \tag{11.16}$$

as we can show directly,

$$\int d\Gamma_N f_N^{\text{eq}} e^{t' H^\times} [AB(t)] = \int d\Gamma_N (e^{-t' H^\times} f_N^{\text{eq}}) AB(t) = \int d\Gamma_N f_N^{\text{eq}} AB(t) \tag{11.17a}$$

To derive this equation it is convenient to begin with

$$\begin{aligned} \langle AH^\times B \rangle &= \int d\Gamma_N f_N^{\text{eq}} AH^\times B \\ &= -\int d\Gamma_N H^\times (f_N^{\text{eq}} A) B \end{aligned} \tag{11.17b}$$

where we employ Eq. (11.7) and integrate by parts. Also,

$$H^\times (f_N^{\text{eq}} A) = f_N^{\text{eq}} (H^\times A) \tag{11.17c}$$

in view of (11.11c). So we find that $H^\times$ is "anti-Hermitian":

$$\langle AH^\times B \rangle = -\langle (H^\times A) B \rangle \tag{11.17d}$$

By using this equation repeatedly with the expansion of $e^{tH^\times}$, as in Eq. (1.82), we obtain

$$\langle A e^{tH^\times} B\rangle = \langle (e^{-tH^\times} A) B\rangle \tag{11.18}$$

From Eq. (11.16) we find

$$(\partial/\partial t)\,\langle B(t)A\rangle = (\partial/\partial t)\,\langle BA(-t)\rangle \tag{11.19a}$$

$$\langle \dot{B}(t)A\rangle = -\langle B\dot{A}(-t)\rangle = -\langle B(t)\dot{A}\rangle \tag{11.19b}$$

$$\langle \dot{B}A\rangle = -\langle B\dot{A}\rangle \tag{11.19c}$$

which reduces to Eq. (11.17d). These equations hold even if $B = A$, so we have

$$\langle \dot{A}A\rangle = 0 \tag{11.20}$$

It is usually assumed that

$$\langle AB(t)\rangle \longrightarrow \langle A\rangle\langle B\rangle \qquad \text{as } t \longrightarrow \infty \tag{11.21}$$

but there are exceptions if $B \neq A$, as in Problem 11.14a. If $\langle A\rangle$ and $\langle B\rangle$ are nonzero, it often is convenient to replace A and B by their fluctuating parts:

$$\delta A \equiv A - \langle A\rangle, \qquad \delta B \equiv B - \langle B\rangle \tag{11.22a}$$

We shall do this in all cases, but suppress the notational change so that, in the equations that follow, A should be read as "the fluctuating part of A" and similarly for B. Thus we now consider some of the properties of the time correlation function

$$C_{BA}(t) \equiv \langle B(t)A\rangle \equiv \langle \delta B(t)\,\delta A\rangle \tag{11.22b}$$

If we define the Fourier–Laplace transform

$$C_{BA}[z] \equiv \int_0^\infty e^{-izt} C_{BA}(t)\,dt \tag{11.23}$$

which has the inverse

$$C_{BA}(t) = \frac{1}{2\pi}\int_{-\infty}^{\infty} e^{izt} C_{BA}[z]\,dz \tag{11.24}$$

where z is the frequency conjugate to t, then we can see from Eqs. (11.24) and (11.8) that we have

$$\frac{1}{2\pi}\int_{-\infty}^{\infty} (iz)^k C_{BA}[z]\,dz = (-)^k \langle A(H^\times)^k B\rangle \tag{11.25}$$

Relations like this, between the integral over the frequency on the left and the thermodynamic coefficient (!) on the right, are called *sum rules,* while

$\int_{-\infty}^{\infty} z^k C_{BA}[z]\,dz$ is called the kth *moment* of the spectrum $C_{BA}[z]$.

If $B \neq A$, then $C_{AB}(t)$ is called the *distinct-* or *joint-* or *cross-correlation function* of A and B. If A and B are thermodynamically independent in the sense that $\langle AB \rangle = 0$, we have

$$C_{AB}(0) = 0 \qquad \text{if } \langle AB \rangle = 0 \tag{11.26a}$$

In the other case, the *auto-* or *self-*correlation function $C_{AA}(t)$, we have [see Eq. (11.20)]

$$\left(\frac{\partial^m C_{AA}(t)}{\partial t^m}\right)_{t=0} = 0 \qquad \text{if } m \text{ is odd} \tag{11.26b}$$

Also, since we have, in view of Eq. (11.16),

$$0 \leqslant \langle [A(t) \pm A]^2 \rangle = 2\langle A^2 \rangle \pm 2\langle A(t)A \rangle \tag{11.26c}$$

it follows that

$$-\langle A^2 \rangle \leqslant C_{AA}(t) \leqslant \langle A^2 \rangle \tag{11.26d}$$

An important characteristic of a time correlation function is the correlation time defined, only for cases in which $C_{AB}(0) \neq 0$, as

$$\begin{aligned} \tau_{AB} &= \int_0^\infty dt \frac{C_{AB}(t)}{C_{AB}(0)} \\ &= \frac{C_{AB}[0]}{C_{AB}(0)} \end{aligned} \tag{11.26e}$$

An autocorrelation may have the simple form

$$C_{AA}(t) = \langle A^2 \rangle e^{-t/\tau}$$

in which case τ is the same as τ_{AA} given in Eq. (11.26e).

Sometimes the variables A and B of interest are complex numbers. Then it is convenient to replace Eq. (11.22) by

$$C_{BA}(t) = \langle B^*(t)A \rangle \tag{11.27a}$$

where * means complex conjugate.

Some examples of time correlation functions are shown in Fig. 11.1. In quantum mechanics, and sometimes in classical mechanics, too, A and B represent noncommuting operators. Then it is convenient to replace the definition in (11.22b) by (11.27a), where now B^* is the adjoint of B (Kubo, 1966).

The equations of Hamiltonian mechanics, whether classical or quantal, describe reversible processes and consequently imply time reversal symmetry, so we have

$$\langle AB(-t) \rangle_{-\mathbf{B}m} = \epsilon_A \epsilon_B \langle AB(t) \rangle_{\mathbf{B}m} \tag{11.27b}$$

Here $\epsilon_A = +1$ or -1 according to whether A does or does not change sign if t is changed to $-t$. For example, if $A(t) = \rho_i(\mathbf{r}, t)$, the density of particle i at location $\mathbf{r}$ at time t, then $\epsilon_A = +1$, while if $A(t) = \dot{\rho}_i(\mathbf{r}, t)$ or if $A(t) = p_i(t)$, the momentum

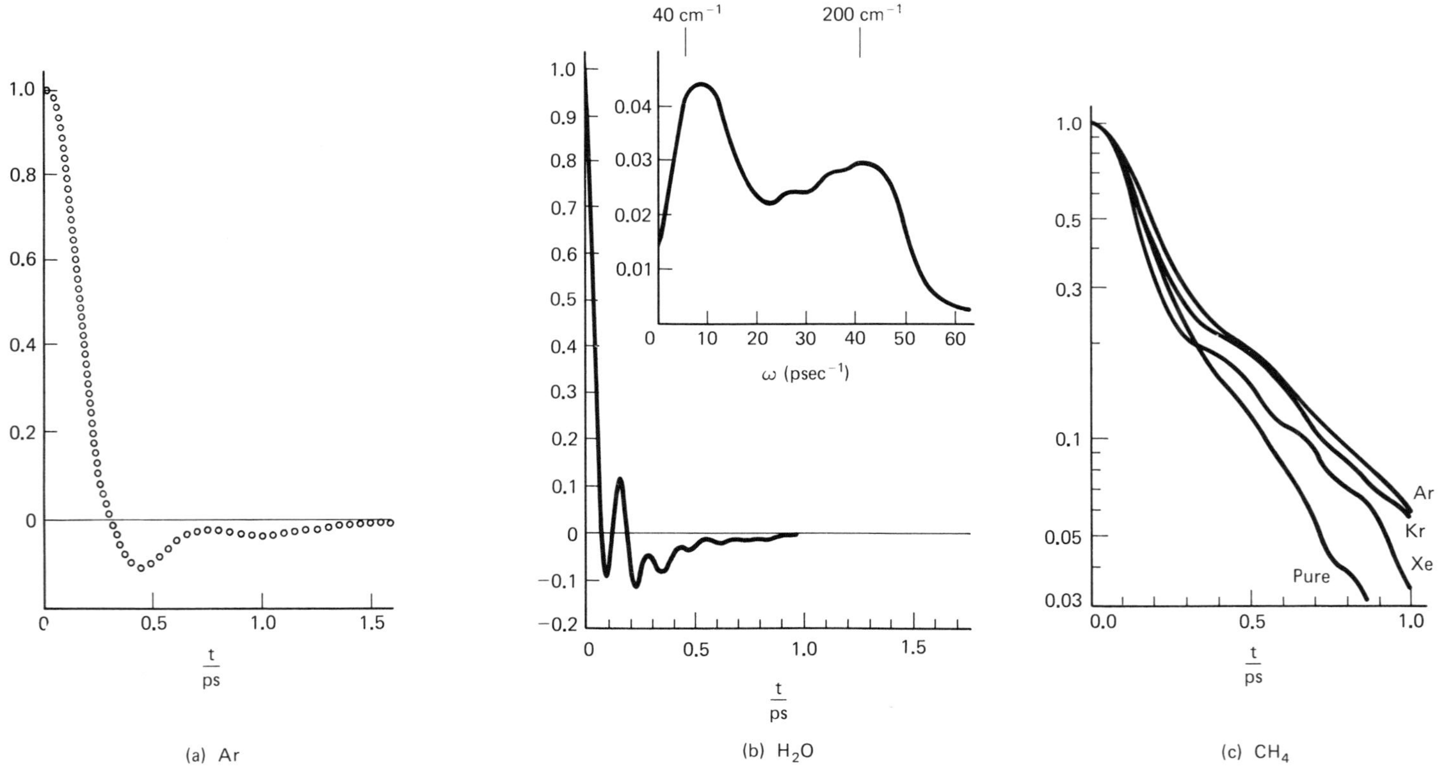

Figure 11.1 (a) The velocity autocorrelation function, $C_{vv}(t)/C_{vv}(0)$ in the notation of Eq. (11.22), for the particles in a 6–12 model for liquid argon at 85.5 K, 1.407 g/cm³ [MD by Nijboer and Rahman (1966)].

(b) $C_{vv}(t)/C_{vv}(0)$ for the center of mass of a water molecule in the ST2 model for the liquid (10°C, 1 g/cm³) obtained by MD by Rahman and Stillinger (1971). The inset shows the corresponding power spectrum, Re $C_{vv}[z]/C_{vv}(0)$ in the notation of Eq. (11.23) (Stillinger, 1975).

(c) The autocorrelation function for molecular reorientation of CH_4 in various liquid media, $\langle \hat{u}(t)\cdot\hat{u}\rangle/\langle\hat{u}\cdot\hat{u}\rangle$, where $\hat{u}$ is a unit vector fixed in the molecule. Obtained from band shapes in the infrared spectra (Steele, 1976).

of particle i at time t, then $\epsilon_A = -1$. Finally, in this equation the subscript $\mathbf{B}_m$ represents an externally applied magnetic field. Unless it is zero, it must be reversed as indicated (Kubo, 1966).

DO PROBLEMS 11.5 TO 11.11.

11.4 LINEAR RESPONSE THEORY

This theory, often called the *Kubo theory,* provides a quite general basis for investigating how the observable transport and relaxation coefficients are determined by intermolecular interactions. The same methods are applicable in classical and quantal systems; the basic paper by Kubo (1966) is concerned primarily with quantal systems. In either case we begin with Fig. 11.2 which shows how f_N changes when a system is subjected to a change in Hamiltonian by an amount $-AK(t)$ where A is a dynamical variable and K is a coefficient which does not depend on the phase.

We assume that before some definite time t_0 we have an ensemble of systems at equilibrium under Hamiltonian H_0; the distribution function is

$$f_N^{\text{eq}}(\Gamma_N) = \frac{e^{-\beta H_0}}{\int d\Gamma_N \, e^{-\beta H_0}} \tag{11.28}$$

On the other hand for $t > t_0$ the Hamiltonian is

$$H_t = H_0 - AK(t) \tag{11.29}$$

Then for $t > t_0$ we may write

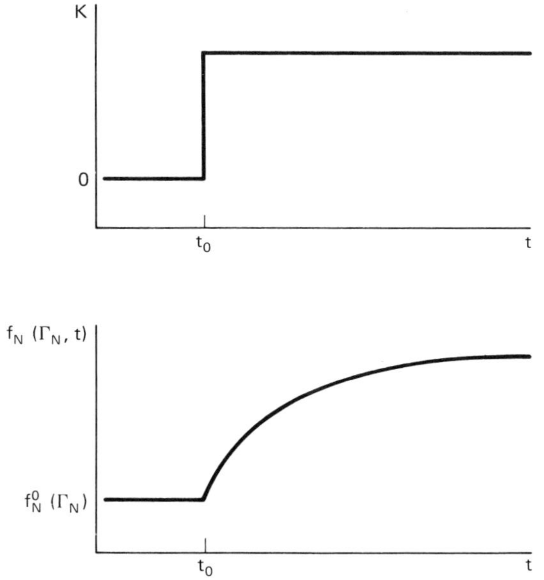

Figure 11.2 Thought experiment for linear response theory. In the case shown here $K(t)$ is constant for $t > t_0$ (i.e., $z = 0$).

$$f_N(\Gamma_N, t) = f_N^{\text{eq}}(\Gamma_N) + \Delta f_N(\Gamma_N, t) + O(K^2) \tag{11.30}$$

Here

$$\Delta f_N = -\int_{t_0}^{t} ds\, e^{-(t-s)H_0^\times} H_s^\times f_N^{\text{eq}} \tag{11.31}$$

is a term that is proportional to K, while terms proportional to K^2 and higher powers are neglected. Rather than derive these equations here, we merely verify their correctness as follows. From Eq. (11.30) we have, in view of Eqs. (11.11) and (11.29),

$$\dot{f}_N = -H_t^\times f_N^{\text{eq}} - H_0^\times \Delta f_N + O(K^2) \tag{11.32a}$$

while from Eqs. (11.30) and (11.31) we have

$$\begin{aligned} \dot{f}_N &= 0 + \frac{\partial}{\partial t} \Delta f_N + O(K^2) \\ &= -H_t^\times f_N^{\text{eq}} - H_0^\times \Delta f_N + O(K^2) \end{aligned} \tag{11.32b}$$

Since Eq. (11.30) gives $\dot{f}_N$ in agreement with the Liouville equation and since it gives the correct f_N at $t = 0$, it must be correct in general for the first-order effect of the perturbation.

Now consider some dynamical variable B that may change due to the perturbation $-AK(t)$. Its response, averaged over the ensemble of systems prepared by perturbing the equilibrium ensemble at $t > t_0$ as described above, is to first order in K,

$$\begin{aligned} \langle \Delta B(t) \rangle_K &= \int d\Gamma_N\, B(\Gamma_N)\, \Delta f_N(\Gamma_N, t) \\ &= -\int_{t_0}^{t} ds \int d\Gamma_N B(\Gamma_N) e^{-(t-s)H_0^\times} H_s^\times f_N^{\text{eq}}(\Gamma_N) \\ &= \int_{t_0}^{t} ds\, K(s) \int d\Gamma_N\, B e^{-(t-s)H_0^\times} A^\times f_N^{\text{eq}} \end{aligned} \tag{11.33}$$

Since $A^\times$ is a linear differential operator we have, in view of Eq. (11.11),

$$\begin{aligned} A^\times f_N^{\text{eq}} &= f_N^{\text{eq}} A^\times(-\beta H_0) \\ &= \beta f_N^{\text{eq}} H_0^\times A \\ &= \beta f_N^{\text{eq}} \dot{A} \end{aligned} \tag{11.34}$$

and Eq. (11.33) becomes

$$\begin{aligned} \langle \Delta B(t) \rangle_K &= \beta \int_{t_0}^{t} ds\, K(s) \int d\Gamma_N\, B e^{-(t-s)H_0^\times} \dot{A}\, f_N^{\text{eq}} \\ &= \beta \int_{t_0}^{t} ds\, K(s) \langle B(t) \dot{A}(s) \rangle \end{aligned} \tag{11.35}$$

where we have used Eq. (11.75) and where the unsubscripted $\langle \cdots \rangle$ refers to an average over an equilibrium ensemble. Equation (11.35) is the linear response theory result; the measurable linear response $\langle \Delta B(t) \rangle_K$ to the perturbation $H_t - H_0$ is determined by the time correlation function for fluctuations of B and $\dot{A}$ in the system at equilibrium. It is a remarkable result because it shows how the calculation of the nonequilibrium quantity $\langle \Delta B(t) \rangle_K$ is reduced to the calculation of the equilibrium ensemble average $\langle B(t)\dot{A}(0)\rangle$.

Finally, we put Eq. (11.35) in a more convenient form. Without loss of generality we take $K(t)$ to be a sum of terms of the form $K_\omega \text{ Re } e^{i(\omega t+\phi_\omega)}$, where K_ω is real and where Re F means "the real part of F." In general, for any complex number F we have $F = \text{Re } F + i \text{ Im } F$ with $\text{Re } F \equiv (F + F^*)/2$ and $\text{Im } F \equiv (F - F^*)/2i$ where F^* is the complex conjugate of F. For example, $\text{Re } e^{i\omega t} = \cos \omega t$. The phase ϕ_ω is not important for us and will be discarded.

We seek the steady-state response to $K(t)$, so we choose $t_0 = -\infty$ and then the response of interest is a component of angular frequency ω, namely

$$\begin{aligned}\langle \Delta B(t) \rangle_{K_\omega} &= \beta K_\omega \text{ Re} \int_{-\infty}^{t} ds \, e^{-i\omega s} \langle B(t)\dot{A}(s)\rangle \\ &= \beta K_\omega \text{ Re } e^{i\omega t} \int_0^\infty dt' \, e^{-i\omega t'} \langle B(t')\dot{A}(0)\rangle\end{aligned} \tag{11.36}$$

where $t' \equiv t - s$.

A phenomenological equation for the same response is

$$\langle \Delta B(t) \rangle_{K_\omega} = \text{Re}(\chi_{BA}(\omega) K_\omega e^{i\omega t}) \tag{11.37}$$

where $\chi_{BA}(\omega)$ is a complex dynamic susceptibility (or admittance, the name depending on the physical nature of A and B). So we see that the phenomenologically defined dynamical susceptibility is proportional to the Fourier–Laplace transform of the relevant time correlation function,

$$\chi_{BA}(\omega) = \beta \int_0^\infty dt \, e^{-i\omega t} \langle B(t)\dot{A}\rangle \tag{11.38}$$

This linear response result is the full expression of the fluctuation-dissipation theorem. The word "fluctuation" refers to fluctuations at equilibrium governed by $\langle B(t)\dot{A}\rangle$, while the word "dissipation" refers to the heat produced by the power $\text{Re } \chi_{B\dot{A}}(\omega) K_\omega^2$ required to maintain the response $\langle \Delta B(t)_{K_\omega}\rangle$.

Equation (11.38) also can be written in the form [see Eq. (11.23)]

$$\chi_{BA}(\omega) = \beta(\langle BA \rangle - i\omega C_{BA}[\omega]) \tag{11.39}$$

One usually writes $\chi' \equiv \text{Re } \chi$ and $\chi'' = \text{Im } \chi$, so $\chi = \chi' + i\chi''$ and we have

$$\chi''_{BA}(\omega) = -\beta\omega \text{ Re } C_{BA}[\omega] \tag{11.40}$$

another form of the fluctuation-dissipation theorem.

DO PROBLEMS 11.11 TO 11.13.

11.5 ELECTRICAL CONDUCTION

As an example of linear response theory, we consider the case in which B is identified with the electric current

$$\mathbf{J} = \sum_{i=1}^{N} j_i \tag{11.41}$$

where j_i is the contribution

$$j_i = \frac{e_i p_i}{m_i} \tag{11.42}$$

to the current due to particle i with electric charge e_i and mass m_i. We suppose that the current is generated by a perturbation [see Eq. (11.29)]

$$-AK(t) = \left(\sum_i e_i \mathbf{r}_i \right) \cdot \mathbf{E}^0(t) \tag{11.43}$$

where $\mathbf{E}^0$ is an externally applied electric field.

As in this example it often is convenient to identify the dynamical variables A and B and the field K as vectors. Alternatively, one could generate a linear response equation for each Cartesian component provided that there were no cross-couplings between distinct components such as one gets for charged particles in a magnetic field.

From Eq. (11.43) we see that

$$\dot{A} = \mathbf{J}$$

in this case, so Eq. (11.36) now becomes

$$\langle \mathbf{J}(t) \rangle_{E_\omega} = \beta e^{i\omega t} \int_0^\infty dt\, e^{-i\omega t} \langle \mathbf{J}(t)\mathbf{J}(0) \rangle \cdot \mathbf{E}^0_\omega \tag{11.44}$$

This result may be compared with the equation

$$\langle \mathbf{J}(t) \rangle_{E_\omega} = \mathrm{Re}\, \mathbf{Y}(\omega) \cdot \mathbf{E}^0_\omega\, e^{i\omega t} V \tag{11.45}$$

where $\mathbf{Y}$ is the complex admittance tensor (or susceptibility or response function). It follows that it has matrix elements

$$Y_{mn}(\omega) = \frac{\beta}{V} \int_0^\infty dt\, e^{-i\omega t} \langle J_m(t) J_n(0) \rangle \tag{11.46}$$

where m and n are indices of Cartesian coordinates. If an electric field is applied to an isotropic system, then $Y_{mn}(\omega) = Y(\omega)\, \delta_{mn}$ and the ordinary conductance is $\sigma(\omega) = \mathrm{Re}\ Y(\omega)$.

The system discussed in this section also may be analyzed by identifying B with the fluctuating polarization

$$\mathbf{P} = \sum_i \mathbf{r}_i e_i - \left\langle \sum_i \mathbf{r}_i e_i \right\rangle \tag{11.47}$$

to obtain an expression for the susceptibility [now identified as the complex dielectric constant $\epsilon(\omega)$] in terms of the polarization fluctuation correlation function $\langle \mathbf{P}(t)\mathbf{P} \rangle$. However, it must be noticed that in either case there is a hidden complication in the linear response theory, namely that the electric field $\mathbf{E}^0(t)$ in Eq. (11.43) is identified as the field due to electric charges that are outside of the N-particle system. On the other hand, the phenomenological coefficient determined in an experiment may be defined by an equation like (11.43) except that $\mathbf{E}^0_\omega$ is replaced by $\mathbf{E}_\omega$, the actual field in the material system. The distinction is not only important but difficult to treat correctly because the relation of $\mathbf{E}^0_\omega$ to $\mathbf{E}_\omega$ depends on the shape of the system. For further analysis of this problem, see Kubo (1966, 1973), Deutch (1977), and Kivelson and Madden (1984).

Another aspect of interest is found in the application of Eq. (11.46) at zero frequency to ionic solutions. Then we find, for a given Cartesian component $m = n$,

$$\sigma_0 = \beta \int_0^\infty dt \sum_{\alpha} \sum_{\beta}_{\text{species}} [\delta_{\alpha\beta}\rho_\alpha \langle j_\alpha(t) j_\alpha(0)\rangle^s + \rho_\alpha \rho_\beta V \langle j_\alpha(t) j_\beta(0)\rangle^d \qquad (11.48)$$

where the sum is over ionic species. A self-correlation function $\langle \cdot \cdot \cdot \rangle^s$ involves velocity self-correlations for one ion of species α, while a joint (or distinct) correlation function $\langle \cdot \cdot \cdot \rangle^d$ involves the cross-correlation between the velocity of one particle of species α and one of species β. Both kinds of correlation function are composition dependent, but it is apparent that in the limit of vanishingly small ion concentrations, the specific conductance σ_0 will be determined solely by the self-correlations for ions moving in the solvent without interacting with each other. Finally, in Eq. (11.48) the thermodynamic limit, $V \longrightarrow \infty$ at fixed $\rho_1, \rho_2, \ldots$ is implied. To explain the appearance of V in Eq. (11.48), which is supposed to give the conductance only in the thermodynamic limit, we notice that the distinct correlation function $\langle j_\alpha(t) j_\beta(0)\rangle^d$ tends to zero in the thermodynamic limit where the chance of some one particle of species α interreacting with a particular particle of species β during a particular time interval becomes vanishingly small. Therefore the product $V\langle \cdot \cdot \cdot \rangle^d$ tends to a definite value, possibly zero, in the thermodynamic limit. Other examples of the different V dependence of self and distinct correlations will be encountered in the following sections.

11.6 VAN HOVE CORRELATION FUNCTIONS

In view of the way the statistical mechanical theory of thermodynamic coefficients and equilibrium structure depends on the spatial correlation function [see Eq. (4.20)],

$$g_{ss'}(\mathbf{R}) \equiv V^2 \langle \delta(\mathbf{r}_{1s} - \mathbf{R}) \delta(\mathbf{r}_{2s'}) \rangle \qquad (11.49)$$

it seems natural to extend this definition to the time-dependent regime. Thus we replace $V^2\langle \cdot \cdot \cdot \rangle$ in Eq. (11.49) by

$$V^2 \langle \delta(\mathbf{r}_{1s}(t + t_0) - \mathbf{R})\, \delta(\mathbf{r}_{2s'}(t_0)) \rangle \equiv \left[\frac{\rho}{\rho_s \rho_{s'}} \right] G^d_{ss'}(\mathbf{R}, t) \qquad (11.50a)$$

as well as by

$$V\langle \delta(\mathbf{r}_{1s}(t+t_0)-\mathbf{R})\,\delta(\mathbf{r}_{1s}(t_0))\rangle \equiv \left[\frac{\rho}{\rho_s}\right] G_s^s(\mathbf{R}, t) \tag{11.50b}$$

In these equations $\mathbf{r}_{1s}(t)$ denotes the location of particle 1 of species s at time t, $\mathbf{r}_{2s'}$ is the location of a distinct particle (even when $s = s'$) and $\rho = \Sigma_s\ \rho_s$ is the total particle number density. Also, $G^d_{ss'}$ and G^s_s are, respectively, the "distinct" and "self" parts of the partial van Hove correlation function

$$G_{ss'}(\mathbf{R}, t) = \delta_{ss'} G_s^s(\mathbf{R}, t) + G^d_{ss'}(\mathbf{R}, t) \tag{11.50c}$$

In the derivation of Eq. (4.20) emphasis was given to the relation of $g_{ss'}$ to the local concentration of s' particles in the field of an s particle. Readers should develop the corresponding interpretations of the self and distinct parts of $G_{ss'}$.

These functions were introduced originally in the theory of the atomic interpretation of neutronscattering (Chapter 13). In that application the species indices s refer to atomic species whether or not the atoms of the system are present in molecular combinations. However, for other applications one may index molecular species and generalize $\mathbf{r}_{1s}$ to include molecular coordinates (Chapter 8). Another aspect of interest is that $\mathbf{r}_{1s}$ may be interpreted as a quantum-mechanical operator that does not commute with other location operators, so that $\langle\cdot\ \cdot\ \cdot\rangle$ can be interpreted as a quantum-mechanical ensemble average. These quantum-mechanical aspects will be suppressed here, so we shall be limited to systems in which classical mechanics is a good approximation.

The partial van Hove function can also be written in the form

$$\rho G_{ss'}(\mathbf{R}, t) = \langle \hat{\rho}_s(\mathbf{R}, t)\hat{\rho}_{s'}(0, 0)\rangle \tag{11.51}$$

where the *local density* (also called a *density field*[1]) introduced in Eq. (4.13) appears in time-dependent form

$$\hat{\rho}_s(\mathbf{R}, t) = \sum_{j=1s}^{Ns} \delta(\mathbf{R} - \mathbf{r}_j(t)) \tag{11.52}$$

where $\mathbf{r}_j(t)$ is the location of particle j at time t. Also, $\hat{\rho}_{s'}(0, 0)$ is the density field of species s at the origin of coordinates at time zero. The translational invariance and stationarity implied in Eq. (11.51) are valid for equilibrium ensembles of isotropic systems. Also valid in these cases is writing $G_{ss'}(R, t)$ in place of $G_{ss'}(\mathbf{R}, t)$, just as the vector $\mathbf{R}$ dependence is lost in the equilibrium function $g_{ss'}(R)$ in isotropic systems. However, we shall often write $G_{ss'}(\mathbf{R}, t)$ solely to make the equations easier to read.

The density–density correlation function represented in Eq. (11.51) is typical of other time correlation functions of field variables of the form

$$\langle A(\mathbf{r}+\mathbf{R}, t+t_0)B(\mathbf{r}, t_0)\rangle \tag{11.53}$$

which may be useful in a variety of problems. An example, the time correlation

[1] A *field* is a function of the spatial coordinates.

TABLE 11.1 SPECIAL VALUES AND ZEROTH MOMENTS OF VAN HOVE CORRELATION FUNCTIONS

	At $t=0$	At $t=\infty$	At $R=\infty$	$\int \cdots d^3R$
$\rho G_s^s(\mathbf{R}, t)$	$\rho_s \delta(\mathbf{R})$	0	0	ρ_s
$\rho G_{ss'}^d(\mathbf{R}, t)$	$\rho_s \rho_{s'} g_{ss'}(\mathbf{R})$	$\rho_s \rho_{s'}$	$\rho_s \rho_{s'}$	$\rho_s \rho_{s'} V$

function for vector current fields, will be encountered shortly. Another important example involves replacing each of the local densities in Eq. (11.51) by its fluctuating part [see Eq. (11.22a)]

$$\hat{\rho}_s(\mathbf{R}, t) \rightarrow \delta\hat{\rho}_s(\mathbf{R}, t) \equiv \hat{\rho}_s(\mathbf{R}, t) - \langle \hat{\rho}_s(\mathbf{R}, t) \rangle \tag{11.54}$$

where the norm $\langle \hat{\rho}_s(\mathbf{R}, t) \rangle$ is actually independent of t. It usually is convenient to define correlation functions in terms of variables with zero norm, and in this respect the density–density correlation functions are no exception, but here we retain the widely used original definitions. Even so, it should be noticed that the notation for the dynamic structure factors (defined below) is only widely accepted up to factors of $\hbar$ and 2π.

Returning to the van Hove correlation functions, we give some of their basic properties in Table 11.1 and some typical forms are shown in Fig. 11.3.

The *intermediate van Hove function* given by[2]

$$\begin{aligned} \rho F_{ss'}(\mathbf{k}, t) &= \int d^3R\, e^{i\mathbf{k}\cdot\mathbf{R}} \rho G_{ss'}(\mathbf{R}, t) \\ &= V^{-1} \langle \rho_s(\mathbf{k}, t) \rho_{s'}(-\mathbf{k}, 0) \rangle \end{aligned} \tag{11.55}$$

where

$$\rho_s(\mathbf{k}, t) = \int d^3R\, e^{i\mathbf{k}\cdot\mathbf{R}} \hat{\rho}_s(\mathbf{R}, t) \tag{11.56}$$

$$= \sum_{j=1_s}^{N_s} e^{i\mathbf{k}\cdot\mathbf{r}_j(t)} \tag{11.57}$$

is a Fourier component of the density field of species s. We also have, by transforming Eq. (11.50c) term by term,

$$F_{ss'}(\mathbf{k}, t) = \delta_{ss'} F_s^s(\mathbf{k}, t) + F_{ss'}^d(\mathbf{k}, t) \tag{11.58}$$

[2] To see the origin of the V^{-1} factor, note that

$$\begin{aligned} \int d^3R\, e^{i\mathbf{k}\cdot\mathbf{R}} \rho(\mathbf{R}, t)\rho(0,0) &= \int d^3R\, e^{i\mathbf{k}\cdot\mathbf{R}} \rho(\mathbf{R}+\mathbf{R}', t)\rho(\mathbf{R}', 0) \\ &= V^{-1} \int d^3\mathbf{R}' \int d^3\mathbf{R}\, \rho(\mathbf{R}+\mathbf{R}', t) e^{i\mathbf{k}\cdot(\mathbf{R}+\mathbf{R}')} \rho(\mathbf{R}', 0) e^{-i\mathbf{k}\cdot\mathbf{R}'} \\ &= V^{-1} \rho(\mathbf{k}, t)\rho(-\mathbf{k}, 0) \end{aligned}$$

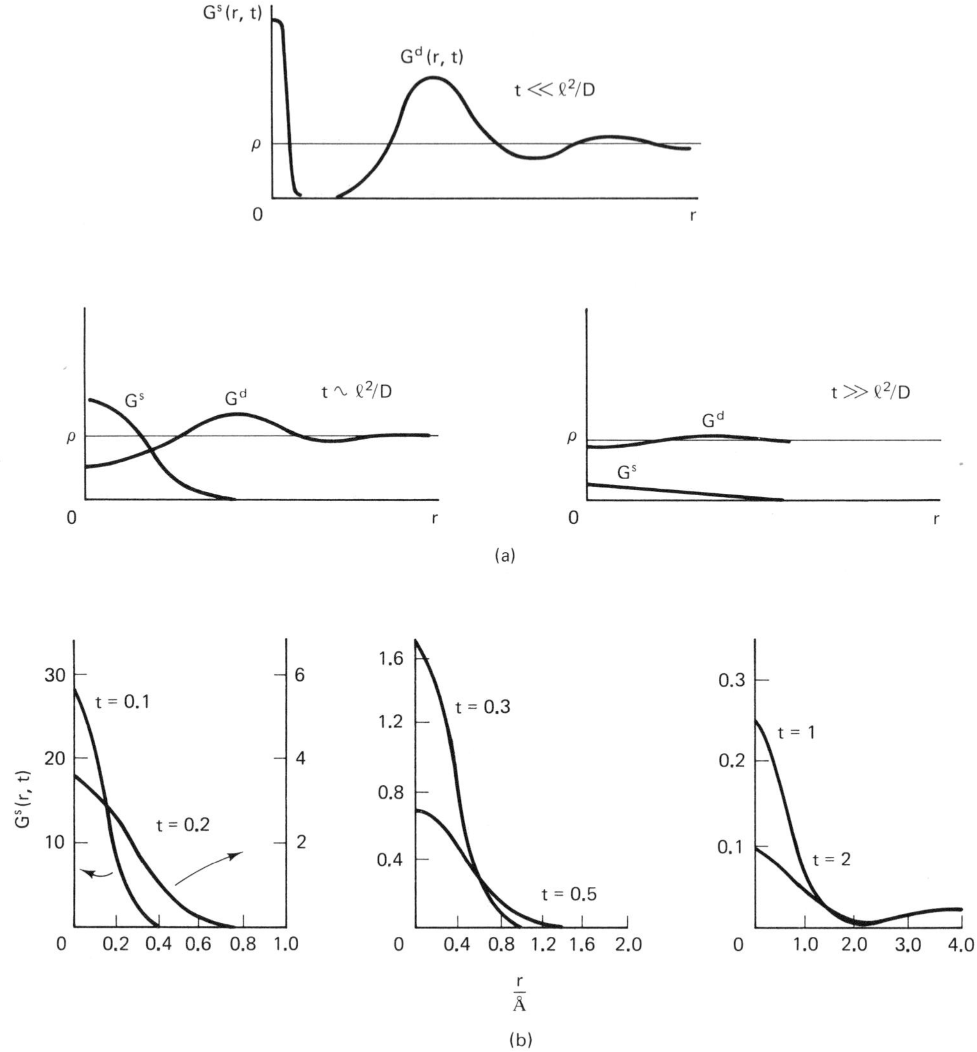

Figure 11.3 Van Hove dynamical correlation functions for a one-component system. (a) Schematic (Van Hove, 1954). In this figure ℓ is the diameter of a particle and D its self-diffusion coefficient. (b) Self-correlations derived from neutron scattering data for liquid argon. Times in picoseconds. Notice that the scales change with t (Dasannacharya and Rao, 1965); see Fig. 10.4. (c) Distinct correlations derived from MD simulations of a 6–12 model for liquid argon at 94.4 K (Rahman, 1964). Notice that G^d seems to change more slowly than G^s as measured by the reduction of the main peak in 1 ps.

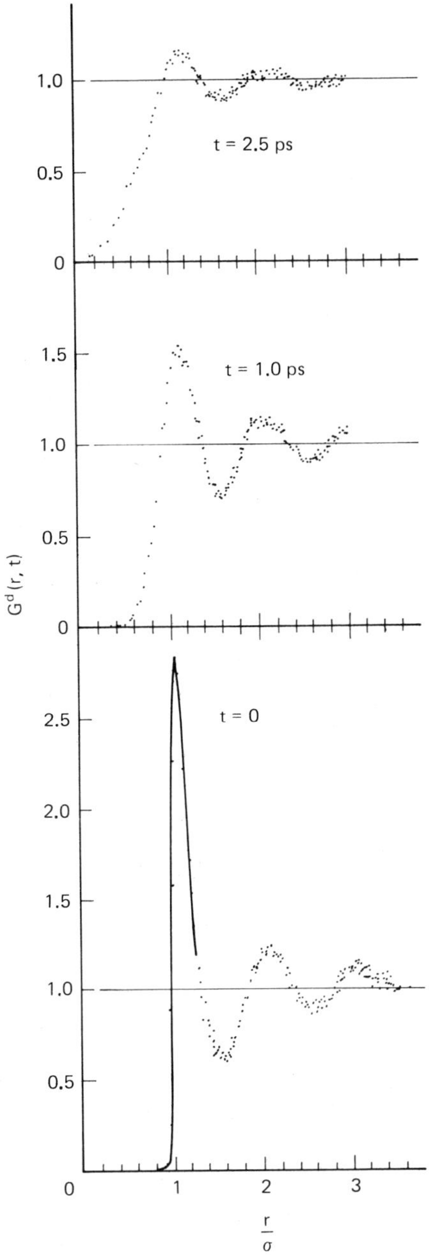

(c)

Current–current correlations are readily generated from the intermediate van Hove functions. Thus we notice that the time derivative of Eq. (11.57) gives

$$\begin{aligned}\dot{\rho}_s(\mathbf{k}, t) &= i\mathbf{k} \cdot \sum_{j=1_s}^{N_s} \dot{\mathbf{r}}_j(t) e^{i\mathbf{k}\cdot\mathbf{r}_j(t)} \\ &\equiv i\mathbf{k} \cdot \mathbf{j}_s(\mathbf{k}, t)\end{aligned} \tag{11.59}$$

where the last member defines the Fourier component of the s-species particle current field[3]

$$\mathbf{j}_s(\mathbf{R}, t) = \sum_{j=1_s}^{N_s} \dot{\mathbf{r}}_j(t)\, \delta[\mathbf{R} - \mathbf{r}_j(t)] \tag{11.60}$$

It may be verified that Eq. (11.59) expresses the conservation of the number of s particles; compare with Eq. (11.10a).

Now it is easy to see that the second time derivative of $F_{ss'}(\mathbf{k}, t)$,

$$\rho \ddot{F}_{ss'}(\mathbf{k}, t) = V^{-1}\mathbf{k} \cdot \langle \mathbf{j}_s(\mathbf{k}, t)\mathbf{j}_{s'}(-\mathbf{k}, 0)\rangle \cdot \mathbf{k} \tag{11.61}$$

generates the ensemble average current–current correlation. It is interesting to discuss this result in a wider context. Thus the basic time correlation function for the two current fields is the tensor

$$\mathbf{J}_{ss'}(\mathbf{R}, t) \equiv \langle \mathbf{j}_s(\mathbf{R}, t)\mathbf{j}_{s'}(0, 0)\rangle \tag{11.62}$$

which, after Fourier transformation, becomes $\mathbf{J}_{ss'}(\mathbf{k}, t)$, which appears in Eq. (11.61) rewritten as

$$\rho \ddot{F}_{ss'}(\mathbf{k}, t) = V^{-1}\mathbf{k} \cdot \mathbf{J}_{ss'}(\mathbf{k}, t) \cdot \mathbf{k} \tag{11.63}$$

Since $i\mathbf{k} \cdot \mathbf{j}_s(k, t)$ is the Fourier transform of $\nabla \mathbf{j}_s(\mathbf{R}, t)$, we also can describe Eq. (11.61) as relating $F_{ss'}$ to the Fourier transform of the time correlation function of two scalar fields, namely

$$\langle \nabla \cdot \mathbf{j}_s(\mathbf{R}, t) \nabla \cdot \mathbf{j}_{s'}(\mathbf{R}', t)\rangle$$

In this connection it is interesting to notice that a vector field like $\mathbf{j}_s(\mathbf{R}, t)$ may be written as a sum

$$\mathbf{j}_s(\mathbf{R}, t) = \mathbf{j}_s^{l}(\mathbf{R}, t) + \mathbf{j}_s^{\text{tr}}(\mathbf{R}, t) \tag{11.64a}$$

of a longitudinal part defined as having no curl,

$$\nabla \times j_s^{l}(\mathbf{R}, t) = 0 \tag{11.64b}$$

and a transverse part defined as having no divergence,

$$\nabla \cdot j_s^{\text{tr}}(\mathbf{R}, t) = 0 \tag{11.64c}$$

Only the longitudinal part of $\mathbf{j}_s$ has a component in the direction of the gradient of $\mathbf{j}_s$, only the longitudinal part contributes to the local density change in Eq. (11.59), and only this part contributes to $\ddot{F}_{ss'}$. Thus we have

[3] Here "current" means the flow of particles, not the electric current.

$$\rho \ddot{F}_{ss'}(\mathbf{k}, t) = k^2 V^{-1} J^l_{ss'}(\mathbf{k}, t) \tag{11.65}$$

where $J^l_{ss'}$ is the correlation function for the longitudinal parts of the vector current fields. The transverse parts are basic to the theory of shear viscosity but do not contribute to the density fluctuations.

Next we further examine the current–current correlation function generated in Eq. (11.61). In the $k = 0$ limit we have, in view of Eq. (11.59),

$$\mathbf{j}_s(\mathbf{k}, t) \longrightarrow \sum_{j=1_s}^{N_s} \dot{\mathbf{r}}_j(t) \qquad \text{as } \mathbf{k} \longrightarrow 0 \tag{11.66}$$

and the longitudinal part of the current–current correlation function in this limit is the sum of the following contributions:

$$\begin{aligned} J^{l,d}_{s,s'}(0, t) &= \langle \hat{k} \cdot \dot{\mathbf{r}}_{1s}(t)\, \hat{k} \cdot \dot{r}_{2s'}(0)\rangle \\ &= \langle \dot{x}_{1s}(t)\dot{x}_{2s'}(0)\rangle N_s N_{s'} \end{aligned} \tag{11.67a}$$

and

$$\begin{aligned} J^{l,s}_{s}(0, t) &= \langle \hat{k} \cdot \dot{\mathbf{r}}_{1s}(t)\, \hat{k} \cdot \dot{\mathbf{r}}_{1s}(0)\rangle N_s \\ &= \langle \dot{x}_{1s}(t)\dot{x}_{1s}(0)\rangle N_s \end{aligned}$$

for the distinct and self-correlations, respectively, and where replacing the projections $\hat{k} \cdot \mathbf{r}(t)$ by $\hat{x} \cdot \mathbf{r}(t) = x(t)$ involves no error in an isotropic system. Finally, we combine these results with Eq. (11.65) to get the velocity cross- and self-correlation functions as functionals of the intermediate van Hove functions.

$$V\langle \dot{x}_{1s}(t)\dot{x}_{2s'}(0)\rangle = \frac{\rho}{\rho_s \rho_{s'}} \lim_{k=0} \ddot{F}^d_{ss'}(\mathbf{k}, t)/k^2 \tag{11.68a}$$

$$\langle \dot{x}_{1s}(t)\dot{x}_{1s}(0)\rangle = \frac{\rho}{\rho_s} \lim_{k=0} \ddot{F}^s_{ss'}(\mathbf{k}, t)/k^2 \tag{11.68b}$$

It should be remarked that in each of these equations the functions equated are finite and generally nonvanishing in the thermodynamic limit; for example,

$$\lim_{V \to \infty} V\langle \dot{x}_{1s}(t)\dot{x}_{2s}(0)\rangle_V$$

$$\rho_1, \rho_2, \ldots \text{ fixed}$$

is not infinite. It is zero for $t = 0$ and $t = \infty$ but not for most values of t in between.

The self-diffusion coefficient of species s, often denoted D^s or $D^\dagger$, given by the integral [Eq. (11.81)]

$$D^s_s = \int_0^\infty dt\, \langle \dot{x}_{1s}(t)\dot{x}_{1s}(0)\rangle \tag{11.69a}$$

is a readily measurable transport coefficient in dilute solutions of species s, and more generally can be measured using tracer methods involving labeling some particles of species s by isotopic substitution or by putting them in particular nuclear spin states. For the theories of mutual diffusion and of Onsager's phenomenological coeffi-

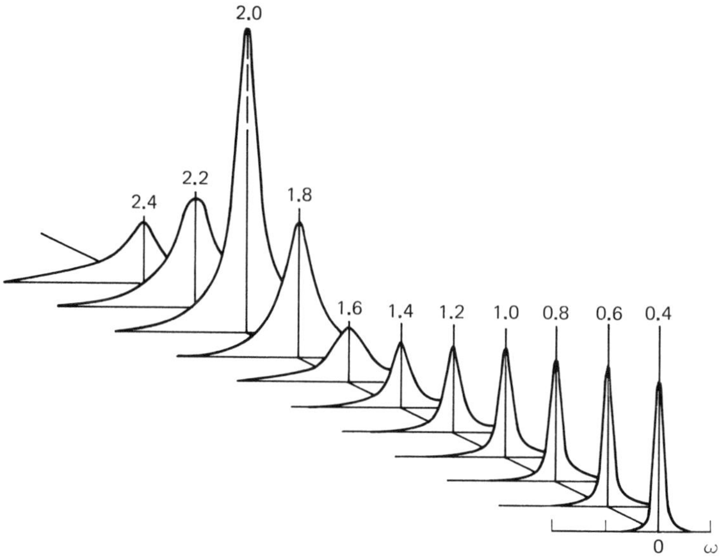

Figure 11.4 $S(k, \omega)$ measured for liquid argon (84.5 K) by inelastic neutron scattering. Each curve is $S(k, \omega)$ for given k [indicated, in Å^{-1}, by the number above the curve, as a function of ω (horizontal axis)]. (From Dasannacharya and Rao, 1965.)

cients for diffusion processes in mixtures, we also need the corresponding cross (or "distinct" or "joint") diffusion coefficients, denoted D^d or D^J,

$$D^d_{ss'} \equiv N\int_0^\infty dt\ \langle \dot{x}_{1s}(t)\dot{x}_{2s'}(0)\rangle \tag{11.69b}$$

where a factor $N = \rho V$ appears rather than V, as in Eq. (11.68a), so D^d will have the same dimensions as D^s. The resulting symmetry appears clearly if we combine Eq. (11.69) with (11.68) to express D^s and D^d in terms of the respective intermediate van Hove functions.[4]

An alternative formulation also may be given in terms of the time transform

$$S_{ss'}(k, \omega) = \frac{1}{2\pi}\int_{-\infty}^{\infty} dt\ e^{i\omega t}F_{ss'}(k, t) \tag{11.70}$$

which is called the *partial dynamical structure factor* because of its role in inelastic neutron scattering (Chapter 13). It follows that

$$F_{ss'} = \delta_{ss'}F^s_s + F^d_{ss'} \tag{11.71a}$$

[4] In general diffusion coefficients depend upon the choice of the reference frame in which the molecular motion is measured, as explained by Kirkwood et al. (1960), DeGroot and Mazur (1962), Miller (1966), and Berne and Pecora (1976). Another distinction arises because, in the terminology introduced by Kirkwood et al. (1960), *practical* diffusion coefficients are defined in terms of constitutive equations relating the flow of species s to the concentration gradients in the system while *fundamental* diffusion coefficients relate the flow of species s to the gradients of chemical potential. It follows from the work of Steele (1969) that the time-correlation function expressions in Eqs. (11.69) give fundamental diffusion coefficients.

implies

$$S_{ss'} = \delta_{ss'} S_s^s + S_{ss'}^d \tag{11.71b}$$

and that we have

$$D_s^s = \frac{\pi \rho}{\rho_s} \lim_{\omega=0} \lim_{k=0} \frac{\omega^2}{k^2} S_s^s(k, \omega) \tag{11.71c}$$

$$D_{ss'}^d = \frac{\pi \rho^2}{\rho_s \rho_{s'}} \lim_{\omega=0} \lim_{k=0} \frac{\omega^2}{k^2} S_{ss'}^d(k, \omega) \tag{11.71d}$$

These expressions stress the behavior of the dynamical structure factors at very small k and very small ω, a region that is not always easily accessible in neutron scattering experiments. An example of $S(k, \omega)$ is given in Fig. 11.4.

11.7 LINEAR RESPONSE THEORY FOR LOCAL FIELDS

In the derivation of the linear response theory, beginning with Eq. (11.33), we did not explicitly consider the case that the dynamical variables A and B may vary from place to place in the system of interest. The example in Section 11.5 illustrates the case in which A and B are defined globally for the system. By construction they are not locally varying. On the other hand, the van Hove correlation functions deal with particle number densities which are locally varying; they are sometimes called *density fields.* Reviewing the derivation of the linear response theory in Section 10.4, we see that it is equally valid if A and B are fields in this sense. Thus we may generalize Eq. (11.35) to find how $\langle \Delta B(\mathbf{r}, t)\rangle_K$ is related to the system's Hamiltonian when the perturbation $-A(\mathbf{r})K(t)$ also is a locally varying field. We find that

$$\langle \Delta B(\mathbf{r}, t)\rangle_K = \frac{\beta}{V} \int_{t_0}^{t} ds \int d^3R \, \langle B(\mathbf{R} + \mathbf{r}, t)\dot{A}(\mathbf{R}, s)\rangle K(s) \tag{11.72a}$$

where the averages are as defined in connection with Eqs. (11.33) and (11.35) and where the integration over $\mathbf{R}$ is appropriate if the equilibrium system at $t < t_0$ is isotropic. This equation is sometimes said to describe response that is nonlocal in space as well as in time; this refers to the fact that the perturbation at $\mathbf{R}$, s gives rise to a response at a different time *and* place.

Now the susceptibility function needs to be generalized accordingly. Proceeding as in the passage from Eq. (11.36) to (11.38), we find that the ratio of the average ΔB response to the local perturbation is, in terms of the angular frequency ω and wave number $\mathbf{k}$,

$$\chi_{BA}(\mathbf{k}, \omega) = \frac{\beta}{V} \int_0^{\infty} dt \, e^{-i\omega t} \int d^3\mathbf{r} \, e^{i\mathbf{k}\cdot\mathbf{r}} \int d^3\mathbf{R} \, \langle B(\mathbf{R} + \mathbf{r}, t) A(\mathbf{R}, 0)\rangle \tag{11.72b}$$

As a one-component-system example, we consider the "experiment" in which we "measure" the average fluctuation in density $\langle \rho_{\mathbf{k}}(t)\rangle_\phi$ due to an externally applied potential $\phi_{\mathbf{k}} e^{i\omega t}$ which couples to the density field. Then

$$\langle \rho_{\mathbf{k}}(\omega)\rangle_{\phi} = \chi_{\rho\rho}(\mathbf{k}, \omega)\phi_{\mathbf{k}} e^{-i\omega t} \tag{11.72c}$$

defines the generalized susceptibility $\chi_{\rho\rho}$. Some of the theory of χ_{AB} is described in Section 14.5.

As already mentioned in Section 1.2, one of the recurring problems in statistical mechanics is to find the exact theoretical formulation of quantities that may be measured experimentally. The linear response theory discussed here provides a systematic way to solve this problem for an important class of measurements. For the coefficients of rate processes, another way is exemplified by the mean-square displacement theory of diffusion coefficients (Problems 11.11 and 11.19), which leads to Eq. (11.69). But it should be noted that even in the linear response regime there are many coefficients whose theory is well known but not developed here. Three examples for a one-component fluid are given in terms of the general form

$$\begin{aligned} K &= \frac{\beta}{V}\int_0^\infty \langle \dot{A}(t)\dot{A}\rangle\, dt \\ &= \frac{\beta}{V}\lim_{t=\infty} \langle [A(t) - A(0)]^2\rangle/2t \end{aligned} \tag{11.73}$$

If A is

$$m \sum_{i=1}^{N} r_{ix}\dot{r}_{iz}$$

then K is the shear viscosity η. If A is

$$m \sum_{i=1}^{N} r_{iz}\dot{r}_{iz}$$

then $K = \zeta + 4\eta/3$, where ζ is the bulk viscosity. If A is

TABLE 11.2 LINEAR RESPONSE THEORY APPLICATIONS

What is measured	$\dot{A}$ in relevant Eq. (11.73)
Rotational diffusion of a molecule changing its α axis	$\hat{u}_\alpha$, a unit vector along α axis
Infrared absorption band shape, transition dipole along α axis of molecule	$\hat{u}_\alpha$
Raman scattering band shape or fluorescence depolarization; transition direction is α axis	Second rank spherical harmonic of $\hat{u}_\alpha$
NMR line shape[a]	**I**, the vector spin operator of the resonating spin
NMR line shape, dipole–dipole or quadrupole relaxation mechanism	$\hat{u}_\alpha$
NMR line shape, spin-rotation relaxation mechanism	$d\hat{u}_\alpha/dt$
Brillouin scattering	$\boldsymbol{\alpha}_j \exp(i\mathbf{k}\cdot\mathbf{r}_j)$, where $\boldsymbol{\alpha}_j$ is the polarizability tensor of molecule j and $\mathbf{r}_j$ its location

[a] The time autocorrelation function of **I** is mostly directly measurable. Depending on the mechanism of coupling **I** to the rest of the system, other time correlation functions are measurable as well.

$$\frac{1}{T}\sum_{i=1}^{N} r_{iz}E_i$$

where E_i is the total energy of particle i, then k is the thermal conductivity λ. Still other examples are indicated in Table 11.2.

DO PROBLEMS 11.15 to 11.20.

THE LITERATURE

The classical equations of motion as described here apply only to structureless particles in the absence of a magnetic field. A complete and general account is given, for example, by Goldstein (1950) and also by Landau and Lifschitz (1957, 1958). The notation $H^\times$ was introduced by Kubo (1962). Most often the Liouville operator is defined with a factor of $i = \sqrt{-1}$ to make it self-adjoint, but here it seems simpler to suppress this factor. In Chester's review (1963) the linear response theory is developed in the context of extensive earlier work. The accounts by Kubo (1966), Berne (1971), and Schofield (1975) develop many aspects of linear response theory that are omitted here, while Schofield (1975), Hansen and McDonald (1976), and Boon and Yip (1980) supplement the theory with many illustrations using experimental data and simulation results. Martin's book (1968) includes the theory of the generalized susceptibility while Helfand (1960) obtains an alternative formulation of linear response theory by generalizing Einstein's theory of Eq. (11.80).

PROBLEMS

11.1 Show that $H^\times H = 0$, $H^\times e^{-\beta H} = 0$, and $H^\times f_N^{\text{eq}}(\Gamma_N) = 0$.

11.2 Show that

$$H^\times \mathbf{q}_i = \frac{\mathbf{p}_i}{m_i} = \dot{\mathbf{q}}_i$$

$$H^\times H^\times \mathbf{q}_i = \frac{\mathbf{F}_i}{m_i}$$

$$H^\times \mathbf{p}_i = F_i$$

11.3. Show that

$$H^\times = H_A^\times + H_B^\times \qquad \text{if } H = H_A + H_B$$

and verify that

$$e^{H^\times t} = e^{H_A^\times t} + \int_0^t e^{H_A^\times (t-s]}\, ds\, H_B^\times e^{H^\times s}$$

Differentiate by dt, and notice the general rule:

$$\frac{\partial}{\partial x}\int_0^x f(x, y)\,dy = f(x, x) + \int_0^x \frac{\partial f(x, y)}{\partial x}\,dy \tag{11.74a}$$

See also Problem 3.10.

11.4. Suppose that P_N is governed by $H^\times$ and $P_{N,A}$ by $H_A^\times$ (see Problem 11.3). Verify that

$$P_N(\Gamma'_N \xleftarrow{t} \Gamma_N) = P_{N,A}(\Gamma'_N \xleftarrow{t} \Gamma_N) + \int_0^t ds \int P_{N,A}(\Gamma'_N \xleftarrow{t-s} \Gamma''_N)\,d\Gamma''_N\, H_B^\times P_N(\Gamma''_N \xleftarrow{s} \Gamma_N) \tag{11.74b}$$

11.5. By repeated application of Eq. (11.17d), prove that

$$\langle Ae^{tH^\times}B\rangle = \langle(e^{-tH^\times}A)B\rangle \tag{11.75}$$

11.6. Prove that the coefficient $\langle A(H^\times)^k A\rangle$ vanishes for odd k.

11.7. Verify that Eq. (11.24) is the inverse of Eq. (11.23).

11.8. Verify the convolution theorem for Fourier–Laplace transforms: If

$$\phi(t) = \int_0^t f(t - t_1)\,dt_1\, g(t_1) \tag{11.76}$$

then

$$\phi[z] = f[z]g[z] \tag{11.77}$$

provided that both $f(t)$ and $g(t)$ vanish for $t < 0$. [Hint: $\phi(t)$ is not changed if the integrand is multiplied by $\Theta(t - t_1)\Theta(t_1)$, where $\Theta(t)$ is the unit step function [see Eq. (7.36)], and the integration extended from $-\infty$ to ∞. Then use the method in Eq. (A1.4).]

11.9. When $K(t)$ in Eq. (11.29) is a component of an electromagnetic wave, the linear response theory gives the line shape function, also called the power spectrum, as a function of the frequency z (see Flygare, 1978):

$$p(z) \equiv \frac{1}{2\pi}\int_{-\infty}^{\infty} dt\, e^{-izt} C_{AA}(t) \tag{11.78a}$$

Show that $p(z)$ is given by

$$p[z] = \frac{1}{2\pi}[C_{AA}[z] + C_{AA}[-z]] = \frac{1}{\pi}\,\mathrm{Re}\, C_{AA}[z] \tag{11.78b}$$

11.10. **(a)** If the Hamiltonian has explicit time dependence, it may be verified that in place of Eq. (11.12) we have

$$f_N(\Gamma_N, t) = \exp\left(-\int_0^t ds\, H_s^\times\right) f_N(\Gamma_N, 0) \tag{11.79a}$$

where H_t is the Hamiltonian at time t. The definition of the exponential in terms of a series now involves "time ordering," so it is not as simple as Eq. (1.82). Without going into all of that, we may follow Eq. (3.56) to verify the following identity (Feynman, 1951) in the case that $H_t^\times = L_t^0 + L_t'$.

$$\exp\left(-\int_0^t ds\, H_s^\times\right) = \exp\left(-\int_0^t ds\, L_s^0\right) - \int_0^t dt' \exp\left(-\int_{t'}^t ds\, L_s^0\right) L'_{t'} \exp\left(-\int_0^{t'} ds\, H_s^\times\right) \tag{11.79b}$$

Verify this equation. Then use it to derive Eq. (11.31) for Δf_N as defined in Eq. (11.30).

(b) Derive Eq. (11.40) from (11.39).

11.11. Einstein (1905) showed that the self-diffusion coefficient of a particle is given by the expression

$$D^s = \lim_{t\to\infty} \frac{\langle [x(t) - x(0)]^2 \rangle}{2t} \tag{11.80}$$

when the limit exists. Here $x(t)$ is the x coordinate of the particle at time t. Using the identity

$$x(t) = x(0) + \int_0^t \dot{x}(u)\, \mathrm{du}$$

show that Eq. (11.80) implies the important result

$$D^s = \int_0^\infty dt\, \langle \dot{x}(t)\dot{x}(0) \rangle \tag{11.81}$$

for the self-diffusion coefficient. You will have to use stationarity and assume that $t \longrightarrow \infty$ in Eq. (11.80) takes one into the range in which Eq. (11.21) is an excellent approximation.

11.12. Einstein (1905) also showed that

$$D^s = \frac{k_B T}{\zeta} \tag{11.82}$$

where ζ is the friction coefficient given by

$$\frac{1}{\zeta} \equiv \lim_{t=\infty} \frac{\langle \dot{x}(t) \rangle_{F_x}}{F_x} \tag{11.83}$$

the ratio of average velocity of a particle to the applied force acting on it in an experiment like that shown in Fig. 11.2. Find the linear response theory expression for $\langle \dot{x} \rangle_{F_x}$ and, using Eq. (11.81), show that Einstein was right about Eq. (11.82).

11.13. In Eq. (11.36) we specialize the linear response theory to the case of a sinusoidally varying perturbation. There are two other very important special cases of Eq. (11.35)

(a) $K(s) = K$, independent of s. Show that one gets the response

$$\langle \Delta B \rangle / \mathrm{K} = \int_0^\infty dt\, \langle B(t)\dot{A}(0) \rangle$$

(b) $K(s) = K\, \delta(s)$. Show that one gets the response

$$\langle \Delta B(t) \rangle / K = \langle B(t)\dot{A}(0) \rangle$$

which is the response to a delta function pulse of unit amplitude at $t = 0$. Notice that the *response function* $\langle B(t)\dot{A}(0) \rangle$ shows how the molecular motions in the equilibrium system determine the *delayed* response at $t > 0$.

11.14. Consider the evaluation of the correlation function formula for the self-diffusion coefficient, which we write in the form

$$D^s = \lim_{t'=\infty} \int_0^{t'} dt \, \langle \dot{x}(t)\dot{x}(0)\rangle_V \tag{11.84}$$

where *V is the volume of the system.*

(a) Derive the equivalent form

$$D^s = \lim_{t'=\infty} \langle x(t')\dot{x}(0)\rangle_V$$

(b) Show that if the origin of the x coordinate is within V and V is finite, then

$$D^s = 0$$

(c) Apparently, Eq. (11.84) is not correct. In view of the derivation from Eq. (11.80), how would you correct it?

(d) Is this aspect generally a problem with time correlation functions?

11.15. **(a)** Write the phase integral expressions (i.e., $\int d\Gamma_N \cdots$) for the ensemble averages $\langle \cdots \rangle$ in Eqs. (11.50a) and (11.50b) [see Eq. (11.15)].

(b) Show that the correlation functions defined in Eq. (11.50a) and (11.50b) are independent of t_0 [see Eq. (11.16)].

(c) Under what conditions is the correlation function in Eq. (11.51) independent of the location at zero time? When is Eq. (11.51) equivalent to the following alternative definition?

$$\rho G_{\varsigma\varsigma'}(\mathbf{R}, t) = \frac{1}{V}\int d^3r \, \langle \rho_\varsigma(\mathbf{R}+\mathbf{r}, t)\rho_{\varsigma'}(\mathbf{r}, 0)\rangle \tag{11.85}$$

(d) Write F^s_ς and $F^d_{\varsigma,\varsigma'}$ in terms of averages over $e^{i\mathbf{k}\cdot\mathbf{r}(t)}$.

11.16. Eq. (11.10) can also be written

$$\dot{c}_\varsigma(\mathbf{R}) = -\nabla \cdot [\dot{\mathbf{R}} c_\varsigma(\mathbf{R})] \tag{11.86}$$

Show that Fourier transformation yields

$$\frac{d}{dt}\tilde{c}_\varsigma(\mathbf{k}) = i\mathbf{k} \cdot \dot{\mathbf{R}}\tilde{c}_\varsigma(\mathbf{k}) \tag{11.87}$$

Then write Eq. (11.59) in its **r**-space form, like Eq. (11.86), which is called a conservation equation or, sometimes, a continuity equation.

11.17. Show that Eq. (4.20) as a definition for $g_{\varsigma\varsigma'}$ is equivalent to

$$g_{\varsigma\varsigma'}(\mathbf{R}'-\mathbf{R}) = \frac{\langle \hat{\rho}_\varsigma(\mathbf{R})\hat{\rho}_{\varsigma'}(\mathbf{R}')\rangle}{\rho_\varsigma\rho_{\varsigma'}} - \frac{\delta_{\varsigma\varsigma'}\delta(\mathbf{R}'-\mathbf{R})}{\rho_\varsigma} \tag{11.88}$$

and that the corresponding definition of $h_{\varsigma\varsigma'}$ is

$$h_{\varsigma\varsigma'}(\mathbf{R}'-\mathbf{R}) = \frac{\langle \delta\hat{\rho}_\varsigma(\mathbf{R})\,\delta\hat{\rho}_{\varsigma'}(\mathbf{R}')\rangle}{\rho_\varsigma\rho_{\varsigma'}} - \frac{\delta_{\varsigma\varsigma'}\,\delta(\mathbf{R}'-\mathbf{R})}{\rho_\varsigma} \tag{11.89}$$

where $\delta\rho_\varsigma$ is defined in Eq. (11.54).

11.18. Verify the following sum rules.

(a)
$$\int_{-\infty}^{\infty} d\omega\ \rho S_{ss'}(\mathbf{k}, \omega) = \rho_s \delta_{ss'} + \rho_s \rho_{s'}[\tilde{h}_{ss'}(\mathbf{k}) + \delta(\mathbf{k})] \equiv \rho S_{ss'}(\mathbf{k}) + \rho_s \rho_{s'} \delta(\mathbf{k}) \tag{11.90}$$

(b)
$$\rho \int_{-\infty}^{\infty} d\omega \omega^2 S_s^s(k, \omega) = \frac{k^2}{V} J_s^{l,s}(k, 0) = \frac{k^2 \rho_s}{\beta m_s} \tag{11.91}$$

(c)
$$\rho \int_{-\infty}^{\infty} d\omega \omega^2 S_{ss'}^d(k, \omega) = \frac{k^2}{V} J_{ss'}^{l,d}(k, 0) = 0 \tag{11.92}$$

11.19. Eq. (11.80) suggests that the distinct diffusion coefficient is given by

$$D_{ss'}^d = N \lim_{t=\infty} \frac{\langle [x_{1s}(t+t_0) - x_{1s}(t_0)][x_{2s'}(t+t_0) - x_{2s'}(t_0)] \rangle}{2t} \tag{11.93}$$

where even if the species are the same ($s = s'$), the two particles are distinct. This definition corresponds to an experiment in which one observes the trajectories of two distinct particles in a sample. Then for various instants t_0 and elapsed time t one forms the indicated product and averages the results for the different t_0. Show that Eq. (11.93) leads to Eq. (11.69b).

11.20. Derive the relation

$$\chi''_{\rho\rho}(k, \omega) = -\pi \beta \rho \omega S(k, \omega) \tag{11.94}$$

12

Langevin's Equation

12.1 INTRODUCTION

Consider any noisy-looking graph of some quantity A as a function of some variable t (Fig. 12.1). The thickness of a wire as a function of position along it or the closing price of some stock market listing as a function of the date are examples. The quantity that is graphed is called a *random process.* It may be characterized by averages like those introduced in Chapter 11. Thus $\langle A \rangle$, the norm, is simply the average. It may be set to zero without loss of generality [see Eq. (11.22)]. The variance $\langle A^2 \rangle - \langle A \rangle^2$ gives a measure of the amplitude and, assuming that $\langle A \rangle = 0$,

$$\tau = \int_0^\infty dt \frac{\langle A(t)A(0) \rangle}{\langle A^2 \rangle}$$

the coherence, is a measure of how big a piece of the t scale is needed to see the fluctuations in A. In the general case for a complete characterization of a random process $A(t)$ in terms of the time correlation functions, we also need three-point $\langle A(t_2)A(t_1)A(t_0) \rangle$ and higher-order correlation functions. However, there is one special class of random processes, called *Gaussian Markov processes,* for which the norm, variance, and coherence tell all. Langevin's equation, to which we turn next, is a kinetic equation that generates Gaussian Markov processes. We may notice that the linear response theory does not involve three-point and higher-order time correlation functions.

Therefore, in studying a dynamical variable $A(t)$ it is interesting to try the point of view in which it is a random process and to approximate it as a Gaussian Markov process. For an introduction to this field, we do not need the whole machinery of the Gaussian Markov processes, much of which is given by Wang and Uhlenbeck (1945); Langevin's equation will do. Treating $A(t)$ as a random process governed

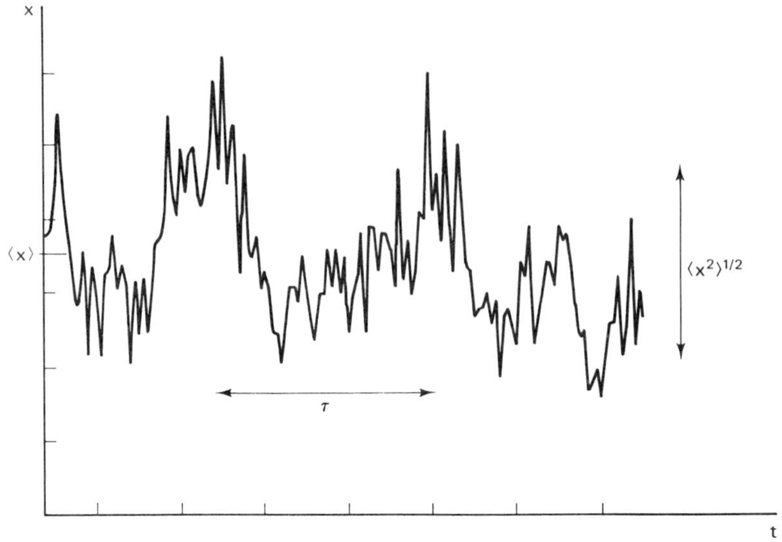

Figure 12.1 Random process $x(t)$ and some of the averages that characterize it.

by Langevin's equation [Eqs. (12.1) below] offers the attractive features that are listed here. To be as clear and explicit as possible, we restrict ourselves to the case in which $A(t)$ is the velocity of a particular particle in the fluid. If the particle has dimensions larger than the wavelength of light, its fluctuating velocity is manifest in the perpetual random displacements which are directly observable as reported by Robert Brown in 1827. So the terms "brownon" or "B particle" seem appropriate to denote a particle of any size, possibly even a labeled molecule of the solvent medium, whose motion is treated as a random process.

1. Langevin's equation gives a detailed description of the motions of the brownon but not of all of the other individual particles; the complete description is usually not practical. The details that are suppressed contribute to the friction force and random force acting on the brownon. The resulting view of the motion of the brownon is often clear in an intuitive way.
2. Time correlation functions derived from Langevin's equation have simple mathematical structures. They serve as idealized examples which illustrate some of the features of more realistic time correlation functions, especially relationships among various time correlation functions.
3. The use of Langevin's equation provides interesting examples of linear response theory and the fluctuation-dissipation theorem.
4. Langevin's equation is fully realistic only in the limit in which the brownon is infinitely massive compared with the other particles in the fluid. However, it often is a reasonably useful approximation even when the brownon is similar to the other particles.
5. A rather similar equation, the generalized Langevin equation, gives the *exact* dynamics for the brownon, regardless of the mass, and is only a little more

difficult mathematically. The generalized Langevin equation is derived and applied in Chapter 14.

12.2 LANGEVIN'S EQUATION

Let $x(t)$ be the x coordinate of a particle of mass m immersed in a fluid medium at equilibrium. Then Langevin's equation is

$$m\ddot{x}(t) = -\frac{m\dot{x}(t)}{\tau} + R(t) \tag{12.1a}$$

The first term on the right is the frictional force on the brownon due to collisions with molecules of the medium; the usual friction coefficient (force per unit velocity) is m/τ, so the parameter τ is smaller the more viscous the medium. The cooling effect of the friction is compensated by the heating effect of the random force $R(t)$, which also represents an effect of collisions with the molecules of the medium. As we shall see in the theory of the generalized Langevin equation, the friction force and the random force are complementary "projections" of the net time-dependent force exerted on the brownon by the medium. We also specify

$$\langle R(t)\dot{x}(0)\rangle = 0 \qquad \text{if } t \geq 0 \tag{12.1b}$$

where $\langle \cdot \cdot \cdot \rangle$ means equilibrium ensemble average. The brownon's velocity does not affect the random force at a later time. We also have

$$\langle R(t)\rangle = 0 \tag{12.1c}$$

$$\langle R(t)R(0)\rangle \sim \delta(t) \tag{12.1d}$$

and a given value of $R(t)$ occurs with probability

$$W(R) = \frac{\exp\left(-R^2/2\langle R^2\rangle\right)}{(2\pi\langle R^2\rangle)^{1/2}} \tag{12.1e}$$

Thus $R(t)$ is a Gaussian random variable with zero norm and zero coherence or correlation time. Finally, in the absence of external fields, the velocity distribution function is the equilibrium Maxwellian function [see Eq. (3.49)]

$$W(\dot{x}) = \frac{\exp\left(-\beta m\dot{x}^2/2\right)}{(2\pi/\beta m)^{1/2}} \tag{12.1f}$$

Langevin's "equation" involves this whole set of six equations.

Equation (12.1d) implies that $\dot{x}(t)$ is a Markov process [Eq. (5.13)], while Eq. (12.1e) implies that $\dot{x}(t)$ is a Gaussian process; $W[x_1(t_1), \ldots, x_n(t_n)]$ is a Gaussian distribution, where $x_1(t_1), \ldots, x_n(t_n)$ is an arbitrary set of observations. (See Wang and Uhlenbeck, 1945.) Sometimes it is convenient to express the same theory in terms of vector quantities such as

$$\mathbf{p} = m(\dot{x}\hat{x} + \dot{y}\hat{y} + \dot{z}\hat{z}) \tag{12.2}$$

where $\hat{x}$, $\hat{y}$, and $\hat{z}$ are Cartesian unit vectors. Then we have

$$\dot{\mathbf{p}} = -\frac{\mathbf{p}}{\tau} + \mathbf{R}(t) \tag{12.3}$$

instead of (12.1a). Also, as we shall see, other dynamical variables may appear in place of $\dot{x}(t)$ in Eq. (12.1a).

DO PROBLEMS 12.1 TO 12.3.

12.3 RELAXATION AND TRANSPORT

To find the velocity time correlation function at equilibrium we multiply Eq. (12.1) by $\dot{x}(0)$ and then average over the equilibrium ensemble to get

$$\langle \ddot{x}(t)\dot{x}(0)\rangle = -\frac{\langle \dot{x}(t)\dot{x}(0)\rangle}{\tau} \tag{12.4}$$

which implies that

$$\langle \dot{x}(t)\dot{x}(0)\rangle = \langle \dot{x}^2\rangle e^{-t/\tau} \tag{12.5}$$

as may be verified, while from Eq. (12.1f)

$$\langle \dot{x}^2\rangle = \frac{1}{\beta m} \tag{12.6}$$

We may confirm the relation of τ to the normalized time correlation function in this case.

$$\tau = \int_0^\infty dt \frac{\langle \dot{x}(t)\dot{x}\rangle}{\langle \dot{x}^2\rangle} \tag{12.7}$$

In the presence of an external force K in the x direction, Eq. (12.1a) becomes

$$m\ddot{x}(t) = -\frac{m\dot{x}(t)}{\tau} + R(t) + K \tag{12.8}$$

Now we average over the nonequilibrium ensemble made up of replica systems brought to a stationary state under the field K. Then we find that

$$0 = -\frac{m}{\tau}\langle \dot{x}\rangle_K + 0 + K \tag{12.9}$$

where $\langle \cdot\cdot\cdot\rangle_K$ specified the nonequilibrium stationary-state ensemble average. The transport coefficient is the ratio of flow to force,

$$\frac{\langle \dot{x}\rangle_K}{K} = \frac{\tau}{m} \tag{12.10}$$

So we see that to measure τ we may either measure the fluctuations in $\dot{x}$ at equilibrium, [Eq. (12.7)], or we may measure the stationary-state transport $\langle \dot{x}\rangle_K$, an illustration of the fluctuation-dissipation theorem (Chapter 11).

The corresponding result for the diffusion coefficient for the brownon may be obtained as follows. We employ the identity, which may easily be verified,

$$\frac{\partial^2}{\partial t^2}\langle[x(t)-x(0)]^2\rangle = 2\langle\dot{x}(t)\dot{x}(0)\rangle \tag{12.11}$$

Then by integrating Eq. (12.11) twice, we get

$$\begin{aligned}\langle[x(t)-x(0)]^2\rangle &= 2\int_0^t dt_1 \int_0^{t_1} dt_2\,\langle\dot{x}(t_2)\dot{x}(0)\rangle \\ &= \frac{2\tau}{m\beta}[t+\tau(e^{-t/\tau}-1)]\end{aligned} \tag{12.12}$$

which may be compared with Einstein's fully general expression for the self-diffusion coefficient

$$D = \lim_{t\to\infty}\frac{\langle[x(t)-x(0)]^2\rangle}{2t} \tag{12.13}$$

to get

$$\begin{aligned}D &= \frac{\tau}{m\beta} \\ &= \frac{k_B T\tau}{m}\end{aligned} \tag{12.14}$$

So we see that a third way to determine τ for a brownon is to measure the self-diffusion coefficient.

A graph of the important equation

$$\frac{\langle[x(t)-x(0)]^2\rangle}{2D\tau} = \frac{t}{\tau} + e^{-t/\tau} - 1 \tag{12.15}$$

derived from Eq. (12.12) is shown in Fig. 12.2. From Eq. (12.12), and $\langle\dot{x}^2\rangle = k_BT/m$ [Eq. (12.6)], we find that

$$\langle[x(t)-x(0)]^2\rangle = \langle\dot{x}^2\rangle t^2 \qquad \text{if } t \ll \tau \tag{12.16}$$

so over sufficiently short intervals the brownon's motion is indistinguishable from the motion of a free particle with the rms thermal velocity $\langle\dot{x}^2\rangle^{1/2}$. Near $t = \tau$ there is a transition to diffusive behavior, and for $t \gg \tau$ we have $\langle[x(t) - x(0)]^2\rangle$ proportional to t rather than to t^2.

It is interesting to notice that for a water molecule or a typical monatomic ion in aqueous solution $\tau \sim 10^{-14}$ s, while for a colloid particle about 1000 Å in diameter $\tau \sim 10^{-8}$ s. We also notice that if the x, y, and z components of $\dot{\mathbf{r}}$ are mutually independent, then Eq. (12.13) is equivalent to the familiar equation

$$D = \lim_{t\to\infty}\frac{\langle r(t)^2\rangle}{6\,Dt} \tag{12.17}$$

where $r(t)^2$ means $|\mathbf{r}(t) - \mathbf{r}(0)|^2$.

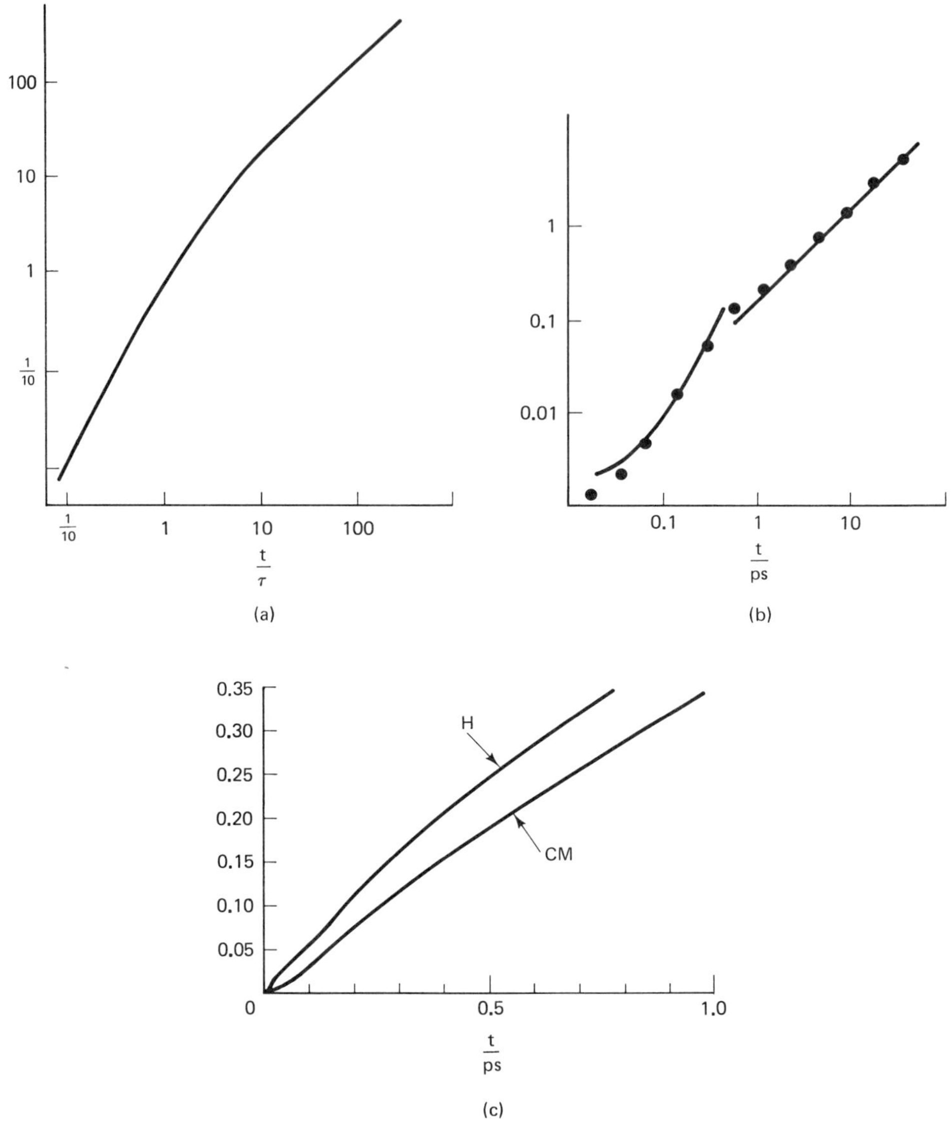

Figure 12.2 Diffusive displacement.

(a) Langevin equation calculation, Eq. (12.15). The ordinate is $\langle [x(t) - x(0)]^2 \rangle / 2D\tau$.

(b) Results of inelastic neutron scattering experiments from liquid argon at 85.2 K. The ordinate is $\langle [\mathbf{r}(t) - \mathbf{r}(0)]^2 \rangle / 6$ in units of Å². The curve at short times is calculated on the basis of a quantal uncertainty effect plus free (i.e., Newtonian) motion, while the line at late times is Dt. (From Sköld et al., 1972.)

(c) Results of MD simulation of a molecular model for liquid water. The ordinate is $\langle [r(t) - r(0)]^2 \rangle / \sigma^2$ for the motion of the H atoms and for the motion of the centers of mass (CM) of the H_2O molecules. (From Rahman and Stillinger, 1971.)

By replacing Eq. (12.8) by

$$m\ddot{x}(t) = -\frac{m\dot{x}(t)}{\tau} + R(t) + Ke^{i\omega t} \tag{12.18}$$

we adapt Langevin's equation to the study of the linear response to a time-dependent external force (see Section 11.4).

DO PROBLEMS 12.4 AND 12.5.

12.4 EVOLUTION OF DISTRIBUTION FUNCTIONS

Langevin's equation describes the evolution of a dynamical variable. In this respect it resembles the Heisenberg equation of dynamics. For some purposes it is convenient to examine the corresponding Liouville equations, giving the evolution of distribution functions. There are three such equations to reckon with, although in common usage the names are not always distinct. Chandrasekhar (1943) has derived all three from Langevin's equation in the one-brownon case. We merely cite his results.

The phase-space distribution function $f(q, p; t)$ for one brownon and one Cartesian coordinate is governed by the Kramers equation,

$$\dot{f} + \hat{c}f = 0 \tag{12.19}$$

$$\hat{c} \equiv \frac{p}{m}\frac{\partial}{\partial q} + K\frac{\partial}{\partial p} - \frac{1}{\tau}\frac{\partial}{\partial p}p - \frac{m}{\beta\tau}\frac{\partial^2}{\partial p^2} \tag{12.20}$$

where K is the external force in Eq. (12.8) and τ the relaxation time in Eq. (12.1a). In using equations such as this one, an operation like

$$\frac{1}{\tau}\frac{\partial}{\partial p}pf \quad \text{is interpreted as} \quad \frac{1}{\tau}\frac{\partial}{\partial p}(pf)$$

Chandrasekhar (1943) also deduced the corresponding propagator, the Green's function of the Kramers equation in the $K = 0$ case. This is the function P such that, with $K = 0$,

$$\dot{P}(q, p \overset{t}{\leftarrow} q_0, p_0) + \hat{c}P(q, p \overset{t}{\leftarrow} q_0, p_0) = 0 \tag{12.21}$$

and

$$P \underset{t=0}{\longrightarrow} \delta(q - q_0)\delta(p - p_0) \tag{12.22}$$

In Eq. (12.21) $\hat{c}$ operates on the "late" phase q, p only. Chandrasekhar's result is

$$P(t \leftarrow 0) \equiv P(q, p \overset{t}{\leftarrow} q_0, p_0) = \frac{1}{\pi\sqrt{D'}}\exp\left(\frac{-GR^2 + 2HRS - FS^2}{D'}\right) \tag{12.23a}$$

where

$$R = q - q_0 - \frac{p_0(1-\Theta)\tau}{m} \qquad G = \frac{2m(1-\Theta^2)}{\beta}$$

$$S = p - \Theta p_0 \qquad H = \frac{2\tau(1-\Theta)^2}{\beta} \tag{12.23b}$$

$$\Theta = e^{-t/\tau} \qquad F = \left(\frac{2\tau^2}{\beta m}\right)\left(\frac{2t}{\tau} - 3 + 4\Theta - \Theta^2\right)$$

$$D' = FG - H^2$$

This is a rare case in which the mathematical form of the propagator is known and not trivial. It is a challenge to verify that this propagator is consistent with Eq. (12.22). The long-time large-displacement limiting form also is remarkably simple,

$$P(t \leftarrow 0) = W(p)\frac{\exp[-(q-q_0)^2/4Dt]}{\sqrt{4\pi Dt}} \qquad \text{if } t \gg \tau \text{ and } \frac{(q-q_0)m}{\tau} \gg p \text{ or } p_0 \tag{12.24}$$

where $W(p)$ follows from Eq. (12.1f).

These equations, written here for displacements q and momenta p on one Cartesian axis in three-dimensional space, are readily generalized to three-dimensional motion in the usual case in which the x, y, and z Cartesian components of the motions are mutually independent.

The preceding Liouville-type equations are fully equivalent to Langevin's equation. On the other hand, the following equations are, respectively, obtained by projecting out location or momentum information.

The velocity distribution function

$$f(p;t) = \int_V dq\, f(q,p;t)$$

of the brownon evolves according to the Fokker–Planck equation:

$$\dot{f}(p;t) + \hat{F}f(p;t) = 0 \tag{12.25}$$

$$\hat{F} \equiv -\frac{1}{\tau}\frac{\partial}{\partial p}p - \frac{m}{\tau\beta}\frac{\partial^2}{\partial p^2} + K\frac{\partial}{\partial p} \tag{12.26}$$

The corresponding propagator for the $K = 0$ case is

$$P_p(t \leftarrow 0) \equiv \int d^3q\, P(q + q_0, p \overset{t}{\leftarrow} q_0, p_0) \tag{12.27}$$

which is found to be

$$P_p(t \leftarrow 0) = \frac{1}{\sqrt{2\pi m(1-\Theta^2)/\beta}} \exp \frac{-\beta(p - \Theta p_0)^2}{2m(1-\Theta^2)} \tag{12.28}$$

From Eq. (12.6) we know that τ is the correlation time for change in momentum. So it is quite consistent that if $t \gg \tau$, then $P_p(t \leftarrow 0)$ reduces to $W(p)$.

The location distribution function $f(q, t) = \int dp\, f(q, p; t)$ of the brownon evolves according to the Smoluchowski equation

$$\dot{f}(q, t) + \hat{S}f(q, t) = 0 \tag{12.29}$$

$$\hat{S} \equiv -\frac{\partial}{\partial q}\left(D\frac{\partial}{\partial q} - \frac{K\tau}{m}\right) \tag{12.30}$$

In the $K = 0$ case the corresponding propagator is deduced from Eq. (12.23).

$$P_q(t \leftarrow 0) = \int dp\, P(q + q_0, p \overset{t}{\leftarrow} q_0, p_0) W(p_0)\, dp_0 \tag{12.31}$$

$$= \frac{e^{-q^2/E}}{\sqrt{\pi E}} \tag{12.31}$$

$$E \equiv 4D[t - \tau(1 - \Theta)] \tag{12.32}$$

The dynamics generated by Langevin's equation is faithfully followed by Smoluchowski's equation only in the time range $t \gg \tau$. This limitation may be illustrated by noting that

$$\dot{P}_q(t \leftarrow 0) + \hat{S}P_q(t \leftarrow 0) = 0 \tag{12.33}$$

is only valid for $t \gg \tau$, while Eq. (12.31) is derived from Eq. (12.23), and hence from Langevin's equation, without further approximation. It follows that $\dot{f}(q, t)$ seen by Smoluchowski's equation can have contributions from slow fluxes that change the local concentrations but not from the instantaneous dynamical velocities. Further development of the Smoluchowski-level theory is given in Section 13.3.

DO PROBLEMS 12.6 AND 12.7.

12.5 ROTATIONAL BROWNIAN MOTION

Assume now that the brownon is a sphere with an axis fixed on it. One may describe the rotational motion of the brownon by noting that the direction of its axis may be expressed in terms of two angles, an azimuthal angle ϕ and a polar angle θ in a coordinate system fixed in the laboratory. The orientational motion of the brownon corresponds to the motion of a representative point on a spherical surface concentric with the particle, the point where an extension from one end of the axis of the particle cuts the spherical surface. If the spherical surface is marked off like a globe, ϕ is the longitude and $\pi/2 - \theta$ the latitude of the representative point. For small changes $d\theta$ and $d\phi$ of the angles, the spatial displacement of the representative point on the spherical surface (of radius R) is

$R\, d\theta$ along an arc of constant longitude.

$R \sin\theta\, d\phi$ along an arc of constant latitude.

Since the latter displacement depends on the θ angle, Langevin's equation for this motion is not just the additive combination of independent Langevin equations in $\dot{\theta}$ and $\dot{\phi}$.

In the face of this difficulty it is fortunate that the treatment of rotational Brownian motion can be based on Smoluchowski's equation for the motion of a representative point on the surface of a sphere. It is simpler to handle than Langevin's equation because the angular momenta do not appear. In this case the Green's function of Smoluchowski's equation has the form

$$P_\Omega(t \leftarrow 0) = P_\Omega(\Omega \overset{t}{\leftarrow} \Omega_0) \tag{12.34}$$

where Ω stands for θ, ϕ. It obeys the fundamental Smoluchowski equation

$$\begin{gathered} \frac{\partial P_\Omega}{\partial t} + \hat{S} P_\Omega = 0 \\ \hat{S} = -\tau^{-1} \frac{\partial^2}{\partial \Omega^2} \end{gathered} \tag{12.35}$$

where τ is a relaxation time and

$$\frac{\partial^2}{\partial \Omega^2} = \frac{1}{\sin\theta} \frac{\partial}{\partial \theta} \sin\theta \frac{\partial}{\partial \theta} + \frac{1}{\sin^2\theta} \frac{\partial^2}{\partial \phi^2}$$

is the Laplacian operator on the surface of the sphere. We also define $d\Omega = \sin\theta \, d\theta \, d\phi$.

A complete set of the basis functions on the surface of the sphere is provided by the spherical harmonics $Y_j^m(\theta, \phi) \equiv Y_j^m(\Omega)$, which obey the orthogonality

$$\int d\Omega \, Y_j^{m*}(\Omega) \, Y_{j'}^{m'}(\Omega) = \delta_{jj'} \delta_{mm'} \qquad \begin{cases} j = 0, 1, 2, \ldots \\ -j \le m \le j \end{cases} \tag{12.36}$$

and completeness

$$\delta(\Omega - \Omega') = \sum_j \sum_m Y_j^{m*}(\Omega) \, Y_j^m(\Omega') \tag{12.37}$$

relations. In view of the completeness of the spherical harmonics, we may write

$$P_\Omega(t \leftarrow 0) = \sum_{j,m} c_j^m(t, \Omega) \, Y_j^m(\Omega_0) \tag{12.38}$$

because the coefficients c_j^m may be calculated by multiplying P_Ω by $Y_j^{m*}(\Omega_0)$ and integrating over Ω_0, according to Eq. (12.36). It is known that

$$\frac{\partial^2}{\partial \Omega^2} Y_j^m(\Omega) = -j(j+1) \, Y_j^m(\Omega) \tag{12.39}$$

Then substituting these equations in Eq. (12.35), we find that

$$\dot{c}_j^m(t, \Omega) = -\tau^{-1} j(j+1) c_j^m(t, \Omega)$$

Using the notation

$$\tau_j \equiv \frac{\tau}{j(j+1)} \tag{12.40}$$

the result of integrating $\dot{c}_j^m$ is

$$c_j^m(t, \Omega) = \exp(-t/\tau_j) c_j^m(0, \Omega) \tag{12.41}$$

and we recover

$$P_\Omega(t \leftarrow 0) \equiv P_\Omega(\Omega \overset{t}{\leftarrow} \Omega_0) = \sum_{j,m} Y_j^m(\Omega_0) c_j^m(0, \Omega) \exp\left(\frac{-t}{\tau_j}\right) \tag{12.42}$$

To evaluate the coefficients $c_j^m(0, \Omega)$ in Eq. (12.42), we note that they are independent of time, so the equation is also valid at $t = 0$, where $e^{-t/\tau_j} = 1$ and where

$$P_\Omega(\Omega \overset{t}{\leftarrow} \Omega_0) = \delta(\Omega - \Omega_0) \tag{12.43}$$

Then comparison with Eq. (12.37) shows that

$$c_j^m(0, \Omega) = Y_j^{m*}(\Omega) \tag{12.44}$$

and so we arrive at the final result, a very important one,

$$P_\Omega(\Omega \overset{t}{\leftarrow} \Omega_0) = \sum_j \sum_m Y_j^m(\Omega_0) Y_j^{m*}(\Omega) e^{-t/\tau_j} \tag{12.45}$$

Notice how much more complicated this is than the corresponding result for P_q, Eq. (12.31), which has only a single correlation time. However, it is to be emphasized that neither result is any better than Langevin's equation, from which they are derived.

Equation (12.45) is limited to the rotational Brownian motion of a spherical particle but often is applied to a molecule with a distinguished axis, as in a dipolar molecule such as CH_3F or to molecules of even lower symmetry. While the correlation times τ_j for a spherical body are all multiples of one basic τ, we expect that for a body with the symmetry of a symmetric top there are two basic rotational correlation times and three for bodies of lower symmetry.

Also, to fix the orientation of a body with less than axial symmetry, one needs the three Eulerian angles (Fig. 8.1) or their equivalent. The more general basis functions required are the Wigner $D^j_{nm}(\alpha, \beta, \gamma)$ described in Appendix 2. Some of these functions do not depend of the γ angle and are proportional to the Y_j^m functions.

$$D^j_{0m}(\alpha, \beta, \gamma) = (-)^m \sqrt{\frac{4\pi}{2j+1}}\, Y_j^m(\beta, \alpha) \tag{12.46}$$

It turns out that for moderate deviations from spherical symmetry of the brownon, the use of Eq. (12.45) does not lead to important errors due to neglect of nonsphericity.

For two dynamical variables $A(\Omega)$ and $B(\Omega)$ that depend on the orientation of the brownon, we have

$$\langle B(t)A^*(0)\rangle = \frac{1}{4\pi}\int d\Omega' \; B(\Omega')P(\Omega' \overset{t}{\leftarrow} \Omega)A^*(\Omega)\, d\Omega \tag{12.47}$$

since in isotropic systems $W(\Omega) = 1/4\pi$ since all orientations are equally possible and $\int d\Omega = 4\pi$. Now if we expand both B and A in spherical harmonics:

$$B(\Omega) = \sum_{j,m} B_j^m Y_j^m(\Omega); \qquad A(\Omega) = \sum_{j,m} A_j^m Y_j^m(\Omega) \tag{12.48}$$

then we find, using Eq. (12.45), that we have

$$\langle B(t)A^*(0)\rangle = \frac{1}{4\pi}\sum_{j,m} B_j^m A_j^{m*} e^{-t/\tau_j} \tag{12.49}$$

with no "cross-terms." In many cases B and A will have nonzero coefficients in Eq. (12.48) only for a single value of j. For example, if B is independent of Ω, then only the $j = 0$ component does not vanish. If B is the electric dipole moment of the brownon, only the $j = 1$ components do not vanish. If B is either the electric quadrupole moment tensor or the electric polarizability tensor, only the $j = 2$ components do not vanish. It is clear from Eq. (12.49) that if B comes in one j value and A in another, the correlation function will vanish. This is a special case of a general law known as *Curie's theorem.* It is also clear from Eq. (12.49) that the various different m terms with the same j have the same time dependence. This is a special case of another general law known as the *Wigner–Eckart theorem.*

In the most common applications of Eq. (12.49), $A = B$ and B has nonvanishing spherical harmonic components for either $j = 1$ or $j = 2$. Sometimes one can measure τ_j for $j = 1$ and $j = 2$ on a given brownon, for example, by measuring the dielectric relaxation ($j = 1$) and the NMR relaxation ($j = 2$) of a water molecule in liquid water. We should find

$$\tau_1 = 3\tau_2 \tag{12.50}$$

according to Langevin's equation. Adherence to Eq. (12.50) is sometimes used as a test of the applicability of Langevin's equation. Surprisingly enough, in the example given, liquid water, it works within the experimental error of $\sim$10%. However, a weak link in this application of Eq. (12.49) is that the dielectric relaxation time cannot in general be identified with τ_1 because of the effect of the difference between the externally applied field $\mathbf{E}_0$ and the internal field $\mathbf{E}$ as discussed following Eq. (11.47). On the other hand, the analysis of certain band shapes in infrared spectra does lead to τ_1.

The reorientation dynamics of a molecule also contributes to the shape of its infrared absorption lines, as illustrated in Fig. 12.3. Reorientation affects the spectrum because the transition dipole moment vector associated with an infrared transition is fixed in the molecule and rotates with it, leading to line shape effects due to a sort of lifetime broadening. Flygare (1978) discusses both the theory of the measurement and the way in which the time correlation function obtained directly by Fourier inversion of the line shape function has to be corrected to remove certain interfering contributions. This procedure is also the source of the data in Fig. 11.1c.

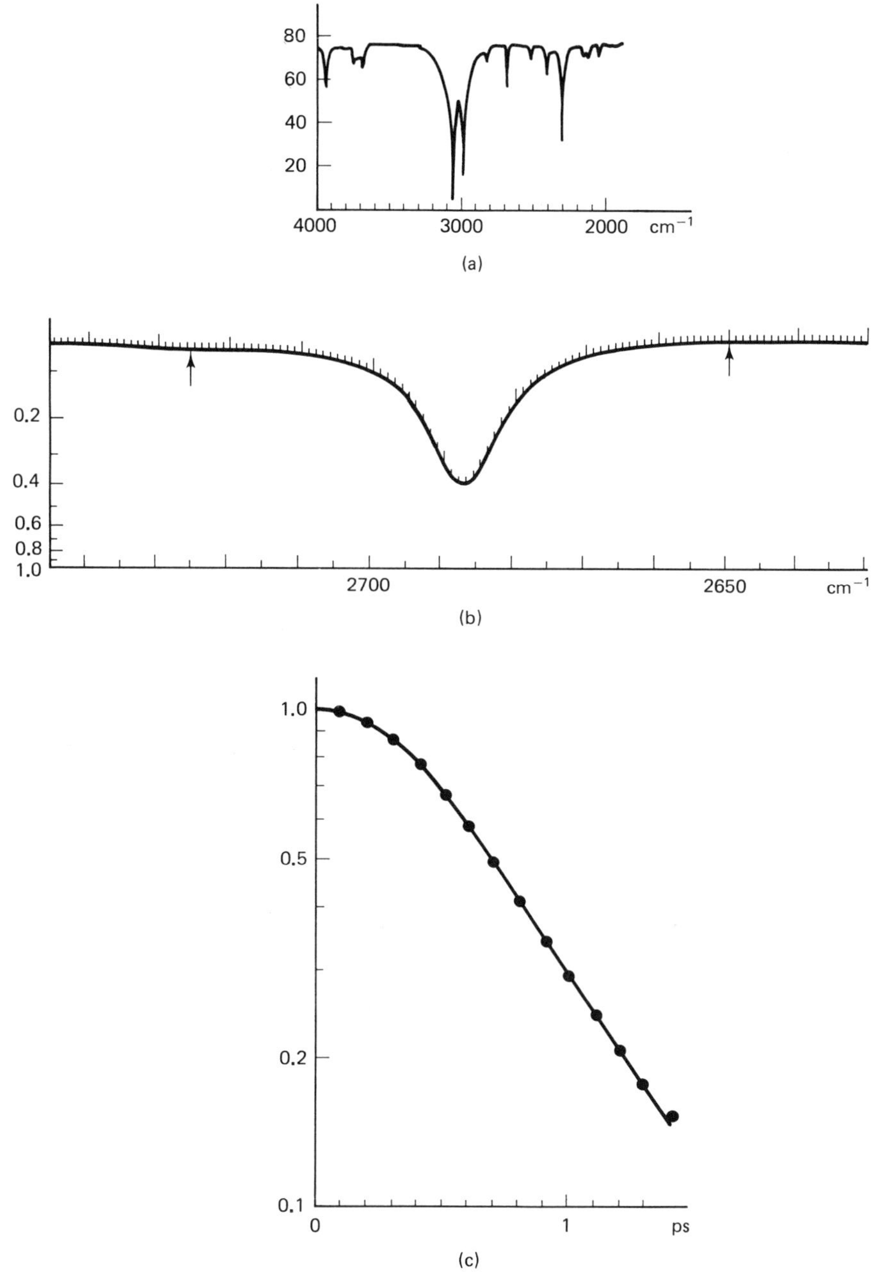

Figure 12.3 Line shape in the infrared. (a) Part of the infrared spectrum of liquid $CH_2C\ell_2$. (b) The line near 2688 cm^{-1} on an expanded scale. Evidently, it is approximately Lorentzian. (c) Time correlation functions for the reorientation of a particular vector (the direction of the transition dipole) fixed in a $CH_2C\ell_2$ molecule, calculated from (b) on the assumption that the only important contribution to the line shape is the reorientational motion of the molecules. If the line shape were Lorentzian, this function would be exponential. (Adapted from Rothschild, 1970; after Flygare, 1978.)

12.6 THE NOISY OSCILLATOR

For a brownon in a field of force that is derived from the simple harmonic oscillator potential

$$U = \frac{1}{2} K x^2 \tag{12.51}$$

the Langevin equation is [see Eq. (12.8)]

$$m\ddot{x}(t) = -\frac{m\dot{x}(t)}{\tau} + R(t) - Kx(t) \tag{12.52}$$

In the zero friction limit $\tau = \infty$ and $R(t) = 0$ and

$$x(t) = \text{Re}\, x_0 e^{i\omega t}, \qquad \text{zero friction} \tag{12.53}$$

where

$$\omega = \sqrt{\frac{K}{m}} \tag{12.54}$$

is the angular frequency (radians per unit time). From the resulting equation

$$\ddot{x}(t) = -\frac{\dot{x}(t)}{\tau} + \frac{R(t)}{m} - \omega^2 x(t) \tag{12.55}$$

various time correlation functions have been calculated by Chandrasekhar (1943) without assuming linear response to K, which would simplify the mathematics at the cost of trivializing the model. However, the theory is much easier and still very informative if we replace the real variable $\dot{x}(t)$ by a complex variable $u(t)$ whose real part is $\dot{x}(t)$ and which obeys the alternative equation

$$\dot{u}(t) = -\frac{u(t)}{\tau} + \frac{R(t)}{m} + i\omega u(t) \tag{12.56}$$

which is fully equivalent to Eq. (12.55) only in the limit of zero friction. Equation (12.56) implies that we have

$$\begin{aligned} C_{uu}(t) &\equiv \frac{\langle u(t)u \rangle}{\langle u^2 \rangle} \\ &= e^{(i\omega - 1/\tau)t} \end{aligned} \tag{12.57}$$

$$C_{uu}[z] = \frac{1}{1/\tau + i(z-\omega)} \tag{12.58}$$

with the power spectrum,

$$\text{Re}\, C_{uu}[z] = \frac{\tau}{1 + \tau^2[z-\omega]^2} \tag{12.59}$$

a Lorentzian function of frequency z (Fig. 12.3). The relation of this calculation to a spectral line shape function may be established by adding a term $[F_0/m]e^{izt}$ on the right of Eq. (12.56) to represent the effect of an external field that exerts a force $F_0 e^{izt}$ on the brownon, for example an electric field acting on a charged brownon. Then construct the ensemble average of each term of the equation to get the observable response, a periodic current of frequency z which is a measure of the energy absorbed from the exciting field [see Eq. (11.36)].

Thus Langevin's equation provides a simple theory of the most commonly observed line shape, namely Lorentzian (Fig. 12.3). However, it should be noted that real line shape functions cannot be Lorentzian; instead, they must decay to zero faster in the $|z - \omega| \gg 1/\tau$ range (i.e., in the "wings" of the line) because of the sum rules in Eq. (11.25).

Two characteristics of the noisy oscillator that are sometimes of interest are difficult to derive from Eq. (12.56) but follow easily from Eq. (12.55). One is the variance of the coordinate

$$\begin{aligned}\langle x^2 \rangle &= \frac{\displaystyle\int_{-\infty}^{\infty} x^2 e^{-\beta K x^2/2}\, dx}{\displaystyle\int_{-\infty}^{\infty} e^{-\beta K x^2/2}\, dx} \\ &= \frac{1}{\beta m \omega^2}\end{aligned} \tag{12.60}$$

and the other is the corresponding correlation time,

$$\begin{aligned}\tau_x &= \int_0^{\infty} \frac{\langle x(t) x \rangle}{\langle x^2 \rangle}\, dt \\ &= \frac{1}{\omega^2 \tau}\end{aligned} \tag{12.61}$$

where the result follows from Chandrasekhar's calculation (1943). Notice that an estimate for τ_x can be gotten from $\langle x^2 \rangle / D$, where $D = \tau/\beta m$ is the diffusion coefficient of the brownon when the potential $U(x)$ is turned off. Thus $\tau_x \sim \langle x^2 \rangle D = 1/\omega^2 \tau$, in agreement with the more detailed calculation cited above.

12.7 MAGNETIC RESONANCE

The Bloch equation for $\mathbf{M}$, the macroscopic magnetization of a system, may be written

$$\dot{\mathbf{M}}(t) = \gamma \mathbf{M}(t) \times \mathbf{B}(t) - [\hat{x}\hat{x} + \hat{y}\hat{y}] \cdot \mathbf{M}(t)/T_2 - \hat{z}[M_z(t) - \langle M_z \rangle]/T_1 \tag{12.62}$$

where $\hat{x}$, $\hat{y}$, and $\hat{z}$ are Cartesian unit vectors and it is assumed that the magnetic field $\mathbf{B}$ fixes the z direction. Also, γ is the gyromagnetic ratio and T_1 and T_2 are the *spin-lattice* and *transverse* relaxation times. To see how these measurable coefficients are related to equilibrium fluctuation in the system, we may proceed as follows.

First we notice that $\mathbf{M} = \Sigma_i \mathbf{m}_i$ is a sum of contributions of all of the particles $1, 2, \ldots, i, \ldots$ of some particular species. (While $\mathbf{M}$ may have contributions from several particle species, the usual magnetic resonance experiment sees the contribution to **M** from one species at a time.) The kinetic equation for the magnetization of any one particle

$$\dot{\mathbf{m}}(t) = \gamma \mathbf{m}(t) \times \mathbf{B}(t) - [\hat{x}\hat{x} + \hat{y}\hat{y}] \cdot \mathbf{m}(t)/T_2 - \hat{z}[m_z(t) - \langle m_z \rangle]/T_1 + \mathbf{R}(t) \qquad (12.63)$$

has a random force term which tends to cancel in the sum over the one particle moments leading from Eq. (12.63) to (12.62).

To uncouple the contributions to Eq. (12.63), one defines the *spherical components* of the **m** vector as follows:

$$m_\pm = m_x \pm i m_y \qquad (12.64a)$$

$$m_0 = m_z - \langle m_z \rangle \qquad (12.64b)$$

where $m_x \equiv \hat{x} \cdot \mathbf{m}$ and so on are Cartesian components. Then $\mathbf{m} = \hat{x}[m_+ + m_-]/2$, $\hat{y}[m_+ - m_-]/2i$, $\hat{z}m_0$ can be substituted in Eq. (12.63) to get

$$\dot{m}_0 = -\frac{m_0(t)}{T_1} + R_0(t) \qquad (12.65a)$$

$$\dot{m}_\pm(t) = \left(\pm\frac{\omega}{i} - \frac{1}{T_2}\right) m_\pm(t) + R_\pm(t) \qquad (12.65b)$$

where $\omega = \gamma|\mathbf{B}|$ is the *Larmor precession frequency.* Now we can easily derive the following fluctuation formulas for the relaxation times:

$$T_1 = \int_0^\infty dt \frac{\langle m_0(t) m_0(0) \rangle}{\langle m_0^2 \rangle} \qquad (12.66a)$$

$$T_2 = \int_0^\infty dt \frac{\langle m_+(t) m_-(0) \rangle e^{-i\omega t}}{\langle m_+ m_- \rangle} \qquad (12.66b)$$

$$T_2 = \int_0^\infty dt \frac{\langle m_-(t) m_+(0) \rangle e^{i\omega t}}{\langle m_+ m_- \rangle} \qquad (12.66c)$$

In the last two equations are examples of time correlation functions for complex quantities, as in Eq. (11.27a). Equations (12.66) are to be contrasted with the following, also derivable from Eqs. (12.65), which illustrate the roles of T_1 and T_2 as transport coefficients

$$\langle m_0(t) m_0(0) \rangle = \langle m_0^2 \rangle e^{-t/T_1} \qquad (12.67a)$$

$$\langle m_+(t) m_-(0) \rangle = \langle m_+ m_- \rangle e^{-t(1/T_2 - i\omega)} \qquad (12.67b)$$

since they show how nonequilibrium states relax.

DO PROBLEMS 12.8 TO 12.12.

12.8 CHEMICAL RATE CONSTANT

The rate of a chemical reaction is often discussed in terms of a diagram showing the free energy as a function of reaction coordinate as in Fig. 12.4a. In the well-

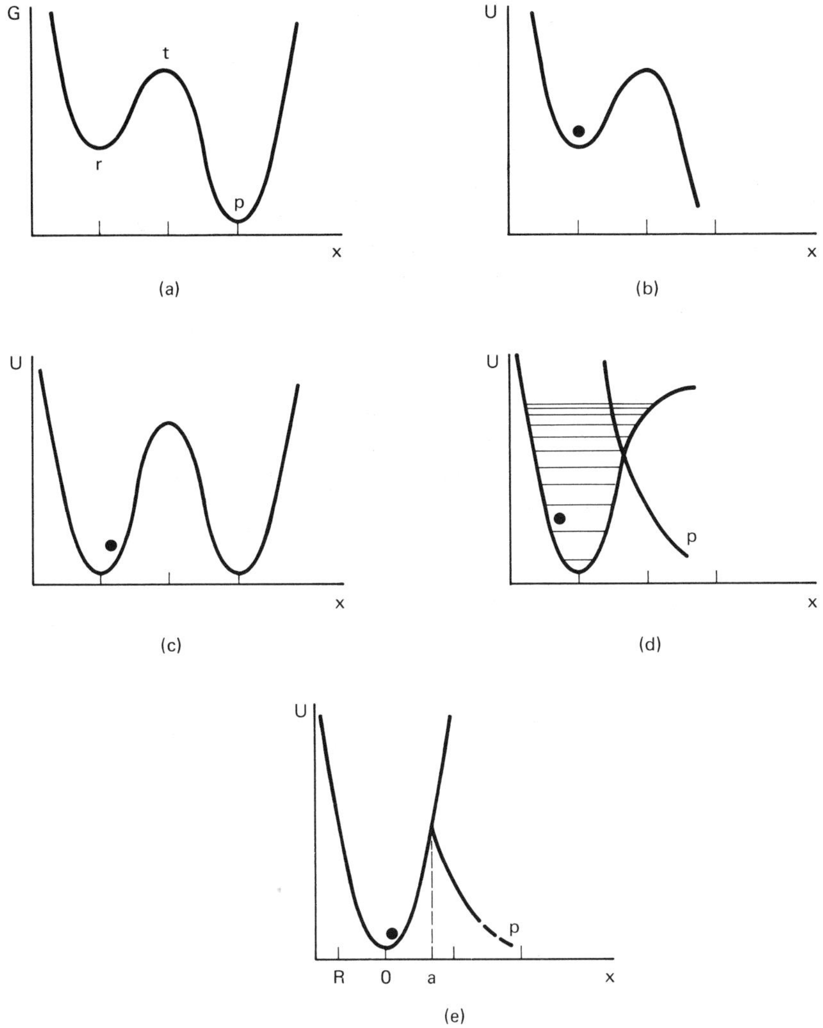

Figure 12.4 Models for chemical reaction.

(a) Usual free-energy diagram of transition-state theory showing free energy G as a function of reaction coordinate x. The reactant state is at r, the transition state at t, and the product at p.

(b) Model introduced by Kramers (1940). The potential is parabolic near each extremum. A mass point (black circle) governed by Langevin's equation represents the state of the reacting system.

(c) Symmetrical version of model (b); model for an isomerization.

(d) A quantum case. The stationary energy levels are shown for the oscillator in the absense of the product state p, and with no coupling to the heat bath.

(e) Model analyzed here. When the mass point is in the well and near $x = a$ it has some probability to pass over to the p state.

known transition state theory (also called the absolute rate theory or the activated complex theory) one focuses on the thermodynamic aspects of the barrier-crossing problem. There are a number of factors, some quantum mechanical, like tunneling through the barrier; and some classical, like the detailed description of the reaction coordinate as a function of many degrees of freedom, which also need to be treated carefully to extend the transition-state theory to allow an accurate description of more interesting chemical reactions. In the case of chemical reactions in solution, one of these additional factors concerns the interaction of the chemical system with a heat reservoir in the barrier-crossing process related to Fig. 12.4a. In terms of the prototype of a chemical reaction,

$$A + B \longrightarrow C + D$$

if there is an energy barrier the requirement that particles A and B have more than typical thermal energies means that the reaction depends on a fluctuation in which the rest of the system, the "heat bath," transfers energy to A and B. Does the process by which the reacting subsystem increases in energy as it climbs the barrier have any other effect than providing a thermalized energy distribution (i.e., a Boltzmann distribution at the temperature T of the heat reservoir) at each value of the reaction coordinate? This factor was first addressed by Kramers (1940) in terms of a mechanical model shown in Fig. 12.4b, where the black circle represents a Brownian particle moving on the potential surface as in Eq. (12.52), but with a more general force "constant" $K(x)$. Here we treat a model of this kind (Fig. 12.4e) using the theory of mean first passage times in a version due to Deutch (1980).

If we have an ensemble of systems like any of those in Fig. 12.4, prepared so that at $t_0 = 0$ all of the brownons are to the left of the barrier, then the fraction of systems $N(t)$ that are unreacted (i.e., have not yet crossed the barrier) at a later time t might possibly have the form

$$N(t) = e^{-t/\tau_e} \tag{12.68}$$

in which case τ_e is the mean first passage time ("mean" because it is determined by the ensemble average). Also, $1/\tau_e$ coincides with the usual definition of the rate constant for this first-order process. Alternatively, even if $N(t)$ is not exponential, we define

$$\tau_{\text{mfpt}} = \int_0^\infty N(t)\, dt \tag{12.69a}$$

Thus τ_{mfpt} is the measure of the decay of unreacted systems that coincides with τ_e when the decay is exponential. Moreover, $k \equiv 1/\tau_{\text{mfpt}}$ may serve as a satisfactory definition of the rate constant even when Eq. (12.68) is not valid, in which case the conventional rate constant, defined in terms of the differential rate of reaction, becomes time dependent.

In terms of a system like one of those in Fig. 12.4 we may write

$$N(t) = \int_{-\infty}^{a} P(x, t)\, dx$$

and

$$\tau_{\mathrm{mfpt}} = \int_0^\infty dt \int_{-\infty}^{a} P(x, t)\, dx \tag{12.69b}$$

where $P(x, t)\, dx$ is the fraction of systems in the ensemble that are in the range between x and $x + dx$ at t and that have not departed from $x < a$ since $t = 0$.

We consider a model in which $P(x, t)$ obeys a Smoluchowski equation

$$\dot{P}(x, t) = -\nabla j(x, t) \tag{12.70}$$

where $\nabla = \partial/\partial x$ and the current field is

$$j(x, t) = -D(x)[\nabla P(x, t) + P(x, t)\beta\nabla U(x)] \tag{12.71}$$

These equations together correspond to Eqs. (12.29) and (12.30). We aim to calculate $P(x, t)$ and τ_{mfpt} for the system in Fig. 12.4e with the boundary conditions

$$j(R, t) = 0 \tag{12.72a}$$

$$\begin{aligned} j(a, t) &= 0 && \text{if } t \le 0 \\ &= \kappa P(a, t) && \text{if } t > 0 \end{aligned} \tag{12.72b}$$

$$\begin{aligned} P(x, t) &= P_0(x) && \text{if } t \le 0 \\ P_0(x) &\equiv \frac{e^{-\beta U(x)}}{\displaystyle\int_R^a e^{-\beta U(x)}\, dx} \end{aligned} \tag{12.72c}$$

Thus $P_0(x)$ is the equilibrium distribution function when there is an impenetrable barrier at $x = a$ as well as at $x = R$. Equation (12.72b) is a *radiation boundary condition* in which $j(a, t)$ is the particle current due to systems passing through the $x = a$ barrier in the direction of the forward reaction (i.e., from $x < a$ to $x > a$). It plays the role of the rate of transformation of transition-state systems ($x \simeq a$) to products in the transition-state theory. Now we seek to find how it contributes to the time dependence of $P(x, t)$ once we start the reaction at $t = 0$.

To facilitate the analysis we introduce the modified functions

$$P'(x, t) \equiv \frac{P(x, t)}{P_0(x)} \tag{12.73a}$$

$$j'(x, t) \equiv \frac{j(x, t)}{P_0(x)} \tag{12.73b}$$

in terms of which the flux equation (12.71) reduces to

$$j'(x, t) = -D(x)\nabla P'(x, t) \tag{12.74}$$

At any time j and p' are continuous functions in the $R \le x \le a$ range:

$$\begin{aligned} j(x, t) &= j(R, t) + \int_R^x \nabla j(u, t)\, du \\ &= 0 - \int_R^x \dot{P}(u, t)\, du \end{aligned} \tag{12.75a}$$

$$\begin{aligned} P'(x, t) &= P'(a, t) + \int_a^x du\, \nabla P'(u, t) \\ &= \frac{j'(a, t)}{\kappa} + \int_x^a du \frac{1}{D(u)} j'(u, t) \end{aligned} \tag{12.75b}$$

Elimination of j' from Eq. (12.75b) with the help of (12.73b) and (12.75a) gives

$$P(x, t) = P_0(x) \left\{ -\frac{1}{\kappa P_0(a)} \int_R^a \dot{P}(u, t)\, du - \int_x^a du \frac{1}{D(u) P_0(u)} \int_R^u dy\, \dot{P}(y, t) \right\} \tag{12.76}$$

Finally, we set $R = -\infty$ and substitute Eq. (12.76) in Eq. (12.69b) to get the result

$$\tau_{\text{mfpt}} = \frac{1}{\kappa P_0(a)} + \int_{-\infty}^a \frac{dx}{D(x) P_0(x)} \left[\int_{-\infty}^x P_0(y)\, dy \right]^2 \tag{12.77}$$

In a model with the potential

$$U(x) = \tfrac{1}{2} K x^2 \tag{12.78}$$

with a constant diffusion coefficient $D(x) = D$ and with a large barrier height $E^\ddagger$,

$$\begin{aligned} \beta U(a) &= \tfrac{1}{2} \beta K a^2 \\ &\equiv \beta E^\ddagger \gg 1 \end{aligned} \tag{12.79}$$

the remaining integrals in (12.77) can be evaluated asymptotically, giving

$$\tau_{\text{mfpt}} = \left[\frac{(\pi \langle x^2 \rangle / 2)^{1/2}}{\kappa} + \frac{\sqrt{\pi}\, \tau_x}{(\beta E^\ddagger)^{1/2}} \right] e^{\beta E^\ddagger} + \cdots \tag{12.80}$$

where the omitted terms grow with $\beta E^\ddagger$ more slowly than those shown. The rate constant is

$$\frac{1}{\tau_{\text{mfpt}}} = k = \frac{e^{-\beta E^\ddagger}}{\pi^{1/2} [\tau_a + \tau_x (\beta E^\ddagger)^{-1/2}]} \tag{12.81}$$

where

$$\tau_a \equiv \frac{(\frac{1}{2} \langle x^2 \rangle)^{1/2}}{\kappa} \tag{12.82}$$

is the mean time for a particle with velocity κ to travel the distance $\frac{1}{2} \langle x^2 \rangle^{1/2}$. To obtain these results we use the expressions for $\langle x^2 \rangle$ and τ_x given in Eqs. (12.60) and (12.61).

In the limit that τ_x / τ_a is sufficiently small, we recover the transition-state theory for this model,

$$k = \frac{e^{-\beta E^\ddagger}}{\pi^{1/2} \tau_a} \tag{12.83}$$

as expected; in this case the depletion of the ratio $P(a, t)/P(0, t)$ due to reaction is not significant. The interaction of the brownon with the heat bath is not so strong that slow motion of the brownon is controlling the rate constant k.

Now if we turn up the interaction of the brownon with the heat bath, for example by increasing the strength of the interaction between the brownon and the degrees of freedom of the heat bath or by increasing the viscosity of the heat bath, we may reach a regime in which τ_x/τ_a is large enough so that $P(a, t)/P(0, t)$ is affected. In this range the size of k is controlled by the diffusion of the brownon as well as by the barrier height. The $\sqrt{\beta E^\ddagger}$ factor that comes into play in this region is also given by some other model calculations. It may be thought of as defining the effective rate constant $1/\tau_{\text{eff}}$ in a chemical kinetics model for the same process

$$\text{reactant} \xrightarrow[1/\tau_{\text{eff}}]{} \text{intermediate} \xrightarrow[1/\tau_a]{} \text{product}$$

that is,

$$\tau_{\text{eff}} = \tau_x[\beta E^\ddagger]^{-1/2}$$

There also is a regime in which the interaction of the brownon with the heat bath is very weak. Thus it is clear that if this interaction vanishes, there is no mechanism for the brownon to get the energy needed to go from $x = 0$ to a. Thus we expect in general that the rate constant will depend on the coupling of the brownon to the heat bath as indicated in Fig. 12.5, first deduced for the model in Fig. 12.4b by Kramers (1940).

The part of Fig. 12.5 with positive slope reflects the slow diffusion of energy to the brownon from the heat bath. The faster part of the energy diffusion involves the kinetic energy diffusion, which is closely related to velocity relaxation with the correlation time τ appearing in Eqs. (12.1) and (12.52). However, as explained in connection with Eq. (12.33), Smoluchowski's equation is not consistent with Langevin's equation for changes in time intervals as short as τ. Consequently, the theory described here is not competent to yield the part of Fig. 12.5 with positive slope.

For the model in Fig. (12.4b) with the curvature

$$\left|\frac{\partial^2 U}{\partial x^2}\right| = \omega^2 m \tag{12.84}$$

in the neighborhood of each extremum, Skinner and Wolynes (1978) derive the rate constant

$$k \simeq \frac{e^{-\beta E^\ddagger}}{\tau(1 + \pi/\omega\tau + 2\pi/\omega^2\tau^2)} \tag{12.85}$$

from an analysis of the large τ and small τ behavior together with an empirically justified interpolation. It is actually this function that is plotted in Fig. 12.5.

Another important contribution is the investigation of a model that is the same but with a non-Langevin equation for the interaction of the brownon with the heat bath. In the alternative model the interaction is by randomly timed collisions that

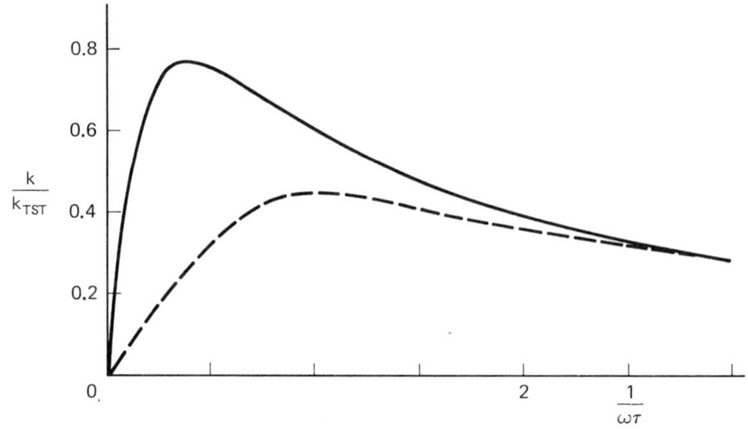

Figure 12.5 Relative rate constant k/k_{TST} for excape over the barrier given the potential function in Fig. 12.4b. Here k is given by Eq. (12.85) and $k_{TST} = (\omega/2\pi)$ exp $(-\beta E^{\ddagger})$ is the transition-state theory approximation for this model. Here $1/\omega\tau$ is a dimensionless measure of the friction (i.e., the coupling to the heat bath). Solid line, Eq. (12.85) (Skinner and Wolynes, 1978); dashed line, BKG impulsive model (Berne et al., 1980).

change the brownon's momentum by finite impulses. The result for a finite impulse model lies below the Langevin equation model except at a large coupling, as shown in Fig. 12.5. The conclusion is that it is reasonable to expect that some chemical reactions in solution will exhibit significant derivations from transition-state theory due to the motion along the reaction coordinate being too weakly or too tightly coupled to the remaining degrees of freedom (i.e., to the motions of the heat bath).

DO PROBLEMS 12.13 TO 12.15.

THE LITERATURE

The mathematical theory of Langevin's equation is developed by Chandrasekhar (1943) in a paper which is reprinted with additional important contributions by Wang and Uhlenbeck (1945) and others in a volume edited by Wax (1954). For more information on some of the applications, see Debye (1929), Abragam (1961), and Steele (1976) for rotational Brownian motion; Kubo (1969) for the noisy oscillator; and Deutch and Oppenheim (1968) for the application to magnetic resonance.

The early work by Kramers (1940) on the statistical mechanical theory of the chemical rate constant has been supplemented by mean first passage time methods (Szabo et al., 1980, 1981; Deutch, 1980), by studies with more general coupling (BKG model) of the particle to the heat bath (Berne et al., 1980), and by application to various kinds of chemical reactions (Grote and Hynes, 1982).

PROBLEMS

12.1. List the last two digits of each of the first 20 telephone numbers on page 52 of the local telephone book.

(a) For this sequence of numbers calculate

$$\langle N \rangle,\ \langle N^2 \rangle,\ \langle \delta N_j \delta N_{j+1} \rangle$$

where $\delta N \equiv N - \langle N \rangle$ and $j = 0, 1, 2, \ldots, 19$ is the index.

(b) Compare with what you expect if the numbers are random integers on the interval 0 to 99.

12.2. Equation (12.1d) specifies merely that $\langle R(t)R \rangle$ is proportional to $\delta(t)$. To learn more we consider a more general sort of random process governed by Eqs. (12.1b) to (12.1e) together with the "generalized Langevin equation" (see Chapter 14)

$$\dot{p}(t) = -\int_0^t M(t-t_1)p(t_1)\,dt_1 + R(t) \tag{12.86}$$

where $p(t) \equiv m\dot{x}(t)$.

(a) Show that Eq. (12.86) implies the following equations [see Eq. (11.19)].

$$-\frac{\partial^2}{\partial t^2}\langle p(t)p \rangle = \int_0^t M(t-t_1)\,dt_1 \frac{\partial}{\partial t_1}\langle p(t_1)p \rangle + \langle R(t)R \rangle \tag{12.87a}$$

$$\frac{\partial}{\partial t}\langle p(t)p \rangle = -\int_0^t M(t_1)\langle p(t-t_1)p \rangle\,dt_1 \tag{12.87b}$$

$$\frac{\partial^2}{\partial t^2}\langle p(t)p \rangle = -M(t)\langle p^2 \rangle - \int_0^t M(t_1)\,dt_1 \frac{\partial}{\partial t}\langle p(t-t_1)p \rangle \tag{12.87c}$$

(b) Eliminate $(\partial^2/\partial t^2)\langle p(t)p \rangle$ between Eqs. (12.87a) and (12.87c) to get the following important result:

$$\langle R(t)R \rangle = \langle p^2 \rangle M(t) \tag{12.88}$$

which is called the second fluctuation-dissipation theorem. It is discussed further in Chapter 14.

12.3. **(a)** Show that if the memory function $M(t)$ in Problem 12.2 has the special form

$$M(t) = \frac{\delta(t)}{\tau} \tag{12.89a}$$

then Langevin's equation is recovered from Eq. (12.86).

(b) Show that in this case the second fluctuation-dissipation theorem reads

$$\begin{aligned}\langle R(t)R \rangle &= \frac{\langle p^2 \rangle}{\tau}\delta(t)\\ &= \frac{k_B T m}{\tau}\delta(t)\end{aligned} \tag{12.89b}$$

12.4. **(a)** Do the differentiation in Eq. (12.11) to see whether you get the result given there.

(b) Verify the last equality in Eq. (12.12).

12.5. Derive the response equation that is related to Eq. (12.18) as Eq. (12.9) is related to Eq. (12.8). How is the response to $K \exp(i\omega t)$ related to $\langle \dot{x}(t)\dot{x} \rangle$ [see Eq. (11.36)]?

12.6. Eqs. (11.10a) and (12.29) are often written as, respectively,

$$\dot{c}_s(\mathbf{R}) = -\nabla_R \cdot \mathbf{j}(\mathbf{R}) \tag{12.90}$$

$$\dot{f}(q, t) = -\frac{\partial}{\partial q} j(q, t) \tag{12.91}$$

(a) Account for the difference in the two current density functions $\mathbf{j}(\mathbf{R})$ and $j(q, t)$.
(b) What are the corresponding current densities in Eq. (11.10b)?
(c) The Fokker–Planck equation (12.25) may be similarly written:

$$\dot{f}(p; t) = -\frac{\partial}{\partial p} j(p; t) \tag{12.92}$$

where $j(p, t)$ is the current density in momentum space. Explain each term in $j(p; t)$. Show that

$$D_p = \frac{k_B T m}{\tau} \tag{12.93}$$

plays the role of diffusion coefficient in momentum space.

12.7. **(a)** Show that $P_q\left(q \xleftarrow{t} q_0\right)$ as generally defined by the integral in Eq. (12.31) is just the self Van Hove correlation function $G^s(q - q_0, t)$.
(b) Show that for displacement in three dimensions, with

$$r = [(x - x_0)^2 + (y - y_0)^2 + (z - z_0)^2]^{1/2} \tag{12.94}$$

as usual, the Langevin equation leads to

$$G^s(\mathbf{r}, t) = \frac{\exp(-r^2/E)}{(\pi E)^{3/2}} \tag{12.95}$$

where E is given in Eq. (12.32).
(c) What is the form of $G^s(\mathbf{r}, t)$ when $t \gg \tau$?
(d) Show that Langevin's equation leads to

$$F^s(\mathbf{k}, t) = e^{-k^2 E/4} \underset{t \gg \tau}{=\!=} e^{-k^2 Dt} \tag{12.96}$$

(e) Show that Langevin's equation leads to

$$S^s(\mathbf{k}, \omega) = \frac{Dk^2}{\pi[\omega^2 + (Dk^2)^2]} \tag{12.97}$$

12.8. For a spherical brownon with a dipole $\mathbf{M}$ embedded at its center the dipole autocorrelation function is, according to Eq. (12.49),

$$\langle \mathbf{M}(t) \cdot \mathbf{M}^* \rangle = \frac{1}{4\pi} (M_1^0)^2 e^{-t/\tau_1} \tag{12.98}$$

where M_1^0 is the magnitude of the dipole moment. The normalized time correlation function is thus

$$C_{\mathbf{MM}}(t) = e^{-t/\tau_1} \tag{12.99}$$

Show that the corresponding result in the general case (i.e., Langevin's equation not assumed) is

$$C_{\mathbf{MM}}(t) = \langle \cos \theta(t) \rangle \tag{12.100}$$

where $\theta(t)$ is the angle between $\mathbf{M}(t)$ and $\mathbf{M}(0)$.

12.9. Show that for an ideal gas of spherical dipolar molecules, $C_{\mathbf{MM}}(t)$ is a *damped* function as a result of averaging over initial rotational velocities.

12.10. **(a)** Using the representation of $\delta(\Omega)$ in Eq. (12.37), prove

$$\int d\Omega\, \delta(\Omega - \Omega')\, Y_j^m(\Omega) = Y_j^m(\Omega') \tag{12.101}$$

(b) Compare Eq. (12.37) with Eq. (3.18) written in the form

$$\delta(x - x') = \frac{1}{2\pi} \int dk\, e^{ikx}(e^{ikx'})^* \tag{12.102}$$

Can you explain why one delta function is represented as a discrete sum while the other is an integral over continuous functions?

12.11. **(a)** Show that for a Lorentzian line shape function, as in (12.59), the nth moment violates a sum rule if n is even and larger than zero.

(b) What is the unrealistic aspect of Eq. (12.57) that is associated with these sum rule violations?

12.12. The imaginary part of Eq. (12.58), called the *dispersion,* is directly observable with some experimental arrangements. Graph both the dispersion and the power spectrum as functions of $z\tau$ for $\omega\tau = \frac{1}{3}$, 1, and 3.

12.13. Use Eqs. (12.31) and (12.32) in the $K = 0$ case to calculate the mean-square displacement in an interval t. Does your result agree with (12.12)? Explain why it should.

12.14. Use Eqs. (12.30) to (12.32) in the $K = 0$ case to calculate $\dot{P}_q$ and $\hat{S}P_q$ explicitly. Show that these quantities obey Eq. (12.33) if and only if $t \gg \tau$.

12.15. If we begin with a number n of brownons very near q_0 at time $t = 0$, the concentration expected at location $q + q_0$ at time t is, in view of Eq. (12.31),

$$\rho(q, t) = nP_q(t \leftarrow 0) \tag{12.103}$$

Show that Smoluchowski's equation gives, for the rate of change of this local concentration,

$$\frac{\partial \rho(q, t)}{\partial t} = -D \frac{\partial^2 \rho(q, t)}{\partial q^2} \tag{12.104}$$

which is also known as *Fick's second law of diffusion.* This problem has been posed for one-dimensional diffusion. Formulate the problem for three dimensions and solve it again, assuming an isotropic system.

12.16. Show how the passage from Eq. (12.77) to (12.79) can be justified by applying the following mathematical relations.

$$\int_{-\infty}^{0} dx\, e^{-x^2} = \int_{0}^{\infty} dx\, e^{-x^2} = \frac{\sqrt{\pi}}{2}$$

$$\int_{y}^{\infty} dx\, e^{-x^2} = -\frac{e^{-x^2}}{2x}\bigg|_{y}^{\infty} - \int_{y}^{\infty} dx \frac{e^{-x^2}}{2x}$$

$$\int_{1}^{y} dx\, e^{x^2} = \frac{e^{x^2}}{2x}\bigg|_{1}^{y} - \int_{1}^{y} dx \frac{e^{x^2}}{2x}$$

13

Simulation and Inelastic Neutron Scattering

13.1 INTRODUCTION

Chapter 5 is concerned largely with simulation methods for calculating thermodynamic coefficients and equilibrium structural functions of Hamiltonian model systems. It also describes the scattering of radiation as an experimental method for determining equilibrium structural functions of real systems. Here both of these themes are extended into the time-dependent regime. In doing so we tend to lose some tools that are powerful for equilibrium properties but which in the time-dependent regime are useful only for the study of slowest processes. Monte Carlo simulation and x-ray scattering are examples, while molecular dynamics (MD) simulation and inelastic neutron scattering (INS) experiments take over as the most general and powerful techniques.

13.2 MOLECULAR DYNAMICS SIMULATION OF RATE PROCESSES

The basic procedure for MD simulation, as described in Chapter 5, leads to a set of states along a trajectory that represents the natural motion of the system in phase space. [Even with algorithms in which only the locations of the particles are stored, the momenta can be recovered trivially as long as the step interval $\hat{t}$ is small compared with the velocity relaxation time; see the discussion of Eq. (5.5).] Given these states along a trajectory representing fluctuations at equilibrium, we can calculate transport coefficients by applying the linear response theory.

The simplest example concerns the self-diffusion coefficient D_s for species s given by Eq. (11.69a). To evaluate $\langle \dot{x}_s(t)\dot{x}_s \rangle$ we notice that the velocity autocorrelation function is the same for all particles of species s and, in an isotropic system, for motion along the x, y, and z Cartesian coordinates, so we have

$$\langle \dot{x}_s(t+t_0)\dot{x}_s(t_0)\rangle \simeq \frac{1}{3N_s}\sum_{j=1_s}^{N_s}\sum_{\alpha=x,y,z} u_{j\alpha}(t+t_0)u_{j\alpha}(t_0) \tag{13.1}$$

where $u_{j\alpha}$ is the α component of the velocity of particle j and N_s is the number of particles of species s in the basic cell. Thus we pick a state i on the trajectory, label it as time t_0, and calculate the average in Eq. (13.1) for a chosen t. The average can be improved by performing the time average over values of t_0, since in a stationary system Eq. (13.1) does not depend on t_0:

$$\begin{aligned}\langle \dot{x}_s(t)\dot{x}_s\rangle &= \frac{1}{t'}\int_0^{t'} \langle \dot{x}_s(t+t_0)\dot{x}_s(t_0)\rangle\, dt_0 \\ &\simeq \frac{1}{3N_s\nu}\sum_{i=1}^{\nu}\sum_{j=1_s}^{N_s}\sum_{\alpha} u_{j\alpha}[(i+k)\hat{t}]u_{j\alpha}(i\hat{t})\end{aligned} \tag{13.2}$$

where $k\hat{t} = t$. If the ν contributions for various t_0 were independent, the improvement in the average would be to reduce the noise by a $\nu^{-1/2}$ factor. However, since $\hat{t}$ is much smaller than the velocity autocorrelation time in a satisfactory MD simulation, we can be sure the contributions in Eq. (13.2) from successive steps $i = 1, 2, 3, \ldots$ are highly correlated. Therefore, an equally good time average can be constructed more cheaply by summing only over the contribution of every lth step, where l is chosen so that $l\hat{t}$ is comparable to the velocity relaxation time $D_s/\langle \dot{x}_s^2\rangle = D_s m_s/k_BT$. Examples of velocity autocorrelation functions obtained by simulation are shown in Figs. 11.1 and 13.1. To obtain D_s from the simulations, one may accumulate $\langle \dot{x}_s(t)\dot{x}_s\rangle$ for enough t values to enable the integral in Eq. (11.69a) to be evaluated. This procedure implies a significant computational effort.

In a simulation, as in a laboratory experiment, we must choose among various ways to determine the coefficient of interest. Thus in the present case we also may extract D_s from mean-square displacements of the s particles, thus exploiting Eq. (11.80). A third method, via Eq. (11.71c), might be attractive if we had other reasons to extract the self-dynamical structure factor from a simulation, say for direct comparison with INS data for a real system. A fourth way to obtain D_s may be derived as follows (Futrelle and McGinty, 1971).

We compute the Fourier–Laplace transform of the velocity autocorrelation function, averaged over starting time t_0, as

$$\begin{aligned}\phi_s(z) &\equiv \int_0^{\infty} dt\, e^{-izt}\langle \dot{x}_s(t)\dot{x}_s\rangle \\ &= \frac{1}{t'}\left\langle \left\{\int_0^{t'} dt_0\, \dot{x}_s(t_0)e^{izt_0}\right\}\left\{\int_0^{\infty} dt\, \dot{x}_s(t+t_0)e^{-iz(t+t_0)}\right\}\right\rangle \\ &= \frac{1}{t'}\left\langle \left\{\int_0^{t'} dt_0\, \dot{x}_s(t_0)e^{izt_0}\right\}\left\{\int_0^{t'} dt_1\, \dot{x}_s(t_1)e^{-izt_1}\right\}\right\rangle\end{aligned} \tag{13.3}$$

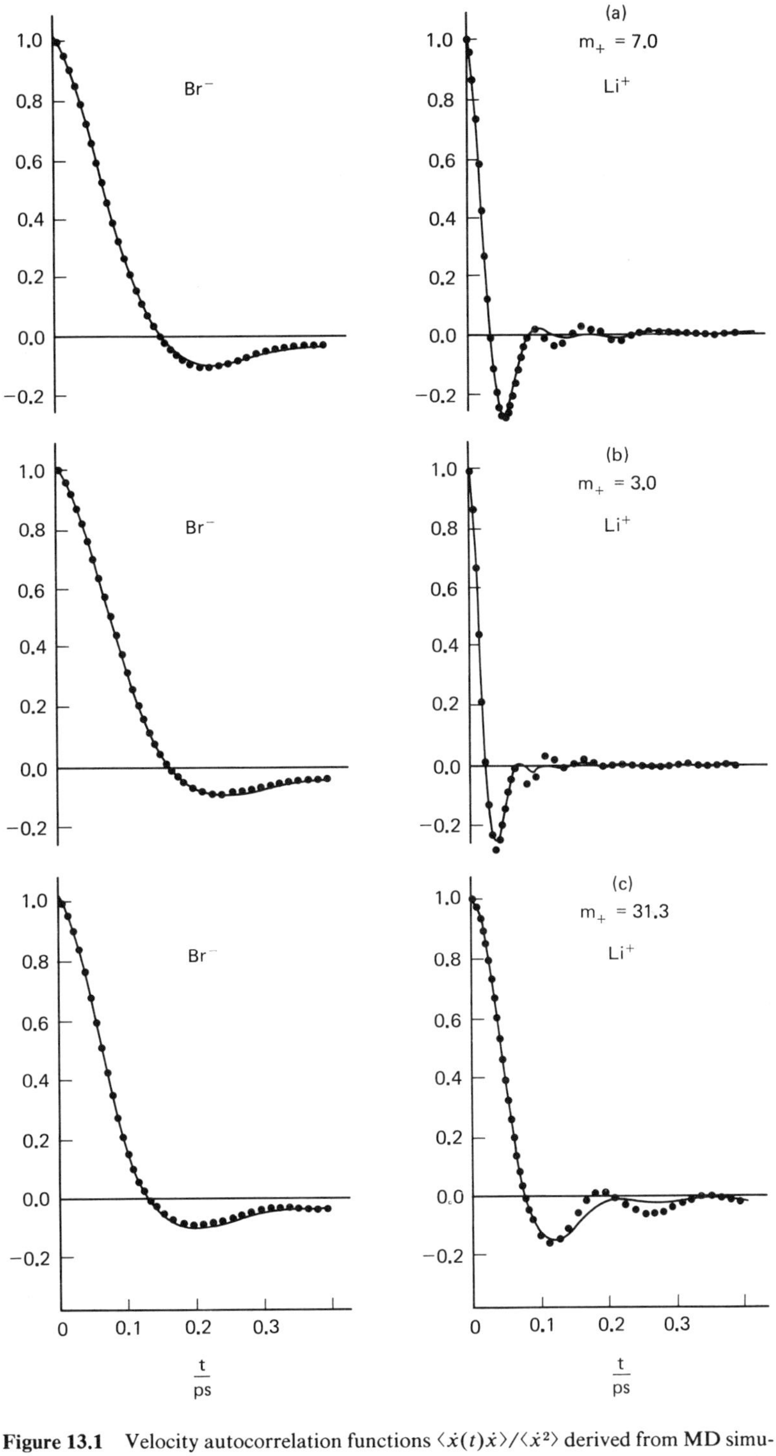

Figure 13.1 Velocity autocorrelation functions $\langle \dot{x}(t)\dot{x}\rangle/\langle \dot{x}^2\rangle$ derived from MD simulations of charged soft-sphere models for liquid LiBr at 1000 K and showing the dependence on the mass of the model cation. The smooth curves are calculated using the memory function model given below in Eq. (14.30a). (From Lantelme et al., 1979.)

where we assume that t' is arbitrarily large and $\dot{x}_s(t)$ is stationary so the integral $\int_0^\infty dt\, f(t + t_0)$ can be replaced by $\int_{-t_0}^\infty dt\, f(t + t_0) = \int_0^\infty dt_1\, f(t_1)$ to obtain the last equality. From this point we proceed as in deriving Eq. (13.2) to obtain

$$\begin{aligned}\phi_s(z) &\simeq \frac{\hat{t}}{3\nu N_s} \sum_{\alpha=x,y,z} \sum_{j=1}^{N_s} \left[\sum_{k=1}^{\nu} \dot{\alpha}_j(k\hat{t}) e^{izk\hat{t}} \right] \left[\sum_{k=1}^{\nu} \dot{\alpha}_j(k\hat{t}) e^{-izk\hat{t}} \right] \\ &= \frac{\hat{t}}{3\nu N_s} \mathbf{A}^*(z) \cdot \mathbf{A}(z)\end{aligned} \tag{13.4}$$

where the approximation is better the larger N_s and ν and where $\mathbf{A}(z)$ is a vector with $3N_s$ elements,

$$A_{j\alpha}(z) = \sum_{k=1}^{\nu} \dot{\alpha}_j(k\hat{t}) e^{-izk\hat{t}} \tag{13.5}$$

During a simulation the vector $\mathbf{A}(z)$ can be accumulated for each z of interest and by Eq. (13.4) trial values of $\phi_s(z)$ can be calculated from time to time to enter on the control chart (Chapter 5) to test convergence.

Since $D_s = \phi_s(0)$, it is interesting to compare Eq. (13.4) in the $z = 0$ limit with Eq. (13.3) in this limit. Do you see the relation?

Still another method for calculating D_s, due to Ciccotti and Jacucci (1975), does not follow so directly from Eq. (11.69a), which depends on fluctuations at equilibrium, as from Eqs. (11.82) and (11.83), which depend on response to an external field. While the self-diffusion coefficient might be derived from Eqs. (11.82) and (11.83) by simulating the motion of a system in which one particle is subjected to an external force F_x, it usually turns out that when F_x is big enough so that the response $\langle \dot{x}(t) \rangle_{F_x}$ [see Eq. (11.83)] can be calculated with significant accuracy, we are outside the regime of linear response. To measure the effect of an external force within the linear response regime, we may proceed as follows.

A trajectory derived from MD simulation of a system at equilibrium under the Hamiltonian H_0 is divided into ν segments, each with M consecutive moves. Then for each segment a modified trajectory is generated as follows. The first phase point (the first state) is the same as the first phase point of the unmodified segment. The rest are generated from the equations of motion under the modified Hamiltonian [see Eq. (11.29)]

$$H_t = H_0 - AK(t) \tag{13.6}$$

where A is (say) the x coordinate of a particle of species s and $K(t)$ is the force acting on the distinguished particle. For simplicity we consider only the case in which $K(t)$ is constant for $t \geqslant t_0$, so it may be identified with F_x in the discussion above. We now apply Eq. (11.33) by identifying the response variable B as $\dot{x}_s$, the velocity of the distinguished particle, and

$$\langle \Delta \dot{x}_s(t) \rangle_K = \frac{1}{\nu} \sum_{l=1}^{\nu} \{[\dot{x}_s(t)]_{H_t} - [\dot{x}_s(t)]_{H_0}\}_l$$

as the average over ν segments of the change in velocity due to the external force.

In the linear regime this response also is given by [see Eq. (11.35)]

$$\begin{aligned}\langle \Delta \dot{x}_s(t)\rangle_K &= \beta K \int_{t_0}^{t} \langle \dot{x}_s(t)\dot{x}_s(t_1)\rangle \, dt_1 \\ &= \beta K \int_0^{t-t_0} dt' \langle \dot{x}_s(t')\dot{x}_s\rangle\end{aligned} \tag{13.7}$$

The self-diffusion coefficient can be extracted in various ways. For example,

$$D_s = \lim_{t\to\infty} \frac{(\langle \Delta \dot{x}_s(t)\rangle_K)}{\beta K} \tag{13.8}$$

That the limit is finite is implied by the known behavior of the integral on the right-hand side of Eq. (13.7) at the $t = \infty$ limit. To extract the full velocity autocorrelation function we proceed as follows, using the notation in Eq. (13.3) and with $t_0 \equiv 0$:

$$\int_0^\infty dt \, [\langle \Delta \dot{x}_s(t)\rangle_K] e^{-izt} = \frac{\beta K \phi_s(z)}{iz} \tag{13.9}$$

provided that the segments are long enough so that

$$\int_0^{M\hat{t}} dt' \; \langle \dot{x}_s(t')\dot{x}_s\rangle \simeq \int_0^\infty dt' \langle \dot{x}_s(t')\dot{x}_s\rangle \tag{13.10}$$

is a good approximation. This condition implies that $M\hat{t}$ is perhaps five velocity correlation times $D_s/\langle \dot{x}^2\rangle = \beta D_s m_s$. The success of this method for determining D_s and $\phi_s(z)$ depends on having sufficient accuracy to distinguish $\langle \Delta \dot{x}_s(t)\rangle_K$ from noise even when t is several correlation times and K is small enough so that linear response is assured. It should be noticed that in the Ciccotti–Jacucci method, unlike the methods based on the fluctuation-dissipation theorem, we are not merely simulating a system at equilibrium; moreover, the particular nonequilibrium states that are generated depend on the transport coefficient being studied.

The methods of determining D_s from simulation data are merely typical of the methods that can be used to determine various transport coefficients and the corresponding time-correlation functions. Still other methods depend on the simulation of systems driven from equilibrium by an external field, as already mentioned. Even when linear response cannot be probed in this way, the method is useful for investigating nonlinear response.

A caveat that must be noted here has to do with the contribution of the "long-time tails" to time correlation functions. For example, it is now known that the ratio $\langle \dot{x}(t)\dot{x}\rangle/t^{3/2}$ approaches a constant as t increases. This remarkable effect, which is discussed in Problem 14.9 and illustrated in Fig. 14.5, was first discovered in a molecular dynamics simulation of a hard-sphere fluid (Alder and Wainwright, 1970). It may be expected that in some cases rather long simulations are necessary to capture the contributions of long-time tails.

DO PROBLEMS 13.1 TO 13.4.

13.3 BROWNIAN DYNAMICS

One frequently encounters systems in which there is a useful role for a simulation method that follows only some of the particles or some of the coordinates. As a simple example, consider simulation of the motion of a particle governed by Langevin's equation

$$m\ddot{x}(t) = -\frac{m\dot{x}(t)}{\tau} + F(x) + R(t) \tag{13.11}$$

where now we think of x as one coordinate of an N-particle system, say the x coordinate of particle 1, and $R(t)$ as the random force along x due to the interactions with all of the other particles of the system. Now $F(x)$ is an externally applied force. Even for such a simple equation of motion, the calculation of the time-correlation functions taxes the known analytical techniques in those cases in which $F(x)$ is complicated and velocity relaxation plays a controlling role, so the methods based on Smoluchowski's equation do not apply. An example may be found in the theory of the barrier-crossing rate constant (Section 12.8) in the weak-friction case. The term *Brownian dynamics* (BD) may be applied to simulations in the spirit of MD but applied to systems governed by stochastic equations of motion such as Langevin's equation.

BD simulation methods are applied to problems in which the MD method would involve impractically large computations, as for the calculation of the conductance of models for dilute electrolyte solutions or the magnetic resonance line shapes for electron spins or nuclear spins in solute particles. It is interesting to note that for the corresponding equilibrium problems one may combine the McMillan–Mayer theory (Chapter 10) with simulation by the MC method using the solvent-averaged equilibrium interaction potentials as input. If one also specifies the relevant solvent-averaged dynamics (i.e., friction coefficients for each solute particle alone in the solvent, hydrodynamic interaction for each pair of solute particles alone in the solvent, etc.), many of the dynamical properties of the MM-level model can be probed by BD simulation based on Smoluchowski's equation.

Since the BD move algorithms are somewhat more complicated than MD algorithms, we consider two interesting examples here. For a simulation based on Langevin's equation for one coordinate, we substitute Eq. (13.11) in the expression for the velocity change in a step of duration $\hat{t}$,

$$\dot{x}(t+\hat{t}) = \dot{x}(t) + \int_0^{\hat{t}} dt_1\, \ddot{x}(t+t_1) \tag{13.12}$$

to get

$$\dot{x}(t+\hat{t}) = \dot{x}(t) + \frac{\hat{t}F(t)}{m} - \frac{[x(t+\hat{t}) - x(t)]}{\tau} + B_1(t, \hat{t}) \tag{13.13}$$

where $F(t)$ means $F(x(t))$ and

$$B_1(t, \hat{t}) \equiv \int_0^{\hat{t}} dt \frac{R(t+t_1)}{m} \tag{13.14}$$

It is assumed that $\hat{t}$ is small enough so that the derivatives of $F(x(t))$ can be neglected (see Chapter 5). Since $R(t)$ has a Gaussian distribution [Eq. (12.1)], so does B_1; moreover, $\langle B_1 \rangle = 0$ and from Eqs. (12.89b) and (13.14),

$$\langle B_1^2 \rangle = 2\langle \dot{x}^2 \rangle \hat{t}/\tau \tag{13.15}$$

Equation (13.13) is awkward because it involves the step in x as well as in $\dot{x}$. However, in place of Eq, (13.13) we can as well obtain

$$\dot{x}(t+\hat{t}) - \dot{x}(t) = \hat{t}\left[\frac{F(t)}{m} - \frac{\dot{x}(t)}{\tau}\right] + B_1(t, \hat{t}) - \int_0^{\hat{t}} dt_1 \frac{\dot{x}(t+t_1) - \dot{x}(t)}{\tau} \tag{13.16}$$

which is an integral equation for the velocity step. Its iterative solution gives the velocity step,

$$\dot{x}(t+\hat{t}) - \dot{x}(t) = (1 - e^{-\hat{t}/\tau})\left[\frac{\tau F(t)}{m} - \dot{x}(t)\right] + B_v(t) \tag{13.17}$$

where we define

$$B_v(t) = \sum_{n=0}^{\infty} (-)^n \int_0^{\hat{t}} dt_1 \cdots \int_0^{t_{n-1}} dt_n\, R(t+t_n)/m \tag{13.18}$$

Since $R(t)$ is Gaussian, so is B_v; we have $\langle B_v \rangle = 0$ and we find that

$$\langle B_v^2 \rangle = \langle \dot{x}^2 \rangle (1 - e^{-2t/\tau}) \tag{13.19}$$

From

$$x(t+\hat{t}) = x(t) + \int_0^{\hat{t}} dt_1\, \dot{x}(t_1) \tag{13.20}$$

and Eq. (13.17), we find

$$x(t+\hat{t}) = x(t) - \dot{x}(t)\tau(1 - e^{-\hat{t}/\tau}) + \frac{(\hat{t}/\tau + e^{-\hat{t}/\tau} - 1)\tau^2 F(t)}{m} + B_x(t, \hat{t}) \tag{13.21}$$

$$B_x(t, \hat{t}) \equiv \int_0^{\hat{t}} dt_1 B_v(t, t_1) \tag{13.22}$$

with $\langle B_x \rangle = 0$ and

$$\langle B_x^2 \rangle = \langle \dot{x}^2 \rangle \tau^2 \left[\frac{2\hat{t}}{\tau} - (1 - e^{-\hat{t}/\tau})^2\right] \tag{13.23}$$

Of course, in a BD simulation based on Eqs. (13.17) and (13.21), the random terms $B_v(t)$ and $B_x(t)$ for given t are not independent since they derive from the same function $R(t + t_1)$ in $0 \leq t_1 \leq \hat{t}$. Indeed, for given t, $\hat{t}$ we have

$$B_x(t, \hat{t}) = B_v(t, \hat{t}) \left(\frac{\langle B_x^2 \rangle}{\langle B_v^2 \rangle}\right)^{1/2} \tag{13.24}$$

to calculate B_x from B_v using (13.19) and (13.23).

It seems straightforward to extend these BD algorithms to systems of interacting brownons. This extension will be illustrated in the example of BD to be described next. It is an algorithm for simulating the Smoluchowski equation (12.29).

At the Smoluchowski level of description, the time correlation function for some dynamical variables $A(q)$ and $B(q)$, which depend on the location q of the brownon, is

$$\langle A(t)B\rangle = \int dq\, Bf^{\text{eq}} e^{t\hat{S}^{\dagger}} A \tag{13.25}$$

where $\hat{S}^{\dagger}$ is the operator such that

$$\dot{A} = (\partial A/\partial q)\dot{q} = \hat{S}^{\dagger} A \tag{13.26}$$

Partial integration gives, if the integrand vanishes at the limits,

$$\langle A(t)B\rangle = \int dq\, A e^{t\hat{S}'}(f^{\text{eq}}B) \tag{13.27}$$

where $\hat{S}'$ is said to be the *adjoint* of operator $\hat{S}^{\dagger}$. Thus with reference to Eq. (11.17d), if $\langle AOB\rangle = \langle (O^{\dagger}A)B\rangle$, then $O^{\dagger}$ is said to be the adjoint of O. Indeed, Eq. (11.17d) shows that $-H^{\times}$ is the adjoint of $H^{\times}$, so for Hamiltonian mechanics we already found that

$$\text{if} \quad \dot{A} = H^{\times}A \quad \text{then} \quad \dot{f}_N = (H^{\times})^{\dagger} f_N \tag{13.28}$$

as seen by comparing Eqs. (11.8) and (11.11). For projected mechanics such as the Kramers, Fokker–Planck, or Smoluchowski mechanics, the relation between the Liouville operator and its adjoint is not so simple.

Now we notice that Eq. (13.27) has the form

$$\langle A(t)B\rangle = \int dq\, Af(q, t) \tag{13.29}$$

where $f(q, t)$ is like a Smoluchowski distribution function which results from the formal solution of Eq. (12.29):

$$\begin{aligned} f(q, t) &= e^{t\hat{S}} f(q, 0) \\ &= e^{t\hat{S}}(f^{\text{eq}}B) \end{aligned} \tag{13.30}$$

where f^{eq} is the equilibrium distribution function ($\hat{S}f^{\text{eq}} = 0$). Then in Eq. (13.27) we may identify $\hat{S}'$ with $\hat{S}$ as given in Eq. (12.30).

To calculate the explicit $\hat{S}^{\dagger}$ we write

$$\frac{\partial}{\partial t}(f^{\text{eq}}B) = f^{\text{eq}}\dot{B} \tag{13.31}$$

which with Eqs. (13.26) and (13.30) become

$$\hat{S}(f^{\text{eq}}B) = f^{\text{eq}}\hat{S}^{\dagger}B \tag{13.32}$$

Now doing the detailed calculation using Eq. (12.30) with $\nabla \equiv \partial/\partial q$ and $\beta D = \tau/m$, and the relations $\hat{S}f^{\text{eq}} = 0$ and $\nabla f^{\text{eq}} = \beta K f^{\text{eq}}$ we find

$$\begin{aligned}\hat{S}(f^{\text{eq}}B) &= f^{\text{eq}}(\nabla - \beta K)\nabla B + f^{\text{eq}}D\nabla^2 B + 2D(\nabla f^{\text{eq}})(\nabla B)\\ &= f^{\text{eq}}(\nabla + \beta K)\end{aligned} \tag{13.33}$$

so in view of Eq. (13.32) we have

$$\hat{S}^{\dagger} = (\nabla D)\nabla + D\nabla\nabla + \beta DK\nabla \tag{13.34}$$

Now using Eq. (13.26) with $A \equiv q$, we find that

$$\dot{q} = \nabla D + 0 + \beta DK \tag{13.35}$$

which must be supplemented by a random velocity term to describe the fluctuations at equilibrium. We first pass to the move algorithm,

$$q(t+\hat{t}) = q(t) + \hat{t}\,\nabla D(t) + \hat{t}\beta D(t)K(t) + Q(t, \hat{t}) \tag{13.36}$$

where $Q(t, \hat{t})$ is the random displacement in an interval $\hat{t}$ beginning at t. In many cases $\nabla D = 0$ and then we find that

$$\begin{aligned}\langle Q\rangle &= 0\\ \langle Q^2\rangle &= \langle[q(t+\hat{t}) - q(t)]^2\rangle_{K=0}\\ &= 2D\hat{t}\end{aligned} \tag{13.37}$$

in which case the move algorithm is completely specified if $Q(t, \hat{t})$ is a Gaussian random process consistent with Eq. (13.37).

This BD algorithm is readily extended to N interacting brownons in three-dimensional space using the well-known result (Ebeling, 1964, 1965; Ermak and McCammon, 1978) for the corresponding Smoluchowski equation (often called the *diffusion equation*)

$$\dot{f}_N(\mathbf{r}_1, \ldots, \mathbf{r}_N) - \hat{S}_N f_N = 0 \tag{13.38}$$

$$\hat{S}_N = \sum_{i,j} \nabla_i \cdot \mathbf{D}_{ij}(\mathbf{r}_i, \mathbf{r}_j) \cdot (\nabla_j - \beta \mathbf{F}_j) \tag{13.39}$$

where $\mathbf{F}_j = -\nabla_j \overline{U}_N + \mathbf{F}_j^{\text{ext}}$ is the sum of the interparticle force derived from the solvent-averaged (MM-level, Chapter 10) potential $\overline{U}_N$ plus the externally applied force, if any. In Eq. (13.39) the sum is over particle labels and $\mathbf{D}_{ij}$ is a diffusion tensor, often assumed to have the form

$$\mathbf{D}_{ij}(\mathbf{r}_i, \mathbf{r}_j) = \delta_{ij}\mathbf{1}D_i^s + \frac{1-\delta_{ij}}{8\pi\eta r_{ij}}(\mathbf{1} + \hat{r}_{ij}\hat{r}_{ij}) \tag{13.40}$$

where we write

$$\mathbf{r}_i - \mathbf{r}_j = r_{ij}\hat{r}_{ij} \tag{13.41}$$

where r_{ij} is the magnitude and $\hat{r}_{ij}$ the unit vector along $\mathbf{r}_i - \mathbf{r}_j$.

Alternative forms of the diffusion tensor are discussed by Ermak and McCammon (1978), Felderhof (1978), and Jones and Burfield (1982). The off-diagonal term shown, called the *Oseen tensor*, has the correct large-r_{ij} asymptotic form. In any case the diagonal terms $\mathbf{D}_{ii}$ with the self-diffusion coefficient D_i^s of the particle in the pure solvent, correspond to the friction term in Eq. (12.1), while the off-diagonal terms $\mathbf{D}_{ij}$ with $i \neq j$ express the effect of the motion of one particle (say i) which makes a flow pattern in the solvent medium which tends to move the other particle (say j). In deriving Eq. (13.40) the medium is assumed to be viscous (with shear viscosity coefficient η) and incompressible. Compressibility would allow a time lag between the motion of i and its effect on j.

By preceeding just as in the one-coordinate case, we find that the adjoint of $\hat{S}_N$ is

$$\hat{S}_N^\dagger = \sum_i \sum_j (\nabla_i \cdot \mathbf{D}_{ij} \cdot \nabla_j) + \beta[\mathbf{D}_{ij} \cdot \mathbf{F}_j] \cdot \nabla_i \tag{13.42}$$

and the corresponding move algorithm is

$$\mathbf{r}_i(t + \hat{t}) = \mathbf{r}_i(t) + \hat{t} \sum_j \nabla_j \cdot \mathbf{D}_{ij}(t) + \hat{t}\beta \sum_j \mathbf{D}_{ij}(t) \cdot \mathbf{F}_j(t) + \mathbf{Q}_i(t, \hat{t}) \tag{13.43}$$

where $\mathbf{D}_{ij}(t)$ means $\mathbf{D}_{ij}(\mathbf{r}_i(t), \mathbf{r}_j(t))$ and $\mathbf{F}_j(t)$ means $\mathbf{F}_j(\mathbf{r}_j(t))$. In the usual case that $\nabla_j \cdot \mathbf{D}_{ij} = 0$ and $\mathbf{Q}_j$ is assumed to be a Gaussian random variable, it is fully characterized by the averages

$$\langle \mathbf{Q}_j(t, \hat{t}) \rangle = 0$$

and (13.44)

$$\langle \mathbf{Q}_i(t, \hat{t}) \mathbf{Q}_j(t, \hat{t}) \rangle = 2\hat{t}\mathbf{D}_{ij}(t)$$

Evidently, in this case the Gaussian distribution from which the $\mathbf{Q}_i(t)$ are drawn changes in each step because the off-diagonal diffusion tensors $\mathbf{D}_{ij}$ depend on the configuration.

DO PROBLEMS 13.5 TO 13.9.

13.4 INELASTIC NEUTRON SCATTERING

In a material body in a beam of neutrons, the important mechanism for scattering the neutrons is through their collisions with atomic nuclei. In cases in which the neutron kinetic energy is low, not much more than $k_B \times 300$ K, these collisions are nearly elastic, relatively little energy being exchanged between neutrons and nuclei. With these conditions inelastic neutron scattering (INS) provides a most powerful experimental tool for the study of the molecular motions whose averages yield the time-relaxed correlation functions for nuclear motions, $\langle[\dot{x}(t) - \dot{x}(0)]^2\rangle$ for example. In this section the theory of the experiment is described at a level that is sufficient to show the relation of the scattering data to the van Hove dynamic structure factor $S_{ss'}(k, \omega)$ but without going into the detail that would be necessary to show how the theory is built up from the quantum-mechanical interaction of a neutron and a nucleus and how the classical $S_{ss'}(k, \omega)$ is obtained in a certain limit (Egelstaff, 1967, and references given therein).

In molecular fluids the collisions of neutrons with nuclei may excite the molecules in which the nuclei are located to higher internal states if the neutrons in the incoming beam are sufficiently energetic. This component of the scattering reflects the same processes as rotational–vibrational optical spectroscopy, but without the selection rules. The following analysis also holds in this range of neutron energies, the basic condition being that the neutron de Broglie wavelength is very large compared to the diameter of a nucleus ($\sim 10^{-5}$ Å). Then the effect of a nucleus on the neutron is like that of a hard sphere with a diameter in the femtometer (10^{-15} m $= 10^{-5}$ Å) range. The diameter of the effective sphere is called the *scattering length* $b(m)$, the m being the quantum number that characterizes the projection of the spin of the neutron–nucleus complex on some particular direction. The neutron spin number is $\frac{1}{2}$, while the nuclear spin number J is an isotopic characteristic. Thus m is allowed to have the values

$$-J-\tfrac{1}{2}, -J+\tfrac{1}{2}, \ldots, J-\tfrac{1}{2}, J+\tfrac{1}{2}$$

But the small size of the nucleus compared to the neutron implies that the only significant scattering is a neutron wave propagating spherically out from the nucleus, for which m may take on only two values, $m = J - \frac{1}{2}, J + \frac{1}{2}$.

Far from the sample (i.e., from S in Fig. 5.4) the superposition of these spherical waves from the numerous neutron–nucleus collisions can be represented in terms of plane waves of scattered radiation proceeding outward from the sample in various directions. Thus instead of Eq. (5.28) our present more detailed analysis leads to the following equation for the wave function of the radiation in the scattered beam [i.e., in $B_s(\mathbf{k}_1)$ in Fig. 5.4], although still with the omission of some factors that are irrelevant to our discussion.

$$A(\mathbf{k}, t) = \sum_{s} \sum_{j=1_s}^{N_s} b_j(m_j) \exp\left[-i\mathbf{k} \cdot \mathbf{r}_j(t)\right] e^{i\omega t} \tag{13.45}$$

This differs from Eq. (5.28) in the specification of $\mathbf{r}_j$ as a function of time; in the inclusion of the $e^{i\omega t}$ factor, where ω is the angular frequency of the waves representing the unperturbed neutron beam and in the replacement of the coherent scattering length f_j^{coh}, which is a certain average quantity, by the fluctuating scattering length $b_j(m_j)$. It is also important to specify that in Eq. (13.45) the sum over s is the sum over each isotopically distinct nuclear species since b_j is determined by the wave function of the nucleus–neutron complex.

The observable quantity of interest, on the basis of the linear response theory described in Section 11.4, is the power spectrum of the time correlation function $\langle A(\mathbf{k}, t + t')A^*(\mathbf{k}, t')\rangle / V$, namely

$$\begin{aligned} C_{AA}(k, z+\omega) \equiv \int_{-\infty}^{\infty} dt\, e^{-it(z+\omega)} \sum_{s} \sum_{s'} \sum_{j=1_s}^{N_s} \sum_{j'=1_{s'}}^{N_{s'}} \langle b_j(m_j) b_{j'}(m_{j'}) \\ \times e^{-i\mathbf{k}\cdot[\mathbf{r}_j(t+t') - \mathbf{r}_{j'}(t')]}\rangle \frac{e^{i\omega t}}{V} \end{aligned} \tag{13.46}$$

where z is the shift in frequency relative to the frequency ω of the incoming neutron beam, B_0 in Fig. 5.4. This formula for the signal that can be measured by a physical detecting instrument is the same as the one that is more usually obtained on the basis of a detailed study of the physics of the scattering process and the detector. It is also similar to the formula for scattering of electromagnetic radiation, the main difference from one case to the next being in the cross section for the scattering, here $b_j b_{j'}$, and the range of momentum change $\hbar k$ and energy change $\hbar\omega$ that can be resolved experimentally.

The ensemble average $\langle \cdots \rangle$ in Eq. (13.46) may be factored because the nuclear spin states m_j and $m_{j'}$ are practically not coupled to each other nor to the locations $\mathbf{r}_j$ and $\mathbf{r}_{j'}$ of the nuclei. Only in the case of the "self" terms, in which $s = s'$ and $j = j'$, do we get coupling between the two scattering length factors. Thus we have, schematically, for the ensemble average in Eq. (13.46),

$$\begin{aligned}\langle \cdots \rangle &= \langle b_j(m_j)^2\rangle \langle \text{function of } \mathbf{r}_j(t+t'), \mathbf{r}_j(t')\rangle \text{ in "self" terms} \\ &\quad + \langle b_j(m_j)\rangle \times \langle b_{j'}(m_{j'})\rangle \langle \text{function of } \mathbf{r}_j(t+t'), \mathbf{r}_{j'}(t')\rangle \text{ in "distinct" terms}\end{aligned} \tag{13.47}$$

In the nomenclature usually encountered, the average over the nuclear spin states

$$f_s^{\text{coh}} \equiv \langle b_j(m_j)\rangle \tag{13.48}$$

is called the *coherent scattering length* of a nucleus of species s. We define

$$f_s = \langle b_j(m_j)^2\rangle^{1/2} \tag{13.49}$$

while

$$f_s^{\text{inc}} = [f_s^2 - (f_s^{\text{coh}})^2]^{1/2} \tag{13.50}$$

is called the *incoherent scattering length.* Then we derive the following results [see Eqs. (11.55), (11.57), and (11.70)]:

$$\begin{aligned}C_{AA}(k, z+\omega) &= \sum_s f_s^2 \int_{-\infty}^{\infty} e^{-izt}\, dt \sum_{j=1_s}^{N_s} \langle \exp[-i\mathbf{k}\cdot(\mathbf{r}_j(t)-\mathbf{r}_j(0))]\rangle \\ &\quad + V^{-1}\sum_s \sum_{s'} f_s^{\text{coh}} f_{s'}^{\text{coh}} \int_{-\infty}^{\infty} e^{-izt}\, dt \sum_{j=1}^{N_s}\sum_{j'=1}^{N_{s'}} \langle \exp[-i\mathbf{k}\cdot(\mathbf{r}_j(t)-\mathbf{r}_{j'}(0)]\rangle\end{aligned} \tag{13.51}$$

$$\begin{aligned}&= \int_{-\infty}^{\infty} e^{-izt}\, dt \Big\{ \sum_s f_s^2 \rho F_s^s(\mathbf{k}, t) \\ &\quad + \sum_s \sum_{s'} f_s^{\text{coh}} f_{s'}^{\text{coh}} \rho F_{ss'}^d(\mathbf{k}, t)\Big\}\end{aligned} \tag{13.52}$$

$$C_{AA}(k, z+\omega) = 2\pi \sum_s f_s^2 \rho S_s^s(\mathbf{k}, z) + 2\pi \sum_{ss'} f_s^{\text{coh}} f_{s'}^{\text{coh}} \rho S_{ss'}^d(\mathbf{k}, z) \tag{13.53}$$

where we define $\rho = N/V$, the total nuclear density, as in Chapter 11, to retain the standard definition of the van Hove functions, and where, it may be recalled, the sum is over nuclear species. Since one expects that in a condensed state, or even a dense gas, both the self and distinct van Hove functions are insensitive to isotopic variations, it is reasonable to reduce Eq. (13.53) to a sum over atomic species provided that f_s and f_s^{coh} are appropriately interpreted as averages over the isotopic species in the system. It must be noticed, however, that quantum corrections to the classical limit involve the nuclear masses [see Eq. (4.36), for example] in a way that makes Eq. (13.53) inaccurate if j, j' denote atomic (rather than nuclear) species and if the system of interest has any isotopic variation. For details, references to earlier work, and applications to ^{36}Ar,^{40}Ar mixtures, see Fredrikze (1983).

In diffraction in crystals there are components (the Bragg peaks) in the scattered beam, due to perfectly elastic scattering, that have no frequency shift z, and the structure is deduced from the k values and intensities of these components. In fluids we recover the corresponding information by integrating over all frequencies

$$\int_{-\infty}^{\infty} C_{AA}(k, z+\omega)\, dz = 2\pi \sum_{s} \left\{ f_s^2 \rho F_s^s(\mathbf{k}, 0) + \sum_{s'} f_s^{\text{coh}} f_{s'}^{\text{coh}} \rho F_{ss'}^d(\mathbf{k}, 0) \right\}$$

$$= 2\pi \sum_{s} \rho_s (f_s)^2 + 2\pi \sum_{ss'} \rho_s \rho_{s'} f_s^{\text{coh}} f_{s'}^{\text{coh}} [\tilde{h}_{ss'}(k) + (2\pi)^3 \delta(\mathbf{k})] \tag{13.54}$$

which agrees with Eq. (5.29) except where f_s appears in place of f_s^{coh}, since the fluctuations of the scattering lengths $b_j(m_j)$ in nuclei with spins were not accounted for in the earlier derivation. The distinction is not important for diffraction, however, since in diffraction the self part F_s^s contributes only a **k**-independent background.

In an inelastic scattering experiment according to Eq. (13.53), we measure a superposition of self and distinct contributions. To resolve them we may vary the isotopic composition if isotopes with suitably different f^{coh} and f^{inc} are available, or we can in principle use the fact that with a polarized incident neutron beam the coherent part of the scattered beam is polarized while the incoherent part is depolarized (Egelstaff, 1967). However, from Eqs. (13.48) and (13.49) it is apparent that we have

$$f_s^2 - (f_s^{\text{coh}})^2 \geq 0 \tag{13.55}$$

so it is possible that the self term contributes much more than the distinct term. This actually happens for hydrogen [$f_H^2/(f_H^{\text{coh}})^2 = 46.1$] and vanadium 51 [$f_V^2/f_V^{\text{coh}})^2 = 159$], but for nuclei without spin, Eq. (13.55) holds with the equal sign.

An example of the exploitation of the fact that $f_H^2/(f_H^{\text{coh}})^2$ is very large is the series of experiments leading to the $S_H^s(k, \omega)$ data in Fig. 13.2; in this case all of the other contributions in Eq. (13.53) are negligible.

As already mentioned, the theory of the use of the inelastic scattering of electromagnetic radiation to probe $S(k, \omega)$ is very similar to the theory of INS, but there are important differences, too; for example, the analogue of f_s^{inc} vanishes for electromagnetic radiation, so all the scattering is coherent. In the case of x-rays the measurement of the frequency change has not been possible, so one measures only $\int d\omega\, S(k, \omega)$, essentially the diffraction, but the advent of synchrotron x-ray sources may relieve this constraint. In the case of light in the visible range, $S(k, \omega)$ is measurable in the k range up to 10^{-3} Å^{-1}, corresponding to the fact that the wavelength of the light is thousands of angstroms, so the scattering process is insensitive to correlations on the few-angstroms scale.

DO PROBLEM 13.10.

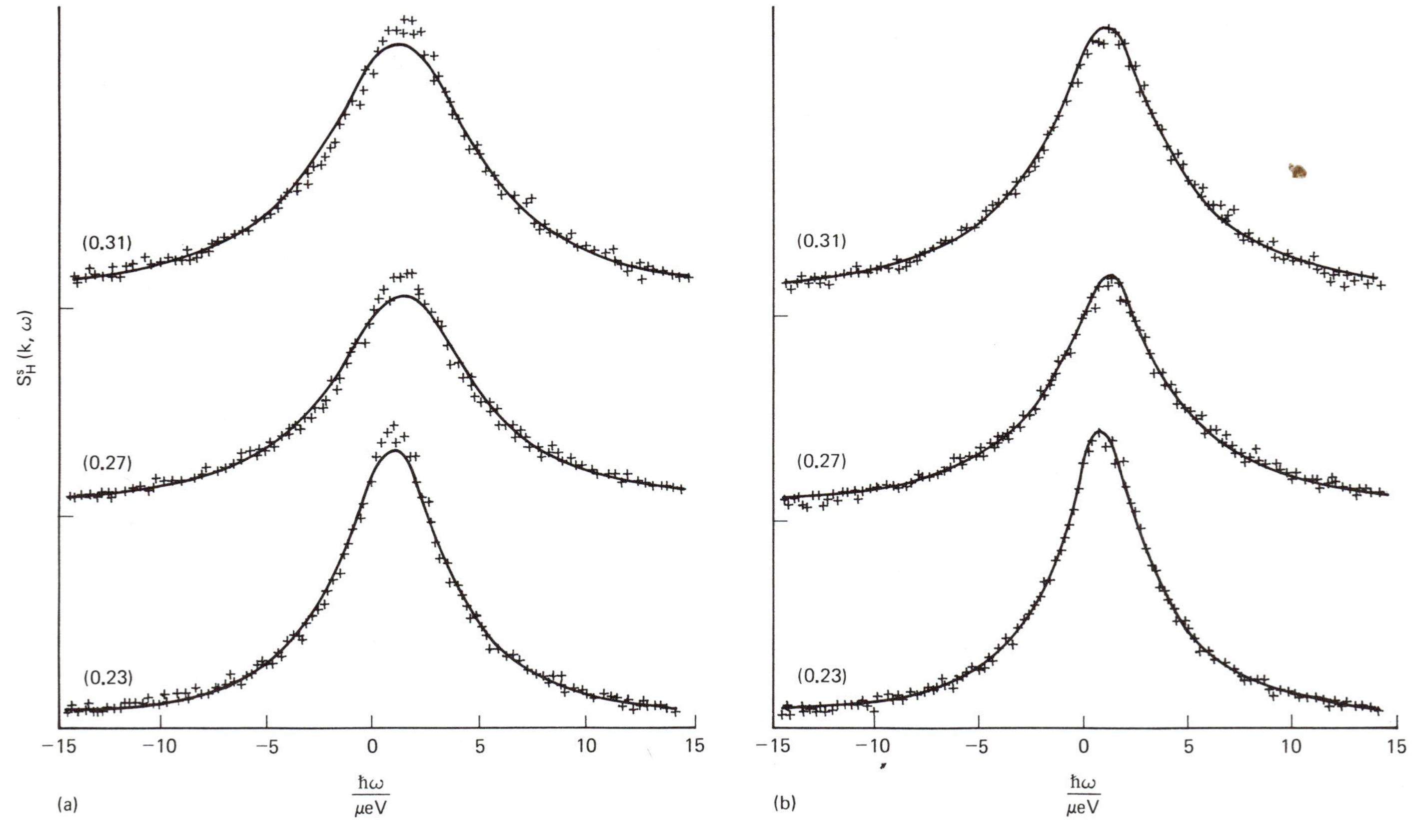

Figure 13.2 Neutron inelastic scattering data for the self van Hove dynamic structure function for protons in $NiCl_2$ solutions in water. The values of k (Å^{-1}, in parentheses) are exceptionally small to enable the diffusive motion of the protons, in intervals long compared to water molecule rotation, to be directly measurable. (a) The smooth curves are best Lorentzian fits to the data at 3 *m* (+). (b) The very similar 2 *m* data are well fit by the sum of two Lorentzians. This shows that in $NiCl_2$ solutions there are two populations of protons which may be distinguished by their diffusion coefficients. Further analysis indicates that one population is in the $Ni(H_2O)_6^{2+}$ solvation complex while the other is comprised of the remaining protons. (From Hewish et al., 1982.)

THE LITERATURE

Most of the references for Chapter 5 are equally relevant to the part of this chapter concerned with molecular dynamics. For an extensive discussion of MD results on simple liquids, see Boon and Yip (1980) as well as Hansen and McDonald (1976) and for new techniques see Ryckaert and Ciccotti (1983) and Hoover (1983).

For Brownian dynamics, see the article by Ermak and McCammon (1978) and references given there as well as Kushick and Berne (1977) and, for spin relaxation problems, Freed (1972). The role of the adjoint of the Smoluchowski operator is discussed very clearly in the context of orientational as well as translational motion by Wolynes and Deutch (1977), while Jones and Burfield (1982) is a source for the theory of the hydrodynamic interaction.

Egelstaff (1967) presents the theory of scattering of radiation, especially neutrons, at a much deeper level than given here, with extensive references to the literature. Other important sources include Martin (1968), Berne and Pecora (1976), and Flygare (1978).

PROBLEMS

13.1. Among the factors that limit the accuracy of the self-diffusion coefficient D_s^s determined by simulation is the distortion of the diffusive trajectories by the periodic boundary conditions. For MD simulations for liquid argon ($D^s \sim 0.16$ Å^2/ps) with 200 atoms in the basic cell, how long does it take for an atom to diffuse about half a box length? In view of Fig. 11.1a, is this factor important in limiting the accuracy of D_s^s in this case?

13.2. Verify the second equality in Eq. (13.7).

13.3. Suppose that the structure in the $\langle v(t)v\rangle$ data in Fig. 13.1 is due to each ion moving in a cage formed by its neighbors. When an ion strikes the cage wall, it is reflected back, leading to the negative portion of $\langle v(t)v\rangle$. Use the data in Fig. 13.1 to estimate the mean distance each ion (Li^+ or Br^-) can move in its cage.

13.4. For a dilute solution of s' in s, one might wish to estimate D_s^s for solvent molecules that are near an s' particle and far from one. Discuss whatever simulation approaches you can think of for this problem that is relevant to the data in Fig. 13.2 as interpreted by Hewish et al. (1982).

13.5. Prove that if R is a Gaussian random variable [Eq. (12.1e)], so is B_1 in Eq. (13.14).

13.6. **(a)** Derive Eq. (13.15) from (13.13) for the $F = 0$ case.

(b) Use the linear response theory to extend the result to nonzero F.

13.7. Work out the details of the passage from Eq. (13.25) to (13.27) assuming, for simplicity, that $\overset{\dagger}{S}$ is given by Eq. (13.34) and D and K are independent of q.

13.8. Show that in the case in which $K(t) = K_0\delta(t - t_0)$, in place of (13.7) we find $\langle \dot{x}_s(t)\rangle_K = \beta K_0\langle \dot{x}_s(t)\dot{x}_s\rangle$ if $t_0 = 0$.

13.9. **(a)** From the definitions in Eq. (11.55) to (11.58), show that the intermediate van Hove functions can be written

$$\rho F_{\mathcal{S}}^{s}(k,\, t) = V^{-1} \sum_{j=1}^{N_{\mathcal{S}}} \langle \exp\{i\mathbf{k} \cdot [\mathbf{r}_j(t) - \mathbf{r}_j(0)]\}\rangle \tag{13.56}$$

$$\rho F_{\mathcal{S}\mathcal{S}'}^{d}(k,\, t) = V^{-1} \sum_{j=1}^{N_{\mathcal{S}}} \sum_{j'=1}^{N_{\mathcal{S}'}} \langle \exp\{i\mathbf{k} \cdot [\mathbf{r}_j(t) - \mathbf{r}_{j'}(0)]\} \rangle [1 - \delta_{\mathcal{S}\mathcal{S}'}\delta_{jj'}] \tag{13.57}$$

(b) Verify that at $t = 0$ we have $\rho F_{\mathcal{S}}^{s}(k,\, t) = \rho_{\mathcal{S}}$, as indicated in Eq. (13.54).

14

Techniques in Linear Response Theory

14.1 INTRODUCTION

In this chapter we describe developments which can, in suitable cases, lead to detailed molecular interpretations of time-correlation functions and transport coefficients in the linear response regime. These developments may well be the first in a battery of approximation methods that parallel those for equilibrium statistical mechanics, with the generalized Langevin equation here playing the pivotal role that the Ornstein–Zernicke equation plays in the equilibrium theory.

The equations in this chapter are written for one-component systems of particles with no internal structure except in Section (14.6), where an even simpler model is assumed. Fortunately, the extension of all these results to more general Hamiltonian models requires little more than either reinterpreting or generalizing the notation.

14.2 THE GENERALIZED LANGEVIN EQUATION

This theory incorporates important developments due to several workers, notably Zwanzig, M. S. Green, and Kubo, as well as Mori (1965), whose account is followed here except that we use the notation appropriate to classical mechanics.

Consider dynamical variables A and B governed by the equations [see Eq. (11.8)]

$$\frac{d}{dt}A(t) = H^{\times}A(t) \tag{14.1a}$$

$$\frac{d}{dt}B(t) = H^{\times}B(t) \tag{14.1b}$$

where the Hamiltonian is the same in both cases. We assume that these variables are so defined that $0 = \langle A \rangle = \langle B \rangle$ [see Eq. (11.22)]. The equilibrium average

$$\langle A^*B \rangle = \frac{\int d\Gamma\, e^{-\beta H} A^* B}{\int d\Gamma\, e^{-\beta H}} \tag{14.2}$$

is called the *projection* of B on A. Here A^* is the adjoint of A; if A is a matrix with elements A_{mn}, then A^* is the matrix whose elements are the complex conjugates of A_{nm} [see Eqs. (12.49) and (12.67)]. In the simplest cases A is a simple variable, $A^* = A$, and the adjoint concept is not needed. We introduce the operator $\mathscr{P}_0$, which acts as follows:

$$\mathscr{P}_0 B = A \frac{\langle A^*B \rangle}{\langle A^*A \rangle} \tag{14.3}$$

an example of a projection operator. In general, a projection operator $\mathscr{P}$ obeys the equations

$$\mathscr{P}[1 - \mathscr{P}] = 0 \tag{14.4a}$$
$$\mathscr{P}\mathscr{P} = \mathscr{P} \tag{14.4b}$$

One often finds this sort of projection operator discussed with reference to a *Hilbert space,* which serves to give $\mathscr{P}_0$ a geometric meaning. In the case of ordinary n-dimensional vectors $\mathbf{A}$ and $\mathbf{B}$, a projection operator $\mathscr{P}$ that obeys Eqs. (14.4) might be given as follows:

$$\mathscr{P}\mathbf{B} = \mathbf{A} \frac{\sum_{k=1}^{n} A_k B_k}{\sum_{k=1}^{n} A_k^2} = \mathbf{A} \frac{\mathbf{A} \cdot \mathbf{B}}{\mathbf{A} \cdot \mathbf{A}}$$

In a slightly more general case, we introduce a weight function f_k such that

$$\sum_{k=1}^{n} f_k = 1$$

and then define the projection operator by the equation

$$\mathscr{P}\mathbf{B} = \mathbf{A} \frac{\sum_{k=1}^{n} A_k B_k f_k}{\sum_{k=1}^{n} A_k^2 f_k} \tag{14.5a}$$

Now passing from the discrete variable k to a continuous variable z in the interval z' to z'', we may define a projection operator that projects an arbitrary function of z onto $A(z)$ by the equation

$$\mathscr{P}B(z) = A(z) \frac{\int_{z'}^{z''} A(z)B(z)f(z)\, dz}{\int_{z'}^{z''} A(z)^2 f(z)\, dz}$$

In this simple case any function $A(z)$ or $B(z)$ or $\cdots$ that is defined on the interval z' to z'' can be thought of as a vector in a space which is the $n = \infty$ limit of the n-dimensional geometrical space in which ordinary vectors **A**, **B**, $\cdots$ are defined. This Hilbert space is associated with the generalized scalar product

$$(C, D) = \int_{z'}^{z''} C(z)D(z)f(z)\,dz \tag{14.6a}$$

for any "vectors" $C(z)$, $D(z)$ in the Hilbert space. In this notation the projection operator $\mathscr{P}$ that projects onto $A(z)$ is given by

$$\mathscr{P}B(z) = A(z)\frac{(A, B)}{(A, A)} \tag{14.6b}$$

The Hilbert space of interest here is of this kind except that the simple variable z is replaced by the phase Γ of an N-particle system, the weight function is the N-particle equilibrium distribution function, and we allow the variables $A(\Gamma)$, $B(\Gamma)$, $\cdots$ to be complex numbers or matrices.

The projection operator $\mathscr{P}_0$ in Eq. (14.3) now may be given a geometric significance with reference to Fig. 14.1. We think of a Hilbert space in which dynamical variables like A and B are vectors. As shown, we may take the x axis as the "direction"

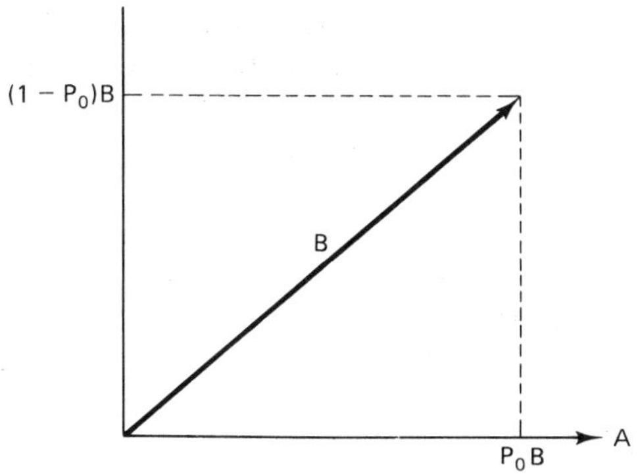

Figure 14.1a $\mathscr{P}_0$ projects B on to A. The complementary part of B is $(1 - \mathscr{P}_0)B$.

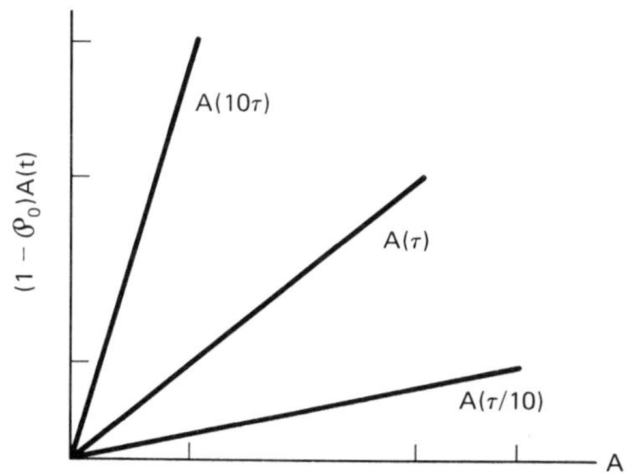

Figure 14.1b As for Fig. 14.1a but for the special case that B is the dynamical variable A(t) which fluctuates with a characteristic time τ.

of dynamical variable A, while B is some other vector. If we use $\mathscr{P}_0$ to project B onto A, the complementary part of B is $(1 - \mathscr{P}_0)B$ which is "perpendicular" to A:

$$\langle A^*(1 - \mathscr{P}_0)B\rangle = \langle A^*B\rangle - \langle A^*B\rangle = 0$$

Now we explore the case in which the second dynamical variable B is identified with the time-displaced A variable

$$A(t) = e^{tH^\times} A \tag{14.7}$$

With the definitions

$$\Sigma_0(t) \equiv \frac{\langle A^*A(t)\rangle}{\langle A^*A\rangle} \tag{14.8}$$

$$\mathrm{A}'(t) \equiv (1 - \mathscr{P}_0)A(t) \tag{14.9}$$

we have

$$A(t) = \Sigma_0(t) \cdot A + A'(t) \tag{14.10}$$

showing how $A(t)$ departs from A in magnitude and direction as t increases from zero. $\Sigma_0(t)$ is the time-correlation function normalized so that $\Sigma_0(0) = 1$.

Equation (14.10) is an equation of motion or kinetic equation for the variable $A(t)$. It exhibits the motion as the sum of a systematic part $\Sigma_0(t) \cdot A$ and a remainder, or "random" part, $A'(t)$. Thus $A(t)$ may be expected, on general grounds, to be a rather noisy function of t. Since the systematic part is free of noise, the noise must all appear in $A'(t)$. The following formal analysis shows how $A'(t)$ can be expressed in terms of $\Sigma_0(t)$ convoluted with a new function $f_1(t)$ that is even less systematic or more random than $A'(t)$.

Noticing that $\langle A(t)^*A'(t_1)\rangle$ vanishes for $t = 0$ but not necessarily for $t_1 > t$, which we would expect if $A'(t)$ were similar to the random force in Langevin's equation, we express $A'(t)$ in terms of a new variable $f_1(t_1)$ which is sure to be orthogonal to $A(t)$ if $t < t_1$ [see Eq. 12.1b)].

We operate on Eq. (14.1a) with $1 - \mathscr{P}_0$ to get

$$\frac{d}{dt}A'(t) = (1 - \mathscr{P}_0)H^\times A(t) \tag{14.11}$$

and combine with (14.10) to get

$$\begin{aligned}\frac{d}{dt}A'(t) &= \Sigma_0(t)(1 - \mathscr{P}_0)H^\times A + (1 - \mathscr{P}_0)H^\times A'(t) \\ &= \Sigma_0(t)f_1 + L_1A'(t)\end{aligned} \tag{14.12a}$$

where we define

$$L_1 \equiv (1 - \mathscr{P}_0)H^\times \tag{14.12b}$$

$$f_1 \equiv L_1A \tag{14.12c}$$

Integrating (14.12a) gives, most easily by using Fourier–Laplace transforms,

$$A'(t) = \int_0^t \Sigma_0(t_1)\, dt_1 f_1(t - t_1) \tag{14.13}$$

where we define

$$f_1(t) \equiv e^{L_1 t} f_1 \qquad \text{if } t \geqslant 0 \tag{14.14}$$

It may be verified that differentiating Eq. (14.13) generates Eq. (14.12a) in view of Eq. (14.14) and also, in view of stationarity, that $\langle A(t)^* f_1(t_1)\rangle = 0$ if $t < t_1$.

These results are of great significance as the basis of two further developments: the generalized Langevin equation and, in the next section, the continued fraction expansion. The generalized Langevin equation is

$$\frac{d}{dt}A(t) = i\omega_0 A(t) - \int_0^t M(t - t_1)\, dt_1\, A(t_1) + R(t) \tag{14.15}$$

where ω_0 is a characteristic frequency, $R(t)$ a random "force," and $M(t)$ is called the memory function [see Eq. (12.86)]. In contrast to Langevin's equation which, like the ideal gas law, is an idealization that is accurate only in certain limiting cases, the generalized Langevin equation is an exact and general description of the evolution of a dynamical variable in the linear response regime. It reduces to Langevin's equation in the limit of short memory [i.e., $M(t) \sim \delta(t)$], as indicated in Problem 12.3. More generally, Eq. (14.15) expresses the fact that $dA(t)/dt$ is influenced by the earlier history of A; this is the "memory" effect. So in the generalized Langevin equation, $A(t)$ is not a Markov process. Neither is it generally a Gaussian process.

While Eq. (14.15) does indeed follow from the preceding equations with no approximations or assumptions, as we show next, it is easy to use only in cases in which the correlation time τ_R for the random force is smaller than the correlation time τ_A for fluctuations of the dynamical variable A. Under these conditions we may hope to describe the evolution of $A(t)$ in terms of the average properties of $R(t)$. On the other hand, if $\tau_A \leqslant \tau_R$ we must expect that $A(t)$ depends on the detailed trajectory $R(t)$.

Before deriving Eq. (14.15) it is convenient to change the notation so that we may exploit the fact that A and f_1 are the zeroth and first elements of an infinite sequence of related variables. Thus the operations of projecting $A(t)$ onto A and expressing the remainder A' in terms of a random "force" $f_1(t)$ as in Eq. (14.13) can be repeated to express $f_1(t)$ in terms of its projection on f_1 and a remainder $f_1'(t)$ which depends, through an equation like (14.13), on a higher-order random "force" $f_2(t)$. While further development of these generalizations is given in the next section, the whole theory is simpler to follow if the notation change is made here.

Accordingly, we now define

$$f_0 \equiv A \tag{14.16a}$$

$$L_0 \equiv H^\times \tag{14.16b}$$

and rewrite some of the foregoing results in the new notation. The results given above are solely for the $j = 0$ cases of the following equations.

$$\begin{aligned} f_j(t) &= \mathscr{P}_j f_j(t) + (1-\mathscr{P}_j) f_j(t) \\ &= \Sigma_j(t)\cdot f_j + f_j'(t) \end{aligned} \tag{14.17a}$$

$$\Sigma_j(t) \equiv \frac{\langle f_j^* f_j(t)\rangle}{\langle f_j^* f_j\rangle} \tag{14.17b}$$

$$f_j'(t) = \int_0^t \Sigma_j(t-t_1)\,dt_1\, f_{j+1}(t_1) \tag{14.17c}$$

$$\mathscr{P}_j B \equiv \frac{f_j\langle f_j^* B\rangle}{\langle f_j^* f_j\rangle} \tag{14.17d}$$

$$f_j(t) \equiv e^{tL_j} L_j f_{j-1} \qquad \text{if } t \geqslant 0 \tag{14.17e}$$

$$L_{j+1} \equiv (1-\mathscr{P}_j)L_j \tag{14.17f}$$

Here B is any dynamical variable.

We also introduce the frequencies ω_j and Δ_j:

$$i\omega_j \equiv \frac{\langle f_j^* \dot{f}_j\rangle}{\langle f_j^* f_j\rangle} \tag{14.18a}$$

$$\Delta_j^2 \equiv \frac{\langle f_j^* f_j\rangle}{\langle f_{j-1}^* f_{j-1}\rangle} \tag{14.18b}$$

As expected on the basis of Eq. (11.20), the frequency ω_j vanishes if f_j is a simple dynamical variable. Examples of nonvanishing ω_j are found in Eqs. (12.56) and (12.66) and in Section 14.4. While the ω_j are frequencies associated with nearly periodic motions in the system, the Δ_j are typically reciprocal decay times.

To derive Eq. (14.15) we begin with the equation

$$f_0(t) = \int_0^t \Sigma_0(t-t_1)\,dt_1\,[f_1(t_1) + f_0\,\delta(t_1)] \tag{14.19}$$

which comes from Eqs. (14.17a) and (14.17c), and differentiate it to get

$$\dot{f}_0(t) = f_1(t) + \int_0^t \dot{\Sigma}_0(t-t_1)\,dt_1\,[f_1(t_1) + f_0\,\delta(t_1)] \tag{14.20}$$

To eliminate $\dot{\Sigma}_0$ we combine the equation

$$\langle f_1^* f_0(t)\rangle = -\langle f_0^* f_0\rangle[\dot{\Sigma}_0(t) - i\omega_0\Sigma_0(t)] \tag{14.21}$$

which follows simply by expanding f_1 according to Eq. (14.12c) with another equation for $\langle f_1{}^* f_0(t)\rangle$, this one from Eq. (14.19),

$$\langle f_1^* f_0(t)\rangle = \int_0^t \Sigma_0(t-t_1)\,dt_1\,\Sigma_1(t_1)\langle f_1^* f_1\rangle \tag{14.22}$$

to get

$$\dot{\Sigma}_0(t) = i\omega_0\Sigma_0(t) + \Delta_1^2\int_0^t \Sigma_0(t-t_1)\,dt_1\,\Sigma_1(t_1) = 0 \tag{14.23}$$

which is used to substitute for $\dot{\Sigma}_0$ in Eq. (14.20). The result is

$$\dot{f}_0(t) = i\omega_0 f_0(t) - \Delta_1^2 \int_0^t \Sigma_1(t-t_1)\, dt_1\, f_0(t_1) + f_1(t) \tag{14.24}$$

which is the same as Eq. (14.15) except for the alternative notation. This important result is somewhat more easily derived by using Fourier Laplace transforms [see Eq. (11.23)] such as

$$\Sigma_0[z] \equiv \int_0^\infty \Sigma_0(t) e^{-izt} dt \tag{14.25}$$

but the derivation in the time space is perhaps marginally more convincing since we do not have to be concerned about the existence of all of the transforms.

Like the strict Langevin equation, the generalized Langevin equation should be considered in the context of a number of auxilliary conditions. In place of Eq. (12.1b), we have

$$\langle f_1(t) f_0(0) \rangle = 0 \qquad \text{if } t \geq 0$$

which is generally valid [see Eq. (12.79d)]. In place of Eq. (12.1c) we have $\langle f_1(t) \rangle = 0$, which also is valid, as one may deduce from Eq. (14.17e) if, as assumed, $\langle f_0(t) \rangle = 0$. In place of Eq. (12.1d) one would have $\Sigma_1(t) \simeq \delta(t)$, which is *not* generally valid, nor is it true in general that f_1 and f_0 are Gaussian random variables as specified in the case of the strict Langevin equations by Eqs. (12.1e) and (12.1f).

The validity of Eq. (14.15) does not depend on $M(t)$ relaxing much faster than $\langle A^* A(t) \rangle$. However, the application of the theory to the molecular interpretation of transport and relaxation data does become more difficult unless $R(t)$ relaxes much more quickly than $A(t)$. In the remainder of this chapter, we describe a few of the techniques that have proved useful in difficult cases.

The theory of Eq. (14.15) as given here is easily extended to the case in which the dynamical variable A is replaced by a set of variables $\mathbf{A} = A_1, A_2, \cdots, A_\nu$. An example is discussed in Section 14.4. Then the correlation function $\langle \mathbf{A}^* \mathbf{A}(t) \rangle$ is an $n \times n$ correlation matrix and we may have nonzero characteristic frequencies $i\omega_{mm',0} = \Sigma_n \langle A_m^* \dot{A}_n \rangle [\langle \mathbf{A}^* \mathbf{A} \rangle^{-1}]_{nm'}$, where $\langle \mathbf{A}^* \mathbf{A} \rangle$ is the matrix of equilibrium correlations $\langle A_m^* A_n \rangle$. However, as already noted, when A is a single simple dynamical variable, then indeed $\omega_0 = 0$. The $\omega_0 \neq 0$ case has been carried through the derivations merely to make it easy to generalize to the correlation matrix case (i.e., the case with $\nu > 1$).

It turns out that a rather simple-looking memory function can give rise to a much more complicated time-correlation function, in rough analogy to the relation of the direct-correlation function $c(r)$ to $g(r)$ in equilibrium theory. For example, the simple form

$$M(t) = \alpha \exp\left(-\frac{t}{t_1}\right) \tag{14.26}$$

in some range of the parameters α and t_1 gives $\langle A(t) A \rangle$, which is oscillatory.

Another interesting feature concerns the approximately additive contribution of different motions. Thus for the self-diffusion coefficient D_{ion} of an ion in solution, we expect an important contribution from the harsh repulsive forces, which control the diffusion in hard-sphere model fluids, as well as from Coulomb forces between the ion and polar solvent molecules. Denoting the respective terms of the random force [Eq. (14.15)] as $R_H(t)$ and $R_C(t)$ and taking $A(t)$ to be the x component of the momentum of an ion, we have

$$D_{\text{ion}} = \int_0^{\infty} \langle A(t)A \rangle \, dt = \frac{\Sigma_0[0]}{m_{\text{ion}}^2} \tag{14.27}$$

and in view of Eq. (14.130),

$$\begin{aligned} \frac{1}{D_{\text{ion}}} &= \int_0^{\infty} dt \, M(t) \\ &= \Delta_1^{-2} \int_0^{\infty} \langle [R_H(t) + R_C(t)][R_H + R_C] \rangle \, dt \\ &\simeq \Delta_1^{-2} \int_0^{\infty} dt \langle R_H(t)R_H + R_C(t)R_C \rangle \end{aligned} \tag{14.28}$$

The approximation in the last step is good if R_H fluctuates much faster than R_C so that the cross terms are relatively small;

$$\langle R_H(t)R_C \rangle \simeq \langle R_C(t)R_H \rangle \simeq 0 \tag{14.29}$$

Beginning in this way, Wolynes (1978) was able to construct a molecular theory of the dielectric drag contribution to D_{ion}.

The memory function for the velocity autocorrelation for a 6–12 fluid modeled to correspond to a liquid state of argon is shown in Fig. 14–2. We may think of the initial sharp fall in $M(t)$ as reflecting a strongly damped periodic motion of an atom in a "cage" of other atoms. The undamped motion would be periodic with the Einstein frequency ω_E, corresponding to the oscillations of an atom about its lattice point in Einstein's theory of the heat capacity of an atomic crystal. The tail on $M(t)$ apparently is controlled by forces which are softer and which have longer relaxation times; they are associated with diffusive motion. So we may model the memory function as

$$M(t) = \omega_E^2 \exp\left(-\frac{t}{\tau_E}\right) + \omega_D^2 \exp\left(-\frac{t}{\tau_D}\right) \tag{14.30a}$$

to show how the memory function may reflect these two quite dissimilar contributions. This sort of model has been very useful in the theory of isotope effects upon diffusion (Lantelme et al., 1979), while a rather similar model

$$M(t) = \omega_E^2 \exp\left(-\frac{t^2}{\tau_E^2}\right) + \omega_A^6 t^4 \exp\left(-\frac{t}{\tau_D}\right) \tag{14.30b}$$

was used by Levesque and Verlet (1970) to fit the data for $M(t)$ for the velocity autocorrelation function for a 6–12 model for liquid argon.

It seems interesting to notice that Fig. 14.2 and Eqs. (14.30) are representations

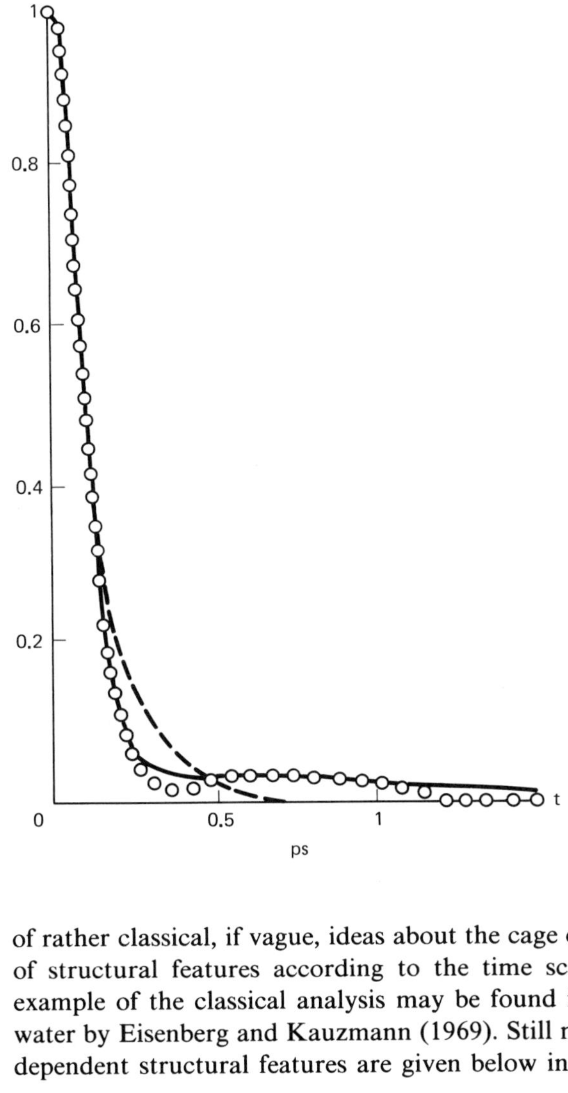

Figure 14.2 Normalized memory function $M(t)/M(0)$ (circles) calculated from the velocity autocorrelation functions for a 6–12 model corresponding to liquid Ar at $T = 94.4$ K and $\rho = 1.374$ g/cm^3 (Rahman, 1964). The exponential approximation for $M(t)$ is represented as a dashed line, while a Gaussian plus exponential approximation is represented as a solid line (Damle and Tillue, 1969; Boon and Yip, 1980).

of rather classical, if vague, ideas about the cage effect in liquids and the classification of structural features according to the time scales on which they persist. A nice example of the classical analysis may be found in the discussion of the structure of water by Eisenberg and Kauzmann (1969). Still more explicit representations of time-dependent structural features are given below in Eqs. (14.40) and (14.65).

DO PROBLEMS 14.1 TO 14.6.

14.3 MORI'S CONTINUED FRACTION THEORY

The analysis of the $j = 0$ case leading to Eq. (14.23) is readily generalized to get

$$\dot{\Sigma}_j(t) - i\omega_j \Sigma_j(t) + \Delta_{j+1}^2 \int_0^t \Sigma_j(t - t_1)\, dt_1 \Sigma_{j+1}(t_1) = 0 \tag{14.31}$$

by combining the equations

$$\begin{aligned}\langle f_{j+1}^* f_j(t)\rangle &= \langle(\mathscr{P}_j L_j f_j)^* f_j(t)\rangle \\ &= \langle f_j^* f_j\rangle[-\dot{\Sigma}_j(t) + i\omega_j \Sigma_j(t)]\end{aligned} \tag{14.32}$$

$$\begin{aligned}\langle f_{j+1}^* f_j(t)\rangle &= \langle f_{j+1}^*[\Sigma_j(t)\cdot f_j + f_j'(t)]\rangle \\ &= \int_0^t \Sigma_j(t-t_1)\, dt_1 \Sigma_{j+1}(t_1)\langle f_{j+1}^* f_{j+1}\rangle\end{aligned} \tag{14.33}$$

in place of Eqs. (14.21) and (14.22). The Fourier–Laplace transform of Eq. (14.31) gives [see Eq. (14.25)]

$$\Sigma_j[z] = \frac{1}{i(z-\omega_j) + \Delta_{j+1}^2\, \Sigma_{j+1}[z]}, \qquad j = 0, 1, 2, \ldots \tag{14.34}$$

so we have the following continued fraction expansion (Mori, 1965) of $\Sigma_0[z]$.

$$\Sigma_0[z] = \cfrac{1}{i(z-\omega_0) + \cfrac{\Delta_1^2}{i(z-\omega_1) + \cfrac{\Delta_2^2}{i(z-\omega_2) + \cfrac{\Delta_3^2}{i(z-\omega_3) + \cdots}}}} \tag{14.35}$$

Since the coefficients ω_j and Δ_j, like the more familiar thermodynamic and equilibrium structural functions, are static ensemble averages with no time-shift operators $e^{L_j t}$ we have here a formally exact expression for a quite general time-correlation function in terms of static ensemble averages. Remarkably, then, the dynamics in the linear response regime is entirely predictable from a *complete* knowledge of the equilibrium thermodynamics and structure.

If explicit formulas for all of the ω_j and Δ_j are known, or if they are connected by recursion relations, one may hope to evaluate the infinite continued fraction computationally (Monroe and Friedman, 1977), if not analytically. More often the only practical way to proceed is to terminate the continued fraction at a low-order memory function, as in

$$\Sigma_0[z] = \cfrac{1}{i(z-\omega_0) + \cfrac{\Delta_1^2}{i(z-\omega_1) + \cdots \cfrac{}{i(z-\omega_{j-1}) + \Delta_j^2 \Sigma_j[z]}}} \tag{14.36}$$

typically with the approximation

$$\Sigma_j[z] = \frac{1}{i(z-\omega_j) + 1/\tau_j} \tag{14.37}$$

which corresponds to an exponentially damped $\Sigma_j(t)$. Some examples have already been mentioned. For example, Eq. (14.26) corresponds to

$$\Sigma_0(z) = \cfrac{1}{iz + \cfrac{\Delta_1^2}{iz + 1/t_1}} \tag{14.38}$$

where $\Delta_1^2 = \alpha$ while Eq. (14.30a) corresponds to

$$\Sigma_0(z) = \cfrac{1}{iz + \cfrac{\Delta_1^2}{iz + \cfrac{\Delta_2^2}{iz + 1/\tau_2}}} \tag{14.39}$$

where Δ_1, Δ_2, and τ_2 can be expressed in terms of the parameters in Eq. (14.30a).

The advantage of the continued fraction expansions over ad hoc memory function approximations is that in the former the parameters ω_j and Δ_j are automatically specified in terms of equilibrium averages according to Eqs. (14.18).

As a further illustration of the use of continued fraction expansions, we consider the case in which f_0 in Eqs. (14.17) is given by

$$\rho_{\mathbf{k}i}(t) = \exp\,[i\mathbf{k} \cdot \mathbf{r}_i(t)]$$

[see Eq. (11.57)]. Then we find that Eq. (14.39) becomes, in view of Eq. (11.58),

$$\tilde{F}^s(k, z) = \int_0^\infty dt\, e^{-izt} F^s(k, t)$$

$$= \cfrac{1}{iz + \cfrac{\omega_0^2}{iz + \cfrac{2\omega_0^2 + \Omega_0^2}{iz + 1/\tau_s(k)}}} \tag{14.40a}$$

with the parameters

$$\omega_0^2 = \frac{k^2}{m\beta} \tag{14.40b}$$

$$\Omega_0^2 = \frac{\rho}{3m} \int d^3r\, g(r)\, \nabla^2 u(r) \tag{14.40c}$$

$$\tau_s(k) \simeq \frac{1}{\beta m D \Omega_0^2} \tag{14.40d}$$

where $u(r)$ is the pair potential and D the diffusion coefficient. This approximation is correct through the fourth moment of $\tilde{F}^s(k, z)$ [see Eq. (11.91)] and, thanks to Eq. (14.40d), also gives the correct small k, small z dependence for $\tilde{F}^s(k, z)$ (i.e., the correct behavior in the hydrodynamic limit). Equation (14.40a) is moderately successful for interpreting the inelastic neutron scattering data for argon (Lovesley, 1973).

DO PROBLEMS 14.7 TO 14.10.

14.4 CORRELATION MATRIX THEORY

Often it is quite natural to analyze relaxation in a system of interest in terms of a set of dynamical variables $A_1(t), A_2(t), \ldots, A_\nu(t)$ which are independent thermodynamically,

$$\langle A_m^*(t)A_{m'}(t)\rangle = \langle A_m^*A_m\rangle\delta_{mm'}$$

but which may be dynamically coupled in the sense that $\langle A_m^*(t_0)A_{m'}(t+t_0)\rangle$ is not necessarily zero even if $m \neq m'$, provided that t is not zero. As an example, if $\nu = 2$, we may have [see Eqs. (11.19) and (11.20)] $A_1 = A$ and $A_2 = \dot{A}$. As a second example, with $\nu = 3$, the A_m might be the projections $S_0 - \langle S_0\rangle$, S_+, S_- of a relaxing spin variable. Or they might be the velocities of ν labeled solute particles, or they might be hydrodynamic modes. In each case it is not wrong to apply the theory of Section 14.2 with f_0 identified as some one A_m, but an analysis in terms of a set of coupled variables may be advantageous, especially if the $A_j(t)$ all fluctuate less rapidly than any of the corresponding random forces.

To proceed we introduce the column vector

$$\mathbf{A}(t) = \text{column}\,[A_1(t), A_2(t), \ldots, A_\nu(t)] \tag{14.41a}$$

and its Hermitian conjugate (or adjoint), the row vector of complex conjugate A_m^*'s:

$$\mathbf{A}^*(t) = \text{row}\,[A_1^*(t), A_2^*(t), \ldots, A_\nu^*(t)] \tag{14.41b}$$

For the first example mentioned above, $A_m^* = A_m$, while for the second the most convenient variables,

$$S_0 \quad \text{and} \quad S_\pm = S_x \pm iS_y$$

include complex numbers or operators (see Section 12.7).

The theory of the generalized Langevin equation in Section 14.2 is easily extended to the case when the single dynamical variable $A(t)$ is replaced by the vector $\mathbf{A}(t)$, the result being

$$\dot{\mathbf{A}}(t) = i\boldsymbol{\omega}\cdot\mathbf{A}(t) - \int_0^t \mathbf{M}(t-t_1)\cdot\mathbf{A}(t_1)\,dt_1 + \mathbf{R}(t) \tag{14.42}$$

with the definitions [see Eqs. (14.15) and (14.24)]

$$i\boldsymbol{\omega} = \langle\mathbf{A}^*\dot{\mathbf{A}}\rangle\cdot(\langle\mathbf{A}^*\mathbf{A}\rangle^{-1}) \tag{14.43a}$$

$$\mathbf{M}(t) = \langle\mathbf{R}^*\mathbf{R}(t)\rangle\cdot(\langle\mathbf{A}^*\mathbf{A}\rangle^{-1}) \tag{14.43b}$$

where $\boldsymbol{\omega}$ and $\mathbf{M}$ are $\nu\times\nu$ matrices and the center dot now means matrix multiplication. For example,

$$\boldsymbol{\omega}\cdot\mathbf{A}(t) = \text{column}\left[\sum_k \omega_{1k}A_k(t),\ \sum_k \omega_{2k}A_k(t)\cdots\sum_k \omega_{\nu k}A_k(t)\right] \tag{14.44a}$$

$$i\omega_{mm'} = \sum_k \langle A_m^*\dot{A}_k\rangle[\langle\mathbf{A}^*\mathbf{A}\rangle^{-1}]_{km'} \tag{14.44b}$$

We also have the usual conditions on the random force,

$$\langle R_i(t)\rangle = 0, \qquad i, = 1, \ldots, \nu \tag{14.45a}$$

$$\langle R_i(t)^* A_j(t_0)\rangle = 0 \quad \text{if } t \geqslant t_0, \qquad i, j = 1, \ldots, \nu \tag{14.45b}$$

Also, introducing the mode–mode normalized time correlation functions

$$C_{mm'}(t) = [\langle \mathbf{A}^*\mathbf{A}(t)\rangle \cdot (\langle \mathbf{A}^*\mathbf{A}\rangle^{-1})]_{mm'} \tag{14.46}$$

and their Fourier–Laplace transforms

$$C_{mm'}[z] = \int_0^\infty dt\, e^{-izt} C_{mm'}(t) \tag{14.47}$$

leads to the matrix form of Eq. (14.34) in the $j = 0$ case, namely[1]

$$\mathbf{C}[z] = [i(z\mathbf{1} - \boldsymbol{\omega}) + \mathbf{M}[z]]^{-1} \tag{14.48}$$

As an example of the use of the correlation matrix theory, we take one of the theories of the van Hove intermediate self-correlation function $F^s(k, t)$ for a one-component system. We assume (Levesque and Verlet, 1970) that

$$\mathbf{A} = \text{column}\{\rho_{\mathbf{k}i}, \dot{\rho}_{\mathbf{k}i}, \sigma_{\mathbf{k}i}\} \tag{14.49}$$

where

$$\rho_{\mathbf{k},i}(t) \equiv \int d^3r\, e^{i\mathbf{k}\cdot\mathbf{r}} \rho_i(\mathbf{r}, t) \tag{14.50}$$

is the transform of the local density of labeled particle i at time t and where

$$\sigma_{\mathbf{k},i}(t) \equiv \ddot{\rho}_{\mathbf{k}i} - \frac{\rho_{\mathbf{k}i}\langle \ddot{\rho}_{\mathbf{k}i}\rho_{-\mathbf{k},i}\rangle}{\langle \rho_{\mathbf{k},i}\rho_{-\mathbf{k},i}\rangle} \tag{14.51}$$

so that $\rho_{\mathbf{k}i}$, $\dot{\rho}_{\mathbf{k}i}$, and $\sigma_{\mathbf{k}i}$ stand in the same relation to each other as the variables f_0, f_1, and f_2 in the Mori theory [Eq. (14.17)]. It follows that $\langle \mathbf{A}^*\mathbf{A}\rangle$ is diagonal [Eq. (14.129a)]. The diagonal elements, evaluated with reference to Section 11.6 and Eqs. (13.56) and (13.57), are

$$\langle \rho_{\mathbf{k}i}\rho_{-\mathbf{k},i}\rangle = F^s(\mathbf{k}, 0) = 1 \tag{14.52a}$$

$$\langle \dot{\rho}_{\mathbf{k}i}\dot{\rho}_{-\mathbf{k},i}\rangle = -\left[\frac{\partial^2}{\partial t^2} F^s(k, t)\right]_{t=0} = \frac{k^2}{m\beta} \equiv \omega_0^2 \tag{14.52b}$$

$$\langle \sigma_{\mathbf{k}i}\sigma_{-\mathbf{k}i}\rangle = 2\omega_0^4 + \frac{\langle F_{iz}^2\rangle k^2}{m^2} \equiv 2\omega_0^4 + \Omega_0^2\omega_0^2 \tag{14.52c}$$

and ω_0 and Ω_0 are the same as in Eq. (14.40). Then we have

$$\langle \mathbf{A}^*\mathbf{A}\rangle = \begin{bmatrix} 1 & 0 & 0 \\ 0 & \omega_0^2 & 0 \\ 0 & 0 & \omega_0^2(2\omega_0^2 + \Omega_0^2) \end{bmatrix} \tag{14.53}$$

[1] Here $\mathbf{1}$ is the unit matrix of the same rank as $\boldsymbol{\omega}$ and $\mathbf{M}$.

Proceeding in the same way, we find the frequency matrix of Eq. (14.42).

$$i\boldsymbol{\omega} = \begin{bmatrix} 0 & 1 & 0 \\ -\omega_0^2 & 0 & 1 \\ 0 & -2\omega_0^2 - \Omega_0^2 & 0 \end{bmatrix} \tag{14.54}$$

Finally, we formulate the memory matrix $\mathbf{M}(t)$ in view of Eq. (14.43). Using the projection operator technique in Section 14.2, we find in the present case, for $m = 1, 2,$ or 3,

$$R_m = (1 - \mathscr{P}_0)H^\times A_m \tag{14.55}$$

where we define

$$\mathscr{P}_0 B = \sum_m A_m \frac{\langle A_m^* B\rangle}{\langle A_m^* A_m\rangle} \tag{14.56}$$

We find, for $\mathbf{A}$ in Eq. (14.49), $R_1 = 0$, $R_2 = 0$, and

$$R_3 = \dddot{\rho}_{\mathbf{k}i} - \dot{\rho}_{\mathbf{k}i}[3\omega_0^2 + \Omega_0^2] \tag{14.57}$$

$$\mathbf{M}(t) = \begin{bmatrix} 0 & 0 & 0 \\ 0 & 0 & 0 \\ 0 & 0 & M_{33}(k, t) \end{bmatrix} \tag{14.58}$$

where M_{33} is proportional to $\langle R_3(k, t)R_3(-k, 0)\rangle$. Combining these results with Eq. (14.48) gives the correlation function matrix

$$\mathbf{C}[z] = \begin{bmatrix} iz & -1 & 0 \\ \omega_0^2 & iz & -1 \\ 0 & 2\omega_0^2 + \Omega_0^2 & iz + \tilde{M}_{33}(k, z) \end{bmatrix}^{-1} \tag{14.59}$$

where $\tilde{M}_{33}(k, z)$ is the Fourier–Laplace transform of $M_{33}(k, t)$. By construction the 1, 1 component of $\mathbf{C}[z]$ is $\tilde{F}^s(k, z)$, the transform of $F^s(k, t)$.

From (14.59) we can recover (14.40a) but with $\tilde{M}_{33}(k, z)$ in place of $1/\tau_s(k)$. However, this difference is not significant, because $1/\tau_s(k)$ is merely an approximation to $\Delta_3^2\Sigma_3(k, z)$, as may be deduced by comparison with Eq. (14.36) for the $j = 3$ case. The same approximation can be made for $\tilde{M}_{33}(k, z)$ in the absence of additional information.

Thus at this point the correlation matrix method based on Eq. (14.49) requires a longer calculation to attain the results more simply deduced from the continued fraction expansion of $F^s(\mathbf{k}, z)$, but it does also at the same time give the associated correlation functions which are the other matrix elements of $\mathbf{C}[z]$. More generally, when the elements of $\mathbf{A}$ do not correspond to a subset of the Mori variables f_0, f_1, . . . , the correlation function theory does not give results which may also be obtained by truncating the memory function continued fraction expansion for a single time-correlation function. Examples of the more general case are $\mathbf{A} = S_0$, S_+S_-, already mentioned, and cases in which **A** is a combination of single particle and product variables, giving what is called the *mode coupling theory.*

DO PROBLEMS 14.11 AND 14.12.

14.5 GENERALIZED HYDRODYNAMICS

Another way to the molecular theory of the time-correlation functions depends on the analytical theory of the suceptibility function $\chi_{BA}(k, \omega)$ [Eq. (11.72)] (Martin, 1968). The real and imaginary parts of $\chi \equiv \chi_{BA}$ obey the Kramers–Kronig relations

$$\chi'(k, \omega) = \frac{1}{\pi} P \int_{-\infty}^{\infty} dz \frac{\chi''(k, z)}{z - \omega} \tag{14.60a}$$

$$\chi''(k, \omega) = -\frac{1}{\pi} P \int_{-\infty}^{\infty} dz \frac{\chi'(k, z)}{z - \omega} \tag{14.60b}$$

where now P means "take the principal part," defined as follows:

$$P \int_{-\infty}^{\infty} \frac{dz\, f(z)}{z - \omega} = \lim_{\epsilon=0} \int_{-\infty}^{\infty} dz \frac{f(z)}{z - \omega - i\epsilon} \tag{14.61}$$

so evidently either χ' or χ'' determines the full χ. For example, one finds that

$$\chi(k, \omega) = \frac{1}{\pi} \int_{-\infty}^{\infty} dz \frac{z\chi''(k, z)}{z^2 - \omega^2} \tag{14.62}$$

and for $\omega = 0$

$$\chi(k) \equiv \chi(k, 0) = \int_{-\infty}^{\infty} dz \frac{\chi''(k, z)}{\pi z} \tag{14.63}$$

For the special case of $\chi_{\rho\rho}$, we have

$$\chi_{\rho\rho}(k) = -\beta \rho S(k) \tag{14.64}$$

where we have used Eqs. (11.90) and (11.94). Then in view of the relation of $G(r, t)$ to the equilibrium structure shown in Fig. 11.3, we may expect $\chi_{\rho\rho}(k, \omega)$ to be expressible in terms of a perturbation, due to dynamic processes, from the equilibrium $\chi_{\rho\rho}(k)$. More generally, $\chi(k, \omega)$ in the small-k, small-ω regime must reflect the hydrodynamic fluxes, such as diffusion and viscous flow. A theory that fulfills this condition but also extends into the regime of large k and ω is called *generalized hydrodynamics.* An example is the following equation for $\chi_{\rho\rho}$, suggested by Kadanoff and Martin (1963) for a one-component fluid:

$$\frac{1}{\chi_{\rho\rho}}(k, z) = \frac{1}{\chi_{\rho\rho}}(k) + \frac{m}{\rho k^2}[z^2 + izk^2 D(k, z)] \tag{14.65}$$

Equations of this kind together with Hamiltonian prescriptions for the new coefficients,

$D(k, z)$ in this case, can also be derived by some of the kinetic theory methods illustrated in Section 14.7.

14.6 PERTURBATION THEORY

In some cases, notably magnetic resonance relaxation, perturbation methods have been useful for interpreting rate data in molecular terms. A very simple mathematical model will be considered here in order to avoid the complexities associated with the physics of spin relaxation while exhibiting the statistical mechanical content of the theory.

We consider the dynamical variable A, with zero norm and unit variance:

$$\langle A \rangle = 0, \qquad \langle A^*A \rangle = 1$$

with the time-correlation function [see Eq. (10.15)]

$$\begin{aligned} C(t) &= \langle A^*A(t) \rangle \\ &= \int d\Gamma\, f^{\text{eq}}(\Gamma) A^*(\Gamma) e^{tH^\times} A(\Gamma) \end{aligned} \tag{14.66}$$

We suppose that we have the Hamiltonian

$$H = H_0 + H_i \tag{14.67}$$

such that

$$H_0^\times A = i\omega A \tag{14.68}$$

so that in the reference system with $H_i = 0$ we have

$$C(t) = e^{i\omega t}, \qquad H_i = 0 \tag{14.69}$$

In the case of spin relaxation this reference system motion is the Larmor precession of the free spin. But more generally we consider a system in which we distinguish a *spin* variable A and call the collection of all of the other variables the *lattice.* Also, $H_0 = H_A + H_L$, where H_A couples only to A and H_L only to the lattice, while only H_i couples the spin to the lattice. We also assume that the equilibrium distribution function $f^{\text{eq}}(\Gamma)$ is very nearly

$$e^{-\beta H_0} \Big/ \int d\Gamma\, e^{-\beta H_0}$$

with negligible dependence on H_i. All of these conditions are met, for example, by a nuclear spin in a particle in a medium if the nucleus with the spin has an electric quadrupole moment which couples it to charged or polar particles in the lattice. The operator of the nuclear spin is A, while all of the other phase variables comprise the lattice.

With nonzero H_i it is convenient to focus on the time-correlation function "in the rotating frame,"

$$C^\ddagger(t) \equiv e^{-i\omega t} C(t) \tag{14.70}$$

which also can be written in the form

$$C^{\ddagger}(t) = \int d\Gamma\, f^{\text{eq}}(\Gamma)(e^{tH_0^\times}A^*)e^{tH^\times}A$$
$$= \int d\Gamma\, f^{\text{eq}}(\Gamma)A^*e^{-tH_0^\times}e^{tH^\times}A \tag{14.71}$$

The last member, which focuses on the incremental motion due to nonzero H_i, is in a form sometimes called an *interaction representation.* It invites the application of the operator identity Eq. (3.56). Thus we have

$$e^{tH^\times} = e^{tH_0^\times} + \int_0^t dt_1 e^{(t-t_1)H_0^\times}H_i^\times e^{t_1H^\times} \tag{14.72}$$

which may equally well be written

$$e^{-tH_0^\times}e^{tH^\times} = 1 + \int_0^t dt_1 L_i(t_1)e^{-t_1H_0^\times}e^{t_1H^\times} \tag{14.73}$$

where we have introduced the operator

$$L_i(t) \equiv e^{-tH_0^\times}H_i^\times e^{tH_0^\times} \tag{14.74}$$

which acts as the Liouville operator coupling the spin to the lattice at time t; both H_A and H_L contribute to its evolution.

Equation (14.73) can be expanded by iteration and substituted in Eq. (14.71) to obtain the *moment expansion* of $C^{\ddagger}(t)$:

$$C^{\ddagger}(t) = 1 + \sum_{n\geq 1}\int_0^t d(1, 2, \ldots, n)\langle A^*L_i(t_1)L_i(t_2)\cdots L_i(t_n)A\rangle \tag{14.75}$$

where we introduce an abbreviation for the *time-ordered* integration

$$\int_0^t d(1\ 2\cdots n)f(t_1, t_2, \ldots, t_n)$$
$$\equiv \int_0^t dt_1\int_0^{t_1} dt_2\cdots\int_0^{t_{n-1}} dt_n f(t_1, t_2, \ldots, t_n) \tag{14.76}$$

This moment expansion, although formally exact, has deplorable properties, rather like the expansion of the grand partition function in Eq. (6.45), and for similar reasons.

To understand this behavior, we at first replace the integrand in any term of Eq. (14.75) by a different function

$$c_{12\cdots n} = c_{12\cdots n}(t_1, t_2, \ldots, t_n) \tag{14.77}$$

that is *connected* in the sense that it vanishes if any interval $t_{j-1} - t_j$ (for $j \leq n$) exceeds some critical value, which we may take to be the longest relaxation time τ_L of a lattice variable. This consideration is prompted by the fact that in the integrands of Eq. (14.75), the operator $L_i(t_{j-1})$ becomes statistically independent of $L_i(t_j)$ if $t_{j-1} - t_j$ is big compared with τ_L. Now the integral

$$\int_0^{t_1} d(2\ 3 \cdots n)c_{12\cdots n}$$

corresponds to the graph in Fig. 14.3a, where the white circles represent the fixed times t_1 and $t = 0$ and the black circles the integrated times, while the bonded circles represent the integrand. As indicated in the figure, this integral is some number independent of t_1 if t_1 is large enough. The integrals in Eq. (14.75) involve one additional integration $\int_0^t dt_1 \cdots$. So with increasing t every term in this equation grows *at least* as fast as t. Notice the analogy to Eqs. (6.45) and (6.64).

As indicated in Fig. 14.3b, if part of an integrand in Eq. (14.75) corresponds to a product of ν functions of the type $c_{1\cdots n}$, the contribution of that part to the integral increases like t^ν at large t. Since the measurable transport and relaxation coefficients are related to

$$\int_0^\infty e^{-izt}C^\ddagger(t)\,dt$$

we must find how terms that grow with t can mutually cancel in order to give a useful perturbation theory.

Next it is necessary to analyze the connectivity, in the sense of Fig. 14.3, of various contributions to the integrands in Eq. (14.75). It is now helpful to think of these integrands as

$$\langle A^*L_i(t_1) \cdots L_i(t_n)A\rangle = \langle A^*\langle L_i(t_1) \cdots L_i(t_n)\rangle_L A\rangle_S \qquad (14.78)$$

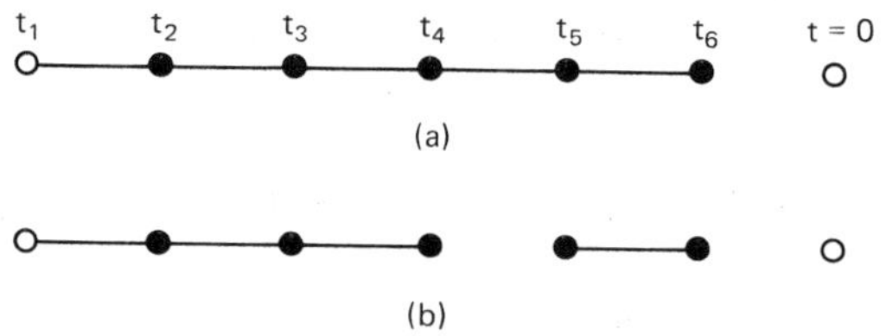

Figure 14.3 (a) Graph representing the integral

$$\int_0^{t_1} d(2\ 3\ 4\ 5\ 6)c_{123456}$$

where the integrand vanishes if any of the bonds (lines) is longer than some critical value. Because there is no bond between the circles at t_6 and $t = 0$, the value of the integral does not depend on t_1 if t_1 is longer than about five critical lengths.

(b) Graph representing the integral

$$\int_0^{t_1} d(2\ 3\ 4\ 5\ 6)c_{1234}c_{56}$$

If t_1 is big enough the integral over $t_1 - t_4$ becomes independent of t_1, the integral over t_5 and t_6 tends to a constant value too because of the growing importance of large t_5 as t_1 increases, but the interval between t_4 and t_5 contributes

$$\int_{t_4}^{t_5} dt = t_5 - t_4$$

so it tends to grow with t_1. The net effect is that at large t_1 this graph grows like t_1.

where $\langle\cdot\cdot\cdot\rangle_L$ means ensemble average of a spin + lattice function over the lattice variables, while $\langle\cdot\cdot\cdot\rangle_S$ means average of the remaining spin function over the spin variables. Then we analyze the lattice average to find the connected and disconnected parts in a purely formal way. For example,

$$\begin{aligned}\langle L_i(t_1)L_i(t_2)\rangle_L &= \{\langle L_i(t_1)L_i(t_2)\rangle_L - \langle L_i(t_1)\rangle_L\langle L_i(t_2)\rangle_L\} \qquad \text{connected} \\ &\quad + \langle L_i(t_1)\rangle_L\langle L_i(t_2)\rangle_L \qquad \text{disconnected}\end{aligned} \tag{14.79}$$

To illustrate other cases we introduce a notation for lattice-averaged moment operators,

$$m_{\mu\nu\cdots\Omega} \equiv \langle L_i(t_\mu)L_i(t_\nu)\cdot\cdot\cdot L_i(t_\Omega)\rangle_L \tag{14.80}$$

In this notation Eq. (13.79) becomes

$$m_{1\,2} = \underset{\text{connected}}{(m_{1\,2} - m_1m_2)} + \underset{\text{disconnected}}{m_1m_2} \tag{14.81}$$

Now the role of the $c_{1\,2}\cdots{}_n$ functions is taken by the connected parts of the moments

$$u_{\mu\nu\cdots\Omega} = \text{connected part of } m_{\mu\nu\cdots\Omega} \tag{14.82}$$

Now we have, for example

$$\begin{aligned} m_1 &= u_1 \\ m_{1\,2} &= u_{1\,2} + u_1u_2 \\ m_{1\,2\,3} &= u_{1\,2\,3} + u_1u_{2\,3} + u_{1\,2}u_3 + u_1u_2u_3 \\ m_{1\,2\,3\,4} &= u_{1\,2\,3\,4} + u_1u_{2\,3\,4} + u_{1\,2\,3}u_4 + u_{1\,2}u_{3\,4} \\ &\quad + u_1u_2u_{3\,4} + u_1u_{2\,3}u_4 + u_{1\,2}u_3u_4 + u_1u_2u_3u_4 \end{aligned} \tag{14.83}$$

These expansions are analogous to the Ursell function expansions (Percus, 1964) of equilibrium statistical mechanics, hence the notation $u_{1\cdots n}$. Equations (14.83) are readily inverted to express the $u_{1\cdots n}$ in terms of the moments, like $u_{1\,2} = m_{1\,2} - m_1m_2$. By construction we are assured that $u_{1\,2\cdots n}$ vanishes if any $t_{j-1} - t_j$, with $j \leqslant n$, is much larger than the greatest lattice correlation time τ_L.

The analogy with Eq. (6.45) of course suggests that one should rearrange the moment expansion of $C^{\ddagger}(t)$ to improve the behavior at large t. We now describe two ways to do this. The first, which has been used by Kubo (1962) and Freed (1972), is along the lines of the passage from Eq. (6.64) to (6.65), called the *cumulant expansion method.*

However, we cannot merely invoke Lemma 1 (Eq. 6.47) in the present case because of the effect of time ordering here. Thus if Lemma 1 were applicable, the expansion of $m_{1\,2\,3}$ in Eq. (14.83) would include $u_{1\,3}u_2$ and the expansion of $m_{1\,2\,3\,4}$ would include $u_2u_{1\,3\,4}$, $u_3u_{1\,2\,4}$, $u_{1\,3}u_{2\,4}$, $u_{1\,4}u_2u_3$, $u_{1\,3}u_2u_4$, and $u_{2\,4}u_1u_3$. But there is no way that lengthening a bond in $m_{1\,2\,3}$ decomposes it into $u_{1\,3}u_2$, and so on, so Lemma 1 does not work here. Nevertheless, the basic operation of taking the logarithm to go from Eq. (6.45) to (6.46) can be extended to apply here.

Thus we define *cumulant functions* by writing

$$C^{\ddagger}(t) = \langle A^*[\exp_O \sum_{n\geqslant 1} K_n(t)]A\rangle_S \tag{14.84}$$

where the subscript O means that we must look to time ordering and where $K_n(t)$ is a cumulant function of the form

$$K_n(t) = \int_0^t d(1\ 2 \cdots n)[L_i(t_1)L_i(t_2) \cdots L_i(t_n)]_c \tag{14.85}$$

where $[\cdots]_c$ means a certain modified lattice average, which also has to be specified. We also introduce, corresponding to $u_{\mu\nu\cdots\Omega}$,

$$k_{\mu\nu\cdots\Omega} \equiv [L_i(t_\mu)L_i(t_\nu) \cdots L_i(t_\Omega)]_c \tag{14.86}$$

To find the prescription for the cumulant lattice average $[\cdots]_c$, we expand the exponential in Eq. (14.84) and compare with the moment expansion to get

$$\begin{aligned}\left\langle A^*\left[1 + \int_0^t d(1)m_1 + \int_0^t d(1\ 2)m_{1\,2} + \int_0^t d(1\ 2\ 3)m_{1\,2\,3} + \cdots\right]A\right\rangle_S \\ = \Big\langle A^*\Big\{1 + K_1(t) + K_2(t) + \cdots \\ + \frac{1}{2!}O[K_1(t) + K_2(t) + \cdots]^2 \\ + \frac{1}{3!}O[K_1(t) + K_2(t) + \cdots]^3 + \cdots\Big\}A\Big\rangle_S\end{aligned} \tag{14.87}$$

Again, O is the time-ordering operator. On the left side of the equation, $L_i(t')$ is on the left of $L_i(t'')$ in each term if $t' > t''$. The same must be true on the right after time ordering. For the products K_nK_m, and so on, the operator works as follows, in terms of the notation in Eq. (14.86).

$$O[K_1(t)K_1(t)] = 2!\int_0^t d(1\ 2)k_1k_2 \tag{14.88a}$$

$$\begin{aligned}O[K_1(t)K_2(t)] &= O[K_2(t)K_1(t)] \\ &= \int_0^t d(1\ 2\ 3)(k_{1\,2}k_3 + k_1k_{2\,3} + k_2k_{1\,3})\end{aligned} \tag{14.88b}$$

Here

$$k_2k_{1\,3} = [L_i(t_2)]_c[L_i(t_1)L_i(t_3)]_c$$

appears although it violates the order of the $L_i(t)$ operators on the left side of Eq. (14.87). This forbidden term must be canceled by another contribution on the right. Evidently, the canceling term is to be found within $O(K_3(t))$. Some other examples of time ordering follow.

$$O[K_1(t)K_1(t)K_1(t)] = 3!\int_0^t d(1\ 2\ 3)k_1k_2k_3 \tag{14.88c}$$

$$O[K_1(t)K_1(t)K_2(t)] = 2!\int_0^t d(1\ 2\ 3\ 4)(k_1k_2k_{3\,4} + k_1k_{2\,3}k_4$$

$$+ k_{1\,2}k_3k_4 + k_1k_3k_{2\,4} + k_2k_3k_{1\,4} + k_2k_4k_{1\,3}) \tag{14.88d}$$

Now we equate terms of Eq. (14.87) with given numbers of $L_i(t)$ factors to find

$$\begin{aligned} m_1 &= k_1 \\ m_{1\,2} &= k_{1\,2} + k_1k_2 \\ m_{1\,2\,3} &= k_{1\,2\,3} + k_1k_{2\,3} + k_{1\,2}k_3 + k_{1\,3}k_2 + k_1k_2k_3 \end{aligned} \tag{14.89}$$

These equations may then be solved for the cumulant lattice averages to obtain

$$\begin{aligned} k_1 &= m_1 \\ k_{1\,2} &= m_{1\,2} - m_1m_2 \\ k_{1\,2\,3} &= m_{1\,2\,3} - m_1m_{2\,3} - m_{1\,2}m_3 - m_{1\,3}m_2 + 2m_1m_2m_3 \end{aligned} \tag{14.90}$$

These results are readily extended to higher order. They already show how the large-t divergences of the moment expansion are suppressed in $\Sigma_n K_n(t)$. In fact, each term of this sum is proportional to t at large t. In spin relaxation the most important term is the $n = 2$ term; truncation of the sum at $n = 2$ is equivalent to Redfield's theory of spin relaxation (Freed, 1972). To this order we have

$$C^{\ddagger}(t) \simeq \left\langle A^*\left(\exp\int_0^t dt_1 \int_0^{t_1} dt_2\, k_{1\,2}\right)A\right\rangle_S \tag{14.91}$$

which, according to the analysis in connection with Fig. 14.3, gives a qualitatively reasonable $C^{\ddagger}(t)$, at least as far as the t dependence is concerned.

The other known way to overcome the difficulties with the moment expansion of $C^{\ddagger}(t)$ is to construct the corresponding expansion of the memory function. Again we content ourselves here with analyzing the first few terms, although the general results are known (Hwang et al., 1975).

The memory function for $C^{\ddagger}(t)$ may be defined by the equation [see Eq. (14.15)].

$$\dot{C}^{\ddagger}(t_1 - t_Z) = \langle A^* m_1 A\rangle C^{\ddagger}(t_1 - t_Z) - \int_{t_Z}^{t_1} M(t_1 - t_Y)\, dt_Y\, C^{\ddagger}(t_Y - t_Z) \tag{14.92}$$

The expansion of $C^{\ddagger}$ in terms of the moment is the left side of Eq. (14.87). The corresponding expansion of $\dot{C}\ddagger$ is

$$\dot{C}^{\ddagger}(t_1) = \left\langle A^*\left\{m_1 + \int_0^{t_1} dt_2 m_{1\,2} + \int_0^{t_1} d(2\;3)\, m_{1\,2\,3} + \cdots\right\}A\right\rangle_S \tag{14.93}$$

Finally, we define the moment expansion of the memory function $M(t_1 - t_Y)$ as follows:

$$-M(t_1 - t_Y) = \left\langle A^*\left\{u_{1Y} + \int_{t_Y}^{t_1} dt_2 u_{1\,2\,Y} + \int_{t_Y}^{t_1} d(2\;3)\, u_{1\,2\,3\,Y} + \cdots\right\}A\right\rangle_S \tag{14.94}$$

where we anticipate that the lattice averages $u_{1\,2\cdots n}$, which are the moments of minus the memory function M, are the functions previously defined in Eq. (14.82).

Now we replace the functions $\dot{C}^{\ddagger}$, $C^{\ddagger}$, and M in Eq. (14.92) by their expansions and equate terms with a given power of $L_i(t)$, just as we did for the cumulant expansion. This exercise indeed leads to the results given earlier in Eq. (14.83).

When $\langle A^* m_1 A\rangle$ is negligible, then Eq. (14.92) may be transformed to give

$$C^{\ddagger}[z] = \frac{1}{iz + M(z)} \tag{14.95}$$

and the Redfield approximation is like

$$M[z] = \int_0^\infty dt\, e^{-izt} \langle A^* u_{1Y}(t_1, t_Y) A\rangle_S \tag{14.96}$$

where $t = t_1 - t_Y$. However, in proper spin relaxation theory we deal with an array of dynamical variables and with correlation matrices and memory matrices, as already mentioned. Nevertheless, the results in Eqs. (14.93) to (14.96) show the simple structure of the statistical-mechanical perturbation theory using the memory function expansion rather than the cumulant expansion. Our demonstration that the moments of the memory function $M(t)$ are the Ursell functions $u_{1\,2\,3\cdots}$, which have the cluster property described following Eq. (14.83) (that is, $u_{1\,2\,3\cdots}$ vanishes if any of the time intervals $t_{j+1} - t_j$ it covers is much bigger than the lattice correlation time) leads to an important conclusion. That is that the terms in the moment expansion of the memory function $M(t)$, unlike those of the moment expansion of the time-correlation function $C^{\ddagger}(t)$, have no secular t dependence. In other words, a perturbation expansion of the memory function is a well-behaved construction, unlike the corresponding expansion of $C^{\ddagger}(t)$. It is amusing to notice that we have already seen an application of this aspect of the memory function in Eq. (14.28).

14.7 KINETIC THEORY

The term *kinetic theory* has been used by various authors to designate different parts of time-dependent statistical mechanics or even the whole field. Here we use it to mean the theory of reduced time-dependent phase-space distribution functions, as in the following simple examples, or the corresponding time-correlation functions, to be defined shortly.

$$f^{(1)}(\mathbf{R}, \mathbf{P}; t) \equiv \int d\Gamma\; D(\Gamma; t) \sum_{i=1}^{N} \delta(\mathbf{R} - \mathbf{r}_i)\, \delta(\mathbf{P} - \mathbf{p}_i) \tag{14.97a}$$

$$\begin{aligned} f^{(2)}(\mathbf{R}, \mathbf{P}, \mathbf{R}'\mathbf{P}'; t) \equiv \int d\Gamma\; D(\Gamma; t) \\ \times \sum_{i,k}^{d} \delta(\mathbf{R} - \mathbf{r}_i)\, \delta(\mathbf{P} - \mathbf{p}_i)\, \delta(\mathbf{R}' - \mathbf{r}_k)\, \delta(\mathbf{P}' - \mathbf{p}_k) \end{aligned} \tag{14.97b}$$

Here the superscript d means the sum over all pairs of distinct particles of the N-particle system and $D(\Gamma;\; t)$ is the phase-space distribution function denoted

$f_N(\Gamma_N, t)$ in Eq. (11.12); the notation $D(\Gamma; t)$ is used here to avoid confusion with other functions which are denoted by f and F in this section.

By applying the operator

$$\int d\Gamma \sum_i \delta(\mathbf{R}-\mathbf{r}_i)\,\delta(\mathbf{P}-\mathbf{p}_i)\cdots$$

to both sides of the Liouville equation

$$\frac{\partial}{\partial t} D(\Gamma; t) = -H^\times D(\Gamma; t)$$

we find, in the absence of external fields,

$$\begin{aligned}\left(\frac{\partial}{\partial t}+\frac{\mathbf{P}}{m}\cdot\frac{\partial}{\partial \mathbf{r}}\right)&f^{(1)}(\mathbf{R}, \mathbf{P}; t)\\ &= -\int d\Gamma \sum_i \delta(\mathbf{R}-\mathbf{r}_i)\,\delta(\mathbf{P}-\mathbf{p}_i) \sum_{k\neq i} \mathbf{F}_{ik}\cdot\frac{\partial}{\partial \mathbf{p}_k} D(\Gamma; t) \\ &= -\int d^3R'\, d^3p'\, \mathbf{F}_{\mathbf{R},\mathbf{R}'}\cdot\frac{\partial}{\partial \mathbf{P}} f^{(2)}(\mathbf{R}, \mathbf{P}, \mathbf{R}', \mathbf{P}'; t)\end{aligned} \tag{14.98}$$

The force $\mathbf{F}_{\mathbf{R},\mathbf{R}'}$ is the force on the particle at $\mathbf{R}$ due to the particle at $\mathbf{R}'$.

It is straightforward to formulate the general reduced distribution function $f^{(n)}(\mathbf{R}, \mathbf{P}, \ldots; t)$ and adapt the steps above to express the evolution of $f^{(n)}$ in terms of an integral over $f^{(n+1)}$; the resultant sequence of equations is called the Bogolyubov, Born, Green, Kirkwood, Yvon hierarchy, or more often, the *BBGKY hierarchy,* the order of the names being nearly in the reverse of the chronological order of their contributions to the role of the hierarchy in statistical mechanics.

Perhaps the simplest kinetic theory result is an approximation to Eq. (14.98) based on the approximation

$$f^{(2)}(\mathbf{R}, \mathbf{P}, \mathbf{R}', \mathbf{P}'; t) \simeq f^{(1)}(\mathbf{R}, \mathbf{P}; t) f^{(1)}(\mathbf{R}', \mathbf{P}'; t) \tag{14.99}$$

which leads to the *Vlasov equation,*

$$\left[\frac{\partial}{\partial t}+\frac{\mathbf{P}}{\mathbf{m}}\cdot\frac{\partial}{\partial \mathbf{R}}+\langle \mathbf{F}(\mathbf{R}; t)\rangle\cdot\frac{\partial}{\partial \mathbf{P}}\right] f^{(1)}(\mathbf{R}, \mathbf{P}; t) = 0 \tag{14.100}$$

in which

$$\langle \mathbf{F}(\mathbf{R}; t)\rangle \equiv \int \mathbf{F}_{\mathbf{R},\mathbf{R}'}\, f^{(1)}(\mathbf{R}', \mathbf{P}'; t)\, d^3\mathbf{R}'\, d^3\mathbf{p}' \tag{14.101}$$

is the average force acting on the particle at $\mathbf{R}$ at time t due to all of the other particles in the system. The Vlasov equation suppresses the dynamical effect of interactions among the particles, so it is often referred to as a collisionless approximation, although some part of the effect of the particle–particle interactions is carried by the average force term. The Vlasov equation is useful only for treating plasmas and other systems in which relatively long-range interactions dominate the picture so that the effect of hard collisions is unimportant.

In the more typical kinetic theory results, the Boltzmann equation and its most important descendant the Enskog equation, versions of which are given below, the effects of hard collisions *are* taken into account and found to be important, as we would expect in view of the importance of the harshly repulsive core parts of the pair potentials in determining the equilibrium properties. Many of these important developments antedated the discovery of the BBGKY hierarchy, and there is a vast literature concerned with their development, application, and reformulation as statistical mechanics progressed from the time of Boltzmann.

It seems most interesting to turn now to see how some of these theories of the evolution of $f^{(1)}(\mathbf{R}, \mathbf{P}; t)$ can be obtained by beginning with the generalized Langevin equation. Compared to the Liouville equation, this starting point makes it easy to get irreversible evolution equations for time-correlation functions and other averages, although, by this procedure, we give up the chance to probe the sources of irreversibility. Another advantage of deriving kinetic theory from the generalized Langevin equation is that this procedure shows the relation of kinetic theory to other approximation methods that can be derived from the same equation.

As the dynamical variable whose evolution is to be generated by a generalized Langevin equation, we choose the phase-space density "operator," a generalization of the operator for the density field in location space introduced in Eq. (4.13).

$$\begin{aligned}\hat{f}(\mathbf{R}, \mathbf{P}; t) &= \sum_{k=1}^{N} \delta[\mathbf{R} - \mathbf{r}_k(t)]\, \delta[\mathbf{P} - \mathbf{p}_k(t)] - \left\langle \sum_{k=1}^{n} \delta[\mathbf{R} - \mathbf{r}_k(t)]\, \delta[\mathbf{P} - \mathbf{p}_k(t)] \right\rangle \\ &= \sum_{k=1}^{N} \delta[\mathbf{R} - \mathbf{r}_k(t)]\, \delta[\mathbf{P} - \mathbf{p}_k(t)] - \rho\Phi(\mathbf{P})\end{aligned} \tag{14.102}$$

where ρ is the equilibrium average density and $\Phi(\mathbf{P})$ is the Maxwellian distribution for the momentum of a particle. The following phase-space correlation function is a generalization of the van Hove density–density correlation function to the full phase space:

$$G(\mathbf{R} - \mathbf{R}', \mathbf{P}, \mathbf{P}'; t) = \langle \hat{f}(\mathbf{R}, \mathbf{P}; t)\hat{f}(\mathbf{R}', \mathbf{P}'; 0)\rangle \tag{14.103a}$$

where an equilibrium ensemble average is indicated.

To make the equations easier to follow, we shall, whenever it does not cause confusion, replace $\hat{f}(\mathbf{R}_1, \mathbf{P}_1; t)$ by $\hat{f}(1; t)$ and similarly for other functions of the phases. In the resulting notation the preceding equation reads

$$G(1\ 2; t) = \langle \hat{f}(1; t)\hat{f}(2; 0)\rangle \tag{14.103b}$$

We also notice that for the equal-time-correlation function we have, denoting $\hat{f}(1; 0) = \hat{f}(1)$,

$$\begin{aligned}G(1\ 2; 0) &= G(1\ 2) = \langle \hat{f}(1)\hat{f}(2)\rangle \\ &= \delta(1\ 2) \sum_{i=1}^{N} \langle \delta(\mathbf{R}_1 - \mathbf{r}_i)\, \delta(\mathbf{P}_1 - \mathbf{p}_i)\rangle\end{aligned}$$

$$+ \sum_{i,j}^{d} \langle \delta(\mathbf{R}_1 - \mathbf{r}_i)\, \delta(\mathbf{P}_1 - \mathbf{p}_i)\, \delta(\mathbf{R}_2 - \mathbf{r}_j)\, \delta(\mathbf{P}_2 - \mathbf{p}_j) \rangle \tag{14.104}$$

$$- \rho^2 \Phi(1)\Phi(2) h(1\ 2)$$

$$= \rho \Phi(1)\Phi(2)\left[\frac{\delta(1\ 2)}{\Phi(2)} + \rho h(1\ 2)\right]$$

The second line of this equation comprises the self-contribution to $G(1\ 2)$, while the third comprises the distinct part. The superscript d on the sum in the third line specifies that only pairs of distinct particles i, j are to be taken. Other notational features are

$$\Phi(1) \equiv \Phi(P_1) \qquad \text{as in Eq. (14.103)} \tag{14.105a}$$

$$h(1\ 2) \equiv h(|\mathbf{R}_1 - \mathbf{R}_2|) \qquad \text{as in Eq. (7.3)} \tag{14.105b}$$

and

$$\delta(1\ 2) \equiv \delta(R_1 - R_2)\, \delta(P_1 - P_2) \tag{14.105c}$$

Also we remark that the van Hove correlation function, Eqs. (11.50c) and (11.51) for a one-component system, which we now write $G^{(vH)}$, is given by

$$G^{(vH)}(\mathbf{R} - \mathbf{R}', t) = \int d^3P d^3P' G(\mathbf{R} - \mathbf{R}', \mathbf{P}, \mathbf{P}'; t) + \rho^2 \tag{14.106}$$

where the term ρ^2 appears as a consequence of the normalization of $\hat{f}$ in Eq. (14.102).

The phase-space density, being a dynamical variable, evolves according to the generalized Langevin equation

$$\frac{\partial}{\partial t} \hat{f}(1; t) = \int i\omega(1\ 2) d(2) \hat{f}(2; t) - \int_0^t ds \int M(1\ 2; t - s) d(2) \hat{f}(2; s) + R(1; t) \tag{14.107}$$

which is like Eq. (14.42) except that now the dynamical variable $\hat{f}$, the frequency $i\omega$, the memory function M, and the random "force" R are given as generalizations (discrete indices $\longrightarrow$ continuous variables) of the arrays $\mathbf{A}$, $i\boldsymbol{\omega}$, $\mathbf{M}$, and $\mathbf{R}$, respectively, in the correlation matrix equation. This transcription is natural if we think of $\hat{f}(1; t)$ as a vector in a Hilbert space (see Section 14.2) since $\mathbf{A}$ in Eq. (14.42) is a vector whose elements are dynamical variables. Thus we have

$$\begin{aligned} i\omega(1\ 2) &= i\omega(\mathbf{r} - \mathbf{r}', \mathbf{p}, \mathbf{p}') \\ &= \int \langle [H^{\times} \hat{f}(\mathbf{r}, \mathbf{p})] \hat{f}\, \mathbf{R}, \mathbf{P}) \rangle\, d^3R\, d^3P\, G^{-1}(\mathbf{R} - \mathbf{r}', \mathbf{P}, \mathbf{p}') \end{aligned} \tag{14.108}$$

where phase 1 is $\mathbf{r}, \mathbf{p}$, while phase 2 is $\mathbf{r}', \mathbf{p}'$. Also, G^{-1} is defined by the equation

$$\int G(1\ 2) d(2) G^{-1}(2\ 3) = \delta(1\ 2) \tag{14.109a}$$

or, in more detailed notation,

$$\int G(\mathbf{r}-\mathbf{R}, \mathbf{p}, \mathbf{P})\, d^3R\, d^3P\, G^{-1}(\mathbf{R}-\mathbf{r}', \mathbf{P}, \mathbf{p}') = \delta(\mathbf{r}-\mathbf{r}')\,\delta(\mathbf{p}-\mathbf{p}') \tag{14.109b}$$

Also, we have

$$M(1\ 3: t) = \int \langle R(1; t)R(2; 0)\rangle d(2)G^{-1}(2\ 3) \tag{14.110a}$$

or

$$M(\mathbf{r}-\mathbf{r}', \mathbf{p}, \mathbf{p}'; t) = \int \langle R(\mathbf{r}, \mathbf{p}; t)R(\mathbf{q}, \mathbf{P}; 0)\rangle\, d^3q\, d^3P\, G^{-1}(\mathbf{q}-\mathbf{r}', \mathbf{P}, \mathbf{p}') \tag{14.110b}$$

and

$$R(1; t) \equiv R(\mathbf{r}, \mathbf{p}, t) = e^{(1-\mathscr{P})tH^\times}(1-\mathscr{P})H^\times \hat{f}(\mathbf{r}, \mathbf{p}) \tag{14.111}$$

which is a particular case of Eq. (14.17e).

By multiplying Eq. (14.107) from the right by $\hat{f}(\mathbf{r}', \mathbf{p}'; 0)$ and then averaging over the equilibrium ensemble, we obtain the equation for the evolution of the time-correlation function

$$\frac{\partial}{\partial t}G(1\ 2; t) = \int i\omega(1\ 3)d(3)G(3\ 2; t) - \int_0^t ds \int M(1\ 3; t-s)d(3)G(3\ 2; s) \tag{14.112}$$

in view of the version

$$\langle R(\mathbf{r}, \mathbf{p}; t)\hat{f}(\mathbf{r}', \mathbf{p}'; 0)\rangle = 0 \tag{14.113}$$

of Eq. (14.45b) that is relevant here.

In traditional kinetic theory, as noted above, the key observable quantity is the reduced one-particle phase-space distribution function

$$\begin{aligned} F(1; t) \equiv F(\mathbf{r}, \mathbf{p}; t) &\equiv \int d\Gamma\, D(\Gamma; 0)\, \hat{f}(\mathbf{r}, \mathbf{p}, t) \\ &= \int d\Gamma\, D(\Gamma; t)\, \hat{f}(\mathbf{r}, \mathbf{p}, 0) \\ &= f^{(1)}(\mathbf{r}, \mathbf{p}; t) - \rho\Phi(p) \end{aligned} \tag{14.114}$$

It is $f^{(1)}$ adjusted to have zero norm. We deduce the following evolution equation for $F(1; t)$ from Eq. (14.107):

$$\frac{\partial}{\partial t}F(1; t) = \int i\omega(1\ 2)d(2)F(2; t)$$

$$-\int_0^t ds \int M(1\ 2;\ t-s)d(2)F(2;\ s) \tag{14.115}$$

$$+\int d\Gamma\ D(\Gamma;\ 0)\ R(1;\ t)$$

where $D(\Gamma;\ 0)$ may be chosen so that the last term vanishes. In this case we see that the time-correlation function $G(1\ 2;\ t)$ and the reduced distribution function $F(1;\ t)$ are governed by the same evolution equation. This will be the basis for our comparison of approximate evolution equations for $G(1\ 2;\ t)$ (derived from the generalized Langevin equation) with approximate evolution equations for $F(1;\ t)$ (taken from the results of kinetic theory).

Equations (14.112) and (14.115) depend on both the frequency function $i\omega$, which represents instantaneous dynamical response, and the memory function M, which contains the delayed response. While $i\omega$ is straightforward to compute, the memory function is usually quite impossible to evaluate with precision. Then the simplest approximation which suggests itself, a "mean field approximation," is to neglect the term in which the memory function M contributes to the evolution. We shall show how this approximation leads to the Zwanzig–Vlasov equation

$$\left(\frac{\partial}{\partial t}+\frac{\mathbf{P}_1}{m}\cdot\nabla_1\right)G(1\ 3;\ t)=\rho\Phi(1)\left(\frac{\mathbf{P}_1}{m}\right)\cdot\nabla_1\int c(1\ 2)d(2)G(2\ 3;\ t) \tag{14.116}$$

where $c(1\ 2)$ is the usual direct correlation function of equilibrium theory (Chapter 7), where $1=\mathbf{R}_1,\ \mathbf{P}_1$ and where $\nabla_1=(\partial/\partial\mathbf{R}_1)$.

First we notice that the corresponding kinetic theory equation may readily be derived from Eqs. (14.100) and (14.114) if we keep only the terms that survive the linearization

$$\langle\mathbf{F}(\mathbf{R};\ t)\rangle\cdot\frac{\partial}{\partial\mathbf{P}}f^1(\mathbf{R},\ \mathbf{P};\ t)\simeq\langle\mathbf{F}(\mathbf{R};\ t)\rangle\cdot\frac{\partial}{\partial\mathbf{P}}\Phi(P) \tag{14.117}$$

The resulting form of the Vlasov equation is

$$\left(\frac{\partial}{\partial t}+\frac{\mathbf{P}_1}{m}\cdot\nabla_1\right)F(1;\ t)=\frac{\rho}{k_BT}\Phi(1)\frac{\mathbf{P}_1}{m}\cdot\nabla_1\int u(1\ 2)d(2)F(2;\ t) \tag{14.118}$$

We see that the pair potential $u(1\ 2)$ in the Vlasov equation is replaced by the "renormalized potential" $-k_BTc\ (1\ 2)$ in the Zwanzig–Vlasov equation. This change indicates that the equation derived from the generalized Langevin equation more nearly incorporates the accurate equilibrium structure of a dense system [see Eq. (7.15)]. However, like the Vlasov equation, the Zwanzig–Vlasov fails to represent the very important effects of collisions involving the strongly repulsive core part of the pair potential.

Now we summarize the technique for the evaluation of $i\omega$ according to Eq. (14.108). The first step is to find the explicit form of G^{-1}. It is easy to verify that

$$G^{-1}(1\ 2)=\frac{\delta(1\ 2)}{\rho\Phi(2)}-c(1\ 2) \tag{14.119}$$

satisfies Eq. (14.109) in view of Eq. (14.104) and the Ornstein–Zernicke equation.

To complete the analysis of ω, we write the Liouville operator as a sum of self and distinct parts

$$H^\times = H_s^\times + H_d^\times \tag{14.120a}$$

$$H_s^\times \equiv \sum_{i=1}^{N} \frac{\mathbf{p}_i}{m_i} \cdot \frac{\partial}{\partial \mathbf{r}_i} \tag{14.120b}$$

$$H_d^\times \equiv \sum_{i,j}^{d} \mathbf{F}_{ij} \cdot \frac{\partial}{\partial p_j} \tag{14.120c}$$

where

$$F_{ij} \equiv -\frac{\partial u(|\mathbf{r}_i - \mathbf{r}_j|)}{\partial r_j} \tag{14.120d}$$

Now we find that

$$\langle [H_s^\times \hat{f}(1)]\hat{f}(2)\rangle = -\frac{\mathbf{P}_1}{m} \cdot \nabla_1 G(1\ 2) \tag{14.121}$$

$$\begin{aligned}\langle [H_d^\times \hat{f}(1)]\hat{f}(2)\rangle &= \frac{1}{k_B T}\,\Phi(1)\Phi(2)\frac{\mathbf{P}_1}{m} \cdot \sum_{ijk} \langle \mathbf{F}_{ij}\, \delta(\mathbf{R}_1 - \mathbf{r}_j)\, \delta(\mathbf{R}_2 - \mathbf{r}_k)\rangle \\ &= \Phi(1)\Phi(2)\frac{\mathbf{P}_1}{m} \cdot \nabla_1 \sum_{ij} \langle \delta(\mathbf{R}_1 - \mathbf{r}_i)\, \delta(\mathbf{R}_2 - \mathbf{r}_j)\rangle \\ &= \rho^2\Phi(1)\Phi(2)\frac{\mathbf{P}_1}{m} \cdot \nabla_1 g(1\ 2)\end{aligned} \tag{14.122}$$

Then by evaluating Eq. (14.108) with these results, we find that

$$\begin{aligned} i\omega(1\ 2) &= \int \left[-\frac{\mathbf{P}_1}{m} \cdot \nabla_1 G(1\ 3)\right] d(3) G^{-1}(3\ 2) \\ &\quad + \int \left[\rho^2\Phi(1)\Phi(3)\frac{\mathbf{P}_1}{m} \cdot \nabla_1 h(1\ 3)\right] d(3) \left[\frac{\delta(3\ 2)}{\rho\Phi(2)} - c(3\ 2)\right] \\ &= -\frac{\mathbf{P}_1}{m} \cdot \nabla_1 [\delta(1\ 2) - \rho\Phi(1) c(1\ 2)]\end{aligned} \tag{14.123}$$

When this formula for $i\omega$ is substituted in Eq. (14.112) and the memory function term is neglected, we recover Eq. (14.116).

To derive some of the more important results of kinetic theory, we supplement the Liouville operator in Eq. (14.120) by a part that represents the action of collisions between particles with hard cores. Thus we suppose that there is some division of the pair potential into soft and hard parts, say as in Fig. 9.2a, and that the hard-core terms have been omitted in Eq. (14.117). Then the complete Liouville operator would be

$$H^\times = H_s^\times + H_d^\times + \sum_{i,j} m^{-1} |\mathbf{p}_{ij} \cdot \mathbf{r}_{ij}| \Theta(\mathbf{p}_{ij} \cdot \mathbf{r}_{ij})\, \delta(\mathbf{r}_{ij} - \sigma)(\hat{b}_{ij} - 1) \tag{14.124}$$

where σ is the diameter of the hard core, where $\mathbf{p}_{ij} = \mathbf{p}_i - \mathbf{p}_j$, $\mathbf{r}_{ij} = r_i - r_j$, Θ is the unit step function, and where $\hat{b}_{ij}$ is the operator whose action is to replace the momenta $\mathbf{p}_i$, $\mathbf{p}_j$ of two particles about to collide by the post-collision momenta:

$$\hat{b}_{ij}\mathbf{p}_i = \mathbf{p}_i{}^* = \mathbf{p}_i - \mathbf{p}_{ij} \cdot \hat{\mathbf{r}}_{ij}\hat{\mathbf{r}}_{ij} \tag{14.125}$$

$$\hat{b}_{ij}\mathbf{p}_j = \mathbf{p}_j{}^* = \mathbf{p}_j + \mathbf{p}_{ij} \cdot \hat{\mathbf{r}}_{ij}\hat{\mathbf{r}}_{ij} \tag{14.126}$$

The new part, the hard-core interaction part, of the Liouville operator is not derived from mechanics in the same way as the rest. Indeed, the standard formulation does not make sense when one has hard-sphere potentials but, as found by Ernst et al. (1969), the form shown here exactly accounts for the discontinuous changes in motion when hard spheres collide. [Jhon and Dahler (1981)]

The frequency $i\omega$ can be calculated for the new Liouville operator as shown by Sung and Dahler (1983). The resulting mean-field approximation to the equation for the evolution of $G(1\ 2;\ t)$ in the case in which the hard cores have diameter σ can be written as follows. Because it is necessary here to give all of the details, we employ the identification

$$G(1\ 2;\ t) = G(\mathbf{r}_1 - \mathbf{r}_2, \mathbf{p}_1, \mathbf{p}_2;\ t)$$

and then we have

$$\begin{aligned}
&\left(\frac{\partial}{\partial t} + \frac{\mathbf{p}}{m} \cdot \frac{\partial}{\partial \mathbf{r}}\right) G(\mathbf{r} - \mathbf{r}', \mathbf{p}, \mathbf{p}';\ t) \\
&= \rho\sigma^2 g(\sigma) \int \left(\hat{\sigma} \cdot \frac{\mathbf{p} - \mathbf{p}''}{m}\right) \Theta[\hat{\sigma} \cdot (\mathbf{p} - \mathbf{p}'')]\ d^3p''\ d^2\hat{\sigma} \\
&\quad \times [G(\mathbf{r} - \mathbf{r}', \mathbf{p}^*, \mathbf{p}';\ t)\Phi(\mathbf{p}''^*) - G(\mathbf{r} - \mathbf{r}', \mathbf{p}, \mathbf{p}';\ t)\Phi(\mathbf{p}'') \\
&\quad + G(\mathbf{r} - \hat{\sigma}\sigma - \mathbf{r}', \mathbf{p}'', \mathbf{p}';\ t)\Phi(\mathbf{p}^*) - G(\mathbf{r} - \sigma\hat{\sigma} - \mathbf{r}', \mathbf{p}'', \mathbf{p}';\ t)\Phi(\mathbf{p})] \\
&\quad + \rho k_B T \Phi(\mathbf{p}) \frac{\mathbf{p}}{m} \cdot \frac{\partial}{\partial \mathbf{r}} \int [c(\mathbf{r} - \mathbf{r}'') - g(\sigma)\{\Theta(r - \sigma) - 1\}]\ d^3r''\ d^3p'' \\
&\quad \times G(\mathbf{r}'' - \mathbf{r}', \mathbf{p}'', \mathbf{p}';\ t)
\end{aligned} \tag{14.127}$$

where $\hat{\sigma}$ is a unit vector along the line of centers of two particles.

In the case in which there are no soft interactions among the particles (i.e., in which the terms that derive from $H_d^\times$ are omitted), this equation is the same as the "generalized" or "renormalized" Enskog equation, which has been derived by Mazenko (1972) and others in various ways. If the low-density limit is taken and if the resultant description resolves only distances that are much greater than σ, we recover the linearized version of Boltzmann's equation for a gas of hard spheres,

$$\begin{aligned}
&\frac{\partial}{\partial t} + \frac{\mathbf{p}}{m} \cdot \frac{\partial}{\partial \mathbf{r}} G(\mathbf{r} - \mathbf{r}', \mathbf{p}, \mathbf{p}';\ t) \\
&\qquad = \rho \int \left\{ \left(\sigma \cdot \frac{\mathbf{p} - \mathbf{p}''}{m}\right) \Theta[\hat{\sigma} \cdot (\mathbf{p} - \mathbf{p}'')] \right\} d^3p''\ d^2\hat{\sigma} \\
&\qquad\quad \times [G(\mathbf{r} - \mathbf{r}', \mathbf{p}^*, \mathbf{p}';\ t)\Phi(\mathbf{p}''^*) - G(\mathbf{r} - \mathbf{r}', \mathbf{p}, \mathbf{p}';\ t)\Phi(\mathbf{p}'') \\
&\qquad\quad + [G(\mathbf{r} - \mathbf{r}', \mathbf{p}, \mathbf{p}';\ t)\Phi(\mathbf{p}^*) - G(\mathbf{r} - \mathbf{r}', \mathbf{p}'', \mathbf{p}';\ t)\Phi(\mathbf{p})]
\end{aligned} \tag{14.128}$$

Equation (14.127) has been solved for a fluid with a pair potential having both hard and soft parts. It is found (Sung and Dahler, 1983) that the soft part enters into the transport coefficients (at zero frequency and wave number) only through its effect on the equilibrium structural coefficient $g(\sigma)$. When applied to a hard-sphere fluid, Eq. (14.127), as already noted, reduces to the renormalized Enskog equation. As indicated in Fig. 14.4, this equation accounts for the dynamical structure factor of the hard-sphere fluid at several densities and over a considerable range of frequency, and for a greater range of wave number than shown in the figure. Apparently, this kind of kinetic theory has a lot of promise for calculating the measurable dynamical properties of models for fluids.

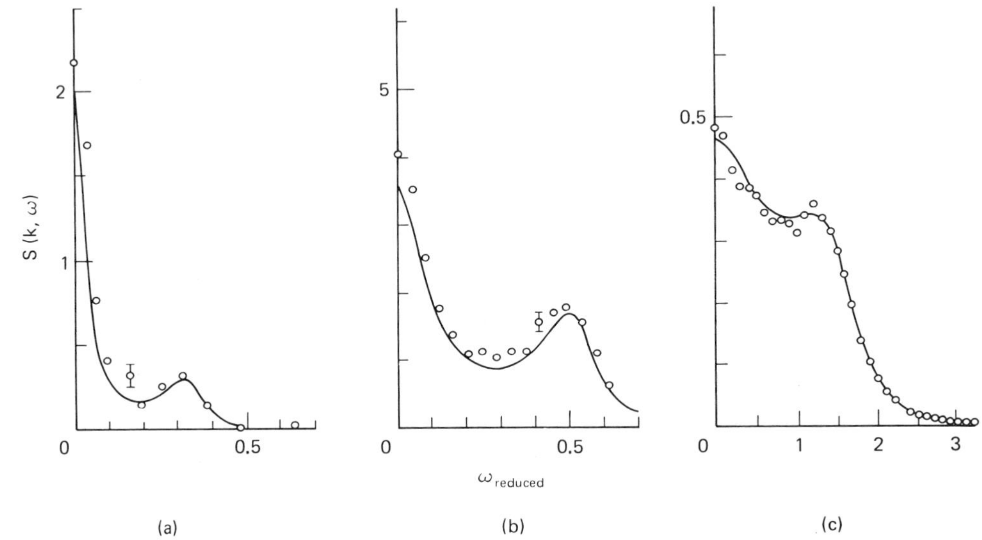

Figure 14.4 Dynamical structure factor $S(k, w)$ for the hard-sphere fluid at the following densities $\rho\sigma^3$ and wave numbers $k\sigma$. (a) 0.883, 0.38. (b) 0.471, 0.44. (c) 0.141, 0.41. The reduced frequency is $\omega\tau_E/k\sigma$, where τ_E is the Enskog mean collision time. The circles are MD simulation results, while the curves are calculated from the renormalized Enskog equation. (From Alley et al., 1983.)

The tendency for the renormalized Enskog equation results to deviate from the simulation data at low frequencies is just what may be expected in view of the fact that the mean-field approximation, from which Eq. (14.127) may be derived, suppresses the effect of the long-time correlations in the memory function and hence is accurate for short times but not long. One way one may hope to obtain a still better theory is through mode coupling techniques, namely the introduction into the dynamical variable of nonlinear terms such as the bilinear term $\hat{f}(1; t)\hat{f}(2; t')$. If it is apt, the resulting frequency array, the generalization of $i\omega(1\ 2)$ in Eq. (14.107), includes more of the collective effects, so the resulting memory effects will be less important. Such a procedure, which is illustrated, for example, in a report by Sung and Dahler (1983), may be viewed as a generalization of the passage from Eq. (14.15) to Eq. (14.42). Thus it illustrates the versatility of approximation techniques based upon the generalized Langevin equation.

DO PROBLEM 14.13.

THE LITERATURE

Most of the topics in this chapter are covered in Hansen and McDonald (1976), where, as in the reviews by Zwanzig (1965), Kubo (1966), and Berne (1971), the citations to the original work of M. S. Green, Zwanzig, Kubo, Mori, Martin, and others may be found. Connections to earlier work in time-dependent statistical mechanics are stressed by Chester (1963), while an account by Hynes and Deutch (1975) describes the various ways in which projection operators may be formulated. Also, the theory of cumulant averages has been treated by Freed (1968) and van Kampen (1974), and mode coupling methods are described by Keyes (1977). Akcasu and Duderstadt (1969) first derived the kinetic theory from the generalized Langevin equation, while other features are discussed by Boon and Yip (1980), Mazenko and Yip (1977), and Dorfman and van Beijeren (1977). Recent work by Evans (1983) indicates that long-time tails may be much more important than has been appreciated.

PROBLEMS

14.1. Verify that $\mathscr{P}_0$ as defined in Eq. (14.3) conforms to the general projection operator equations (14.4).

14.2. If $\mathbf{Q}$ and $\mathbf{R}$ are ordinary vectors in three-dimensional space, and $\mathbf{R} = |\mathbf{R}|\hat{\mathbf{R}}$ defines the unit vector $\hat{R}$, show that the operator $\mathscr{P}_R$ defined by the equation

$$\mathscr{P}_R\mathbf{Q} = \frac{\hat{R}\mathbf{Q}\cdot\mathbf{R}}{|\mathbf{R}|}$$

is a projection operator. [Notice that it looks like Eq. (14.3) if we write $\mathscr{P}_R\mathbf{Q} = \mathbf{R}\mathbf{Q}\cdot\mathbf{R}/\mathbf{R}\cdot\mathbf{R}$.]

14.3. Beginning with the definitions in Eqs. (14.17d)–(14.17f), derive the following rules.

(a) $\langle f_k^* f_j\rangle = \delta_{jk}\langle f_j^* f_j\rangle$ (14.129a)

(b) $L_j = (1 - \mathscr{P}_0 - \mathscr{P}_1 - \cdots - \mathscr{P}_{j-1})L_0$ (14.129b)

(c) For an arbitrary function ϕ

$\langle f_k^* L_j \phi\rangle = 0 \quad \text{if } k < j$ (14.129c)

(d) $\langle f_k^* f_j(t)\rangle = 0 \quad \text{if } t > 0 \text{ and } k < j$ (14.129d)

14.4. In the case in which $f_0 = A = m\dot{x}$, the x component of a particle's momentum in a system of N particles of the same species, find explicit equations for f_1, f_2, and f_3 in terms of the Hamiltonian. Verify that these random forces are mutually orthogonal. Find the explicit formulas for $\langle f_j^* f_j\rangle$ for $j = 0, 1, 2$. Notice that $\langle f_1^* f_1\rangle = \langle \dot{f}_0^2\rangle$. Is this true for any f_0?

14.5. **(a)** Calculate $\langle \dot{A}(t)A \rangle$ for the generalized Langevin equation, Eq. (14.15), and compare your result with Eq. (14.23). Do they agree after allowing for differences in notation?

(b) From your result in part (a), show that the generalized Langevin equation implies the following relation among Fourier–Laplace transforms:

$$\Sigma_0[z] = \frac{1}{i(z - \omega_0) + M[z]} \tag{14.130}$$

From this find the limiting conditions in which $\langle A(t)A \rangle$ exhibits simple exponential decay.

14.6. Find the conditions on the parameters α and t_1 in Eq. (14.26) that are required for $\langle A(t)A \rangle$ to have an oscillatory component when $A = A^*$ and $\omega_0 = 0$.

14.7. Show that if Eq. (14.26) is valid, then $\langle A\dot{A} \rangle = 0$. Compare with the strict Langevin equation and compare with Problem 11.3.

14.8. **(a)** Show that Eq. (11.25) implies that all of the moments of the power spectrum for a process are finite.

(b) Show that if $A(t)$ is governed by the strict Langevin equation, the zeroth moment of $\langle A(t)A \rangle$ is finite but all of the higher, even moments are infinite.

(c) Show that if $A(t)$ is governed by the generalized Langevin equation with the memory function given by Eq. (14.26), the zeroth and second moments are finite but the moments of order 4, 6, 8, . . . are all infinite.

14.9. For a sphere of radius R moving in a viscous, incompressible medium, the linear relation between the velocity response $\langle u(t) \rangle_F$ and the external force $F(t)$ on the sphere has been calculated by Landau and Lifschitz (1959). At any particular frequency ω their result may be expressed as follows. The force

$$F(t) = u_\omega e^{-i\omega t} \zeta_\omega \tag{14.131}$$

along a given coordinate gives the velocity

$$\langle u(t) \rangle_F = u_\omega e^{-i\omega t} \tag{14.132}$$

along the same coordinate; the friction coefficient ζ_ω has the complex conjugate

$$\zeta_\omega^* = [A + (i\omega)^{1/2} B + i\omega C] m \tag{14.133}$$

where

$$A \equiv \frac{6\pi\nu\rho R}{m} \tag{14.134a}$$

$$B \equiv \frac{3\pi\rho R^2 (\nu/2)^{1/2}}{m} \tag{14.134b}$$

$$C \equiv \frac{2\pi\rho R^3}{3m} \tag{14.134c}$$

in terms of the kinematic viscosity $\nu = \eta/\rho$ and the density (ρ = mass/volume) of the solvent and the mass m of the sphere. You may notice that the A term alone gives Stoke's law.

(a) Using the linear response theory show that in this case the memory function in Eq. (14.15) has the Fourier–Laplace transform

$$M[\omega] = A + (i\omega)^{1/2} B + i\omega C \tag{14.135}$$

(b) Show that the Fourier–Laplace transform of $C_{uu}(t)$, the normalized velocity autocorrelation function, has the low-frequency behavior

$$C_{uu}[\omega] = \frac{1}{A} - \frac{B(i\omega)^{1/2}}{A^2} + O(\omega) \tag{14.136}$$

(c) Show that these equations also imply

$$C_{uu}(t) \longrightarrow \frac{1}{\sqrt{\pi} A^2 t^{3/2}} \qquad \text{as } t \longrightarrow \infty \tag{14.137}$$

This behavior, called a long-time tail, is found in other studies of velocity autocorrelation functions (e.g., Fig. 14.5). [*Note*: You may find it easier to transform $C_{uu}(t)$, including the $t^{-3/2}$ term, to get Eq. (14.104) rather than inverting the latter.]

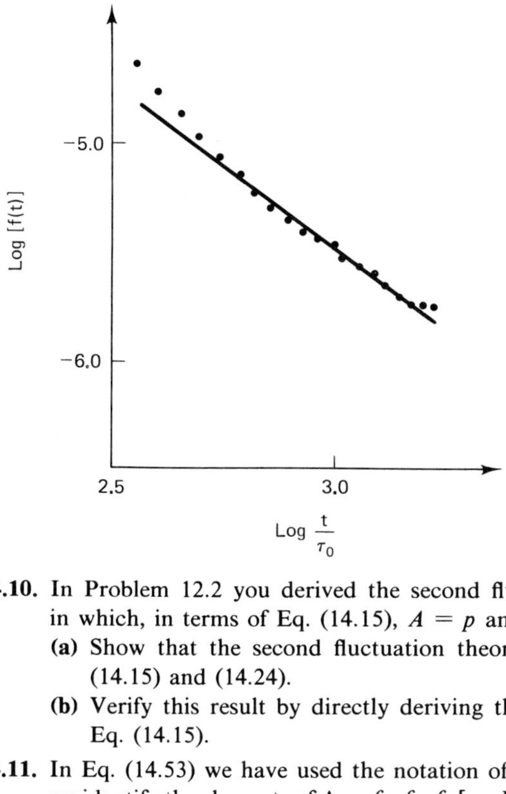

Figure 14.5 Log-log plot of $f(t) = \langle \dot{x}(t)\dot{x}\rangle / \langle \dot{x}^2 \rangle$ (dots) as a function of t obtained by MD simulation of a very large (4000-atom) 6–12 model showing the long-time tail of the velocity autocorrelation function. The state has reduced density $\rho\sigma^3 = 0.45$ and the reduced temperature $k_B T/e_0 = 2.17$; it corresponds to argon at roughly the critical density and twice the critical temperature. The line has slope $-\frac{3}{2}$, so the simulation shows rather clearly that at large t the velocity autocorrelation function decays like $t^{-3/2}$, not like $e^{-\alpha t}$. (From Levesque and Ashurst, 1974.)

14.10. In Problem 12.2 you derived the second fluctuation-dissipation theorem for the case in which, in terms of Eq. (14.15), $A = p$ and $\omega_0 = 0$.

(a) Show that the second fluctuation theorem is generally valid by comparing Eqs. (14.15) and (14.24).

(b) Verify this result by directly deriving the relation $\langle R^* R(t) \rangle = M(t) \langle A^* A \rangle$ from Eq. (14.15).

14.11. In Eq. (14.53) we have used the notation of Levesque and Verlet (1970). Show that if we identify the elements of **A** as f_0, f_1, f_2 [see Eq. (14.49)], then in terms of the frequencies in Eqs. (14.18), we have

$$\Delta_1 = \omega_0 \tag{14.138a}$$

$$\Delta_2 = (2\omega_0^2 + \Omega_0^2)^{1/2} \tag{14.138b}$$

14.12. Show that Ω_0, as defined in Eq. (14.52c), is also given by

$$\Omega_0^2 = \frac{\langle \partial F_{iz} / \partial z \rangle}{m} \tag{14.139}$$

in view of Eq. (3.36), so Ω_0 has the character of the average Einstein frequency for a particle in a cage formed by its neighbors.

14.13. How do

$$\int_0^t d(1\ 2\ 3\ 4)u_{1\ 2}u_{3\ 4}$$

$$\int_0^t d(1\ 2\ 3\ 4)u_{1\ 2}u_{2\ 3}u_{3\ 4}$$

grow with large t in the case where

$$u_{mn} = e^{-bt} \quad ?$$

Appendix 1

Fourier Transforms in Three Dimensions

Only the essential definitions and results are listed here, with a few examples. Suppose that $f(\mathbf{r})$ is any function of a location vector $\mathbf{r}$. [Example: $f(\mathbf{r}) = e^{-\alpha r}$ where $r = |\mathbf{r}|$.] Then the Fourier transform of f is defined as

$$\tilde{f}(\mathbf{k}) = \int d^3r\, e^{i\mathbf{k}\cdot\mathbf{r}} f(\mathbf{r}) \tag{A1.1}$$

The three-dimensional variables $\mathbf{r}$ and $\mathbf{k}$ are said to be conjugate to each other. Next in the light of Eq. (3.18) we look at the transform of $\tilde{f}$. It is called the *inverse transform*.

$$\begin{aligned}(2\pi)^{-3}\int d^3k\, e^{-i\mathbf{k}\cdot\mathbf{r}'}\tilde{f}(\mathbf{k}) &= (2\pi)^{-3}\int d^3r\, f(\mathbf{r})\int d^3k\, e^{i\mathbf{k}\cdot(\mathbf{r}-\mathbf{r}')} \\ &= \int d^3r\, f(\mathbf{r})\,\delta(\mathbf{r}-\mathbf{r}') = f(\mathbf{r}')\end{aligned} \tag{A1.2}$$

Thus the Fourier transform process applied twice yields the original function.

Suppose that

$$\phi(r') = \int d^3r\, f(\mathbf{r}) g(\mathbf{r}' - \mathbf{r}) \tag{A1.3}$$

This kind of integral is called a *convolution* (or folding or faltung or chain integral). It also is represented by the graph

if the distance between the white circles is fixed at r′ and the circles are all 1-circles in the nomenclature of Section 6.5. We examine its Fourier transform:

$$\begin{aligned}\tilde{\phi} &= \int d^3r' e^{i\mathbf{k}\cdot\mathbf{r}'} \int d^3r\, f(\mathbf{r}) g(\mathbf{r}' - \mathbf{r}) \\ &= \iint d^3r'\, d^3r\, e^{i\mathbf{k}\cdot\mathbf{r}} f(\mathbf{r})\, e^{i\mathbf{k}\cdot(\mathbf{r}'-\mathbf{r})} g(\mathbf{r}' - \mathbf{r}) \\ &\underset{\mathbf{u}=\mathbf{r}'-\mathbf{r}}{=} \int d^3r\, e^{i\mathbf{k}\cdot\mathbf{r}} f(\mathbf{r}) \int d^3u\, e^{i\mathbf{k}\cdot\mathbf{u}} g(\mathbf{u}) = \tilde{f}(\mathbf{k})\tilde{g}(\mathbf{k})\end{aligned} \tag{A1.4}$$

This result for a chain of two links is readily extended to a chain of any length. Thus the Fourier transform of a convolution is the product of the transforms of the functions of the chain.

In the case of a spherical function, we have

$$f(\mathbf{r}) \rightarrow f(|\mathbf{r}|) = f(r)$$

and then we find that the transform calculated by Eq. (A1.1) depends only on k = $|\mathbf{k}|$. In this case, which is especially important for us, Eqs. (A1.1) and (A1.2) can be reduced to integrals over the magnitudes r and k.

$$\tilde{f}(k) = \int_0^\infty dr\, r^2 f(r) \int_0^{2\pi} d\phi \int_0^\pi d\theta\, \sin\theta\, e^{ikr\cos\theta} \underset{x=\cos\theta}{=} 2\pi \int_0^\infty dr\, r^2 f(r) \int_{-1}^{1} dx\, e^{ikrx}$$

$$\tilde{f}(k) = 4\pi \int_0^\infty dr\, r^2 \frac{\sin kr}{kr} f(r) \tag{A1.5}$$

and

$$f(r) = (2\pi^2)^{-1} \int_0^\infty dk\, k^2 \frac{\sin kr}{kr} \tilde{f}(k) \tag{A1.6}$$

Now we examine a few examples.

Gaussian functions

$$f(r) = \left(\frac{n}{\pi}\right)^{3/2} e^{-nr^2} \tag{A1.7}$$

Note that $\int d^3r\, f(r) = 1$. Using Eq. (A1.1), it is easy to find

$$\tilde{f}(k) = \exp\left(\frac{-k^2}{4n}\right) \tag{A1.8}$$

We expect the same result from Eq. (A1.5), but that is hard to show.

To test the inversion formula in this case, write $v = 1/4n$, $\tilde{f}(k) = e^{-vk^2}$, and get

$$f(r') = (2\pi)^{-3} \int d\mathbf{k}\, e^{-i\mathbf{k}\cdot\mathbf{r}'} e^{-vk^2} = \left(\frac{1}{2\pi}\sqrt{\frac{\pi}{v}}\right)^3 e^{-r'^2/4v} = (n/\pi)^{3/2} e^{-nr'^2} \tag{A1.9}$$

A very important aspect of the Fourier transform operation can be learned from the Gaussian example. See Fig. A.1. Thus the transform operation converts a flat function into a peaked one, and a peaked one into a flat one, but leaves the normal Gaussian function unchanged. Although it is *only* the $n = \frac{1}{2}$ Gaussian function which is unchanged by Fourier-transforming, the qualitative aspect of these conclusions holds for any function at all.

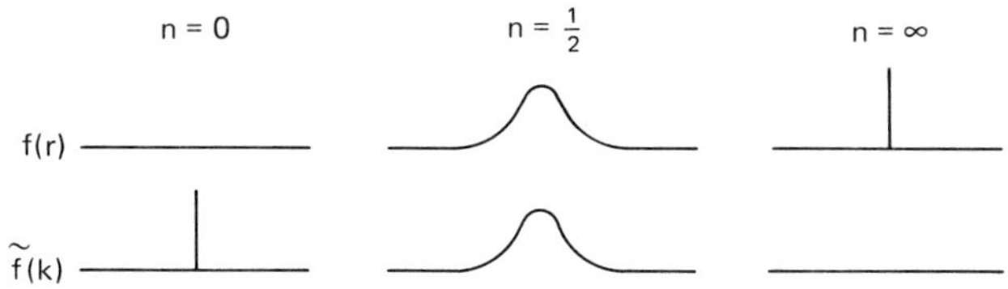

Debye or Yukawa potential

$$f(r) = \frac{e^{-sr}}{r} \tag{A1.10}$$

Using Eq. (A1.5):

$$\tilde{f}(k) = \left(\frac{4\pi}{k}\right) \int_0^\infty dr\, e^{-sr} \sin kr = \left(\frac{4\pi}{k}\right) \mathrm{Im} \int_0^\infty dr\, e^{-r(s-ik)} \tag{A1.11a}$$

Here we recall that $e^{iz} = \cos(z) + i \sin(z)$, so $\sin(z) = \mathrm{Im}\, e^{iz}$, where Im means "the imaginary part of." For any complex number C we have $\mathrm{Im}\, C = (C - C^*)/2i$, where * means the complex conjugate (i.e., change every i to $-i$). See discussion leading to Eq. (11.36). The integral is elementary and we find that

$$\tilde{f}(k) = \frac{4\pi}{k} \mathrm{Im} \frac{1}{s - ik} = \frac{4\pi}{s^2 + k^2} \tag{A1.11b}$$

This $\tilde{f}(k)$ is a *Lorentzian* function of k. Conversely, the Debye potential is the (3D) Fourier transform of the Lorentzian function.

THE LITERATURE

A good introductory account is given by Matthews and Walker (1965). The close relation of Fourier transforms and Dirac delta functions is developed by Lighthill (1959). Very extensive tables of integrals like those in (A1.8) and (A1.11b) are given by Erdelyi (1954).

PROBLEMS

A1.1. The function $f(r) = e^{-sr}$ has the transform

$$\tilde{f}(k) = \frac{8\pi s}{s^2 + k^2} \tag{A1.12}$$

Verify this result by applying the operator $(-\partial/\partial s)$ to $\tilde{f}(k)$ in Eq. (A1.11).

A1.2. The function $f(r) = e^{-sr}/r^2$ has the transform $\tilde{f}(k) = (4\pi/k) \tan^{-1}(k/s)$. Verify this result by applying the operator $\int_s^\infty ds$ to $\tilde{f}(k)$ in Eq. (A1.11).

A1.3. Verify that Eq. (A1.5) is equivalent to the pair of equations

$$\tilde{f}(k) = 4\pi \int_0^\infty dr \cos(kr)\, F(r) \tag{A1.13}$$

$$F(r) \equiv \int_r^\infty dR\, Rf(R) \tag{A1.14}$$

A1.4. Verify the following result:

$$\begin{aligned} &\lim_{|\mathbf{r}_1 - \mathbf{r}_n|=0} \int f_1(r_{1\,2})\, d^3r_2 f_2(r_{2\,3})\, d^3r_3 \cdots d^3r_{n-1} f_{n-1}(r_{n-1,n}) \\ &= \frac{1}{2\pi^2} \int_0^\infty dk\, k^2 \tilde{f}_1(k)\, \tilde{f}_2(k) \cdots \tilde{f}_{n-1}(k) \end{aligned} \tag{A1.15}$$

The r-space integral in the $\mathbf{r}_1 = \mathbf{r}_n$ limit is called a *ring integral.* Identify the ring integrals in Eqs. (6.42).

Appendix 2

Generalized Fourier Transforms for Orientation-Dependent Interactions

An invariant pairwise function $F(1\ 2)$ of the locations and orientations of two molecules, one at the location and orientation specified by 1 and the other at the location and orientation specified by 2, has been defined as a function that is unchanged if the pair of molecules is translated or rotated as a whole (Fig. 8.2a). The molecular pair correlation function $h_{ab}(1\ 2)$ and direct correlation function $c_{ab}(1\ 2)$ in isotropic fluids are examples of invariant pairwise functions. For such functions a convolution integral, like that in the Ornstein–Zernike equation, can be reduced to a matrix product of certain generalized Fourier transforms.

If the convolution integral is defined as

$$F(1\ 3) = \int G(1\ 2)d(2)\mathrm{H}(2\ 3) \tag{A2.1}$$

and G and H are invariant pairwise functions, then the convolution theorem is

$$\hat{F}(k)^{l_1 l_3}_{m_1 m_3 n} = \sum_{l_2, m_2} \hat{G}(k)^{l_1 l_2}_{m_1 m_2 n} \hat{H}(k)^{l_2 l_3}_{m_2 m_3 n} \tag{A2.2}$$

where we define the integral transform for F, and similarly for G and H,

$$\hat{F}(k)^{ll'}_{mm'n} \equiv \sum_L 4\pi i^L \int_0^\infty r^2\, dr\, j_L(kr) \int d\Omega_1\, d\Omega_2\, \Psi^{ll'L}_{mm'n}(1\ 2)F(1\ 2) \tag{A2.3}$$

where j_L is the spherical Bessel function of order L and $\Psi^{ll'L}_{mm'n}$ is a generalized spherical harmonic

$$\Psi^{ll'L}_{mm'n} \equiv \frac{1}{64\pi^4} \sum_s (-)^{n+s+L}[(2l+1)(2l'+1)]^{1/2}(2L+1) \times \begin{pmatrix} l & l' & L \\ -n & n & 0 \end{pmatrix} \begin{pmatrix} l & l' & L \\ -s & s & 0 \end{pmatrix} D^l_{ms}(\Omega_1)\, D^{*l'}_{m's}(\Omega_2) \tag{A2.4}$$

expressed in terms of Wigner $3j$ coefficients $(: \; : \; :)$ and Wigner D functions (see below). In these equations Ω_1 and Ω_2 are the rotations specified in Fig. 8.2a.

We note that the range of the variables in the summations is controlled by the range of indices in the nonzero $3j$ coefficients and, moreover, is the same range as one encounters for the quantum numbers related to angular momentum in quantum mechanics. Finally, we note that the transformation that is inverse to Eq. (A2.3) is

$$F(1\,2) = \sum_{\substack{ll'L \\ mm'n}} \Psi^{*ll'L}_{mm'n}(1\,2) \frac{1}{2\pi^2 i^L} \int_0^\infty k^2\, dk\, j_L(kr)\hat{F}(k)^{ll'}_{mm'n} \tag{A2.5}$$

The Blum–Torruella invariant expansion of an invariant pairwise function $F(1\,2)$ is

$$F(1\,2) = \sum_{\substack{ll'L \\ mm'}} F(r)^{ll'L}_{mm'}\, \Phi^{ll'L}_{mm'}(1\,2) \tag{A2.6}$$

The rotational invariants are, with a particular normalization, usually given by the equation

$$\Phi^{ll'L}_{mm'}(1\,2) = \frac{L!}{\begin{pmatrix} l & l' & L \\ 0 & 0 & 0 \end{pmatrix}} \sum_{nn'\nu} \begin{pmatrix} l & l' & L \\ n & n' & \nu \end{pmatrix} D^l_{mn}(\Omega_1)\, D^{l'}_{m'n'}(\Omega_2)\, D^L_{0\nu}(\hat{r}_{1\,2}) \tag{A2.7a}$$

where the rotations Ω_1 and Ω_2 and the direction $\hat{r}_{1\,2}$ are those specified in Fig. 8.2b. In this equation the rotational invariant *apparently* depends on eight angles. However, a simple transformation yields the equivalent equation

$$\Phi^{ll'L}_{mm'}(1\,2) = \frac{L!}{\begin{pmatrix} l & l' & L \\ 0 & 0 & 0 \end{pmatrix}} \sum_s \begin{pmatrix} l & l' & L \\ -s & s & 0 \end{pmatrix} D^l_{m-s}(\Omega_1) D^{l'}_{m's}(\Omega_2) \tag{A2.7b}$$

where the rotations Ω_1 and Ω_2 now are those specified in Fig. 8.2a. Now we see the proper dependence of the rotational invariant on five independent angles.

The rotational invariants are mutually orthogonal:

$$\int d\Omega_1\, d\Omega_2\, \Phi^{ljL}_{mn}(1\,2)\Phi^{*l'j'L'}_{m'n'}(1\,2) = \delta_{ll'}\delta_{jj'}\delta_{LL'}\delta_{mm'}\delta_{nn'} \left[\frac{8\pi^2 L!}{\begin{pmatrix} l & j & L \\ 0 & 0 & 0 \end{pmatrix}}\right]^2 \frac{64\pi^2}{(2l+1)(2j+1)(2L+1)} \tag{A2.8}$$

The Cartesian forms of the first few rotational invariants are given in Table 8.1.

The orthogonality between the rotational invariants and the generalized spherical harmonics defined in Eq. (A2.4) is expressed by

$$\int d\Omega_1\, d\Omega_2 \Psi^{ll'L}_{mm'0}(1\,2)\, \Phi^{jj'J}_{kk'}(1\,2) \qquad \text{(A2.9)}$$
$$= \delta_{lj}\delta_{l'j'}\delta_{LJ}\delta_{m,-k}\delta_{m'k}L!(-)^{L+m}[(2l+1)(2l'+1)]^{-1/2}$$

Now by substituting the invariant expansion (A2.6) for $F(1\,2)$ in Eq. (A2.3) we find, with the help of (A2.9), the generalized Fourier transforms $\hat{F}$ in relation to the coefficients $\tilde{F}$ in the invariant expansion.

$$\hat{F}(k)^{ll'}_{mm'0} = \sum_L \{L!(-)^{L+m}[(2l+1)(2l'+1)]^{-1/2}\}\tilde{F}(k)^{ll'L}_{mm'} \qquad \text{(A2.10)}$$

The coefficients on the right are the Hankel transforms

$$\tilde{F}(k)^{ll'L}_{mm'} = 4\pi i^L \int_0^\infty r^2\, dr\, j_L(kr)\, F(r)^{ll'L}_{mm'} \qquad \text{(A2.11)}$$

of the coefficients in the invariant expansion [see Eq. (8.5a).

As an example of the application of Eq. (A2.10), we have

$$\hat{c}_{ww'}(k)^{11}_{000} = \tfrac{1}{3}\,[\tilde{c}_{ww'}(k)^{110}_{00} - i\tilde{c}_{ww'}(k)^{111}_{00} + 2\tilde{c}_{ww'}(k)^{112}_{00}] \qquad \text{(A2.12)}$$

It is easily shown that if an invariant expansion coefficient $F(r)^{ll'L}_{mm'}$ is short range in the sense that the integral in (A2.11) converges even if the j_L function is left out, then its Hankel transform vanishes at $k = 0$ unless $L = 0$. Furthermore, if $L = 0$, the $3j$ coefficients vanish unless $l = l'$, so the only nonvanishing transforms at $k = 0$ that derive from short-range functions in r-space have the structure $\hat{F}(k)^{l\ \ l}_{m\,-m\,0}$.

The functions $D^l_{mn}(\Omega) = D^l_{mn}(\alpha, \beta, \gamma) = D^l_{mn}(A \leftarrow B)$ appearing in Eqs. (A2.4) and (A2.7) are generalized spherical harmonics called Wigner D functions. Here the rotation $\Omega = \alpha, \beta, \gamma$ (see Fig. 8.1) is the rotation of a solid object to some orientation A from orientation B. They are defined for $-l \leq m \leq l$ and $-l \leq n \leq l$. Their most important equations are listed here. The * specifies the complex conjugate.

$$D^0_{00}(\Omega) = 1 \qquad \text{(A2.13a)}$$

$$\int d\Omega\, D^l_{mn}(\Omega)\, D^{*\lambda}_{\mu\nu}(\Omega) = \frac{8\pi^2}{2l+1}\,\delta_{l\lambda}\,\delta_{m\mu}\,\delta_{n\nu} \qquad \text{(A2.13b)}$$

$$\begin{aligned} D^l_{mn}(A \leftarrow B) &= D^{*l}_{nm}(B \leftarrow A) \\ &= \{-\}^{m+n} D^l_{-n,-m}(B \leftarrow A) \qquad \text{(A2.13c)} \\ &= \{-\}^{m+n} D^{*l}_{-m,-n}(A \leftarrow B) \end{aligned}$$

$$D^l_{mn}(A \leftarrow C) = \sum_s D^l_{ms}(A \leftarrow B)\, D^l_{sn}(B \leftarrow C) \qquad \text{(A2.13d)}$$

$$\int d\Omega\, D^l_{mn}(\Omega)\, D^{l'}_{m'n'}(\Omega)\, D^L_{MN}(\Omega) = 8\pi^2 \begin{pmatrix} l & l' & L \\ m & m' & M \end{pmatrix} \begin{pmatrix} l & l' & L \\ n & n' & N \end{pmatrix} \qquad \text{(A2.13e)}$$

In these equations

$$\int d\Omega \;\ldots \qquad \text{means} \qquad \int_0^{2\pi} d\alpha \int_0^{\pi} \sin\beta \, d\beta \int_0^{2\pi} d\gamma \;\ldots$$

The number $\begin{pmatrix} l & l' & L \\ m & m' & M \end{pmatrix}$ is called a Wigner $3j$ coefficient. The indices l, l', and so on, are integers or half-integers but only integer indices are encountered in classical mechanics. Some properties are given here.

$$\begin{pmatrix} l & l' & L \\ m & m' & M \end{pmatrix} = \{-\}^{l+l'+L} \begin{pmatrix} l & l' & L \\ -m & -m' & -M \end{pmatrix} = \{-\}^{l+l'+L} \begin{pmatrix} l' & l & L \\ m' & m & M \end{pmatrix} \qquad \text{(A2.14a)}$$

$$\begin{pmatrix} l & l' & L \\ m & m' & M \end{pmatrix} = 0 \text{ unless } l,\ l',\ L \text{ can form a triangle } \textit{and } m + m' + M = 0 \qquad \text{(A2.14b)}$$

(for example, 1, 1, 1 or 1, 1, 2 can form a triangle but 1, 1, 3 cannot).

$$\begin{pmatrix} l & l' & L \\ 0 & 0 & 0 \end{pmatrix} = 0 \qquad \text{Unless } l + l' + L \text{ is even} \qquad \text{(A2.14c)}$$

Some examples of $3j$ coefficients:

$$\begin{pmatrix} l & l & 0 \\ m & -m & 0 \end{pmatrix} = \frac{\{-\}^{l+m}}{(2l+1)^{1/2}} \qquad \text{(A2.15a)}$$

$$\begin{pmatrix} 1 & 1 & 1 \\ m & -m & 0 \end{pmatrix} = \frac{m}{6^{1/2}} \qquad \text{(A2.15b)}$$

$$\begin{pmatrix} 1 & 1 & 2 \\ m & -m & 0 \end{pmatrix} = 2^{-|m|} \left(\frac{2}{15} \right)^{1/2} \qquad \text{(A2.15c)}$$

In other cases tables of the $3j$ coefficients may be used.

The spherical Bessel functions $j_m(x)$ in Eqs. (A2.3), (A2.5), and (A2.11) have the simplest members

$$j_0(x) = \sin x / x \qquad \text{(A2.16a)}$$

$$j_1(x) = \frac{\cos x}{x} - \frac{\sin x}{x^2} \qquad \text{(A2.16b)}$$

and the recursion relation

$$j_{m-1}(x) + j_{m+1}(x) = \frac{2m+1}{x} j_m(x), \qquad m \geq 1 \qquad \text{(A2.16c)}$$

They also obey certain orthogonality relations such as

$$\int_0^{\infty} j_m(rt) j_m(st) t^2 \, dt = \frac{\pi \, \delta(r-s)}{2rs} \qquad \text{(A2.17a)}$$

$$\int_0^{\infty} j_m(t) j_n(t) \, dt = \frac{\pi \, \delta_{mn}}{2(2n+1)} \qquad \text{(A2.17b)}$$

THE LITERATURE

These generalized Fourier transforms and similar transforms were derived by Jepsen and Friedman (1963) and Blum and Torruella (1972). The presentation here follows Friedman (1982). See also Gray and Gubbins (1982). Edmonds (1960) and Wigner (1959) give the complete theory of the D^l_{mn} and $3j$ coefficients. The theory of the spherical Bessel functions is given in Abramowitz and Stegun (1965). For extensive tables of the $3j$ coefficients, see Rotenberg et al. (1960), while more convenient formulas like Eq. (A2.15) for special cases are given by Brink and Satchler (1975).

PROBLEMS

A2.1. The spherical Bessel functions may be defined by the equation

$$j_m(z) = (1/2i^m)\int_{-1}^{1} e^{izu}P_m(u)\,du \tag{A2.18}$$

where P_m is the Legendre polynomial. Using

$$P_0(u) = 1, \qquad P_1(u) = u$$

calculate j_0 and j_1 and compare with Eq. (A2.16).

A2.2. Using the equation

$$P_m(\cos\beta) = D^m_{00}(\alpha, \beta, \gamma) \tag{A2.19}$$

and Eq. (A2.13b), verify the Neumann expansion

$$e^{i\mathbf{k}\cdot\mathbf{r}} = e^{ikr\cos\beta} = \sum_{n=0}^{\infty} \mathrm{i}^n[2n+1]j_n(\mathrm{kr})D^n_{00}(\alpha\beta\gamma) \tag{A2.20}$$

[Equation (A2.20) explains the association of spherical Bessel functions with Wigner D functions in the generalized Fourier transforms.]

A2.3. Use the Neumann expansion to derive Eq. (A1.5) from (A1.1) in the case that $f(\mathbf{r}) = f(r)$.

REFERENCES

ABRAGAM, A. (1961), *Nuclear Magnetic Resonance,* Clarendon Press, Oxford.

ABRAMOWITZ, M., and STEGUN, I. A. (1965), *Handbook of Mathematical Functions,* Dover, New York.

ADELMAN, S. A. (1976a), *J. Chem. Phys. 64,* 724.

ADELMAN, S. A. (1976b), *Chem. Phys. Lett. 38,* 567.

ADELMAN, S. A., and DEUTCH, J. M. (1975), *Adv. Chem. Phys. 31,* 103.

AKCASU, A. Z., and DUDERSTADT, J. J. (1969), *Phys. Rev. A2,* 962.

ALDER, B. J., and POLLOCK, E. L. (1981), *Annu. Rev. Phys. Chem. 32,* 311.

ALDER, B. J., and WAINRIGHT, T. E. (1957), *J. Chem. Phys. 27,* 1208.

ALDER, B. J., and WAINRIGHT, T. E. (1959), *J. Chem. Phys. 31,* 459.

ALDER, B. J., and WAINRIGHT, T. E. (1970), *Phys. Rev. A1,* 18.

ALLEY, W. E., ALDER, B. J., and YIP, S. (1983), *Phys. Rev. A27,* 3174.

ANDERSEN, H. C. (1975), *Annu. Rev. Phys. Chem. 26,* 145.

ANDERSEN, H. C. (1977), in *Modern Theoretical Chemistry* (B. J. Berne, ed.), Vol. 5: *Statistical Mechanics,* Part A, p. 1, Plenum Press, New York.

ANDERSEN, H. C., CHANDLER, D., and WEEKS, J. D. (1971), *Phys. Rev. A4,* 1597.

ANDERSEN, H. C., CHANDLER, D., and WEEKS, J. D. (1976), *Adv. Chem. Phys. 34,* 105.

BACON, G. E. (1975), *Neutron Diffraction,* Oxford University Press, London.

BARKER, J. A., and HENDERSON, D. (1971), *Mol. Phys. 21,* 187.

BARKER, J. A., and HENDERSON, D. (1976), *Rev. Mod. Phys. 48,* 587.

BARKER, J. A., FISHER, R. A., and WATTS, R. O. (1971), *Mol. Phys. 21,* 657.

BARTSCH, G. P., and EBELING, W. (1971), *Beitr. Plasmaphys. 11,* 393.

BAXTER, J. R. (1971), in *Physical Chemistry, An Advanced Treatise* (H. EYRING, D. HENDERSON, AND W. JOST, eds.), Vol. 8A, Chapter 4, Academic Press, New York.

BEEBY, J. L. (1973), *J. Phys. C 6,* 2262.

BERNE, B. J. (1977a), *Modern Theoretical Chemistry* (B. J. Berne, ed.), Vol. 5: *Statistical Mechanics,* Part A, Plenum Press, New York.

BERNE, B. J. (1977b), in *Modern Theoretical Chemistry* (B. J. Berne, ed.), Vol. 6: *Statistical Mechanics,* Part B, p. 233, Plenum Press, New York.

BERNE, B. J. (1971), in *Physical Chemistry* (D. Henderson, ed.), Vol. 8B, p. 540, Academic Press, New York.

BERNE, B. J., and PECORA, R. (1976), *Dynamic Light Scattering,* Wiley, New York.

BERNE, B. J., SKINNER, J. L., and WOLYNES, P. G. (1980), *J. Chem. Phys. 73,* 4314.

BLUM, L. (1980), in *Theoretical Chemistry* (D. Henderson, ed.), Vol. 5, Chapter 1, Academic Press, New York.

BLUM, L., and NARTEN, A. H. (1976), *Adv. Chem. Phys. 34,* 203.

BLUM, L., and TORRUELLA, A. J. (1972), *J. Chem. Phys. 56,* 303.

BOON, J. P., and YIP, S. (1980), *Molecular Hydrodynamics,* McGraw-Hill, New York.

BRINK, D. M., and SATCHLER, G. R. (1975), *Angular Momentum,* Clarendon Press, Oxford.

BUFF, F. P., and BROUT, R. (1955), *J. Chem. Phys. 23,* 458.

CEPERLEY, D. M., and CHESTER, G. V. (1977), *Phys. Rev. A15,* 755.

CHANDLER, D. (1977), *J. Chem. Phys. 67,* 1113.

CHANDLER, D., and ANDERSEN, H. C. (1972), *J. Chem. Phys. 57,* 1930.

CHANDLER, D., and PRATT, L. R. (1976), *J. Chem. Phys. 65,* 2925.

CARNAHAN, N. F., and STARLING, K. E. (1969), *J. Chem. Phys. 51,* 63.

CHANDRASEKHAR, S. (1943), *Rev. Mod. Phys. 15,* 1.

CHESTER, G. V. (1963), *Rep. Prog. Phys. 26,* 411.

CICCOTTI, G., and JACUCCI, J. (1975), *Phys. Rev. Lett. 35,* 789.

CUMMINGS, P. T., and STELL, G. (1982), *Mol. Phys. 46,* 383.

DAMLE, P. S., and TILLU, A. D. (1969), *Indian J. Pure Appl. Phys. 7,* 539.

DASANNACHARYA, B. A., and RAO, K. R. (1965), *Phys. Rev. 137,* A417.

DAVIDSON, N. (1962), *Statistical Mechanics,* McGraw-Hill, New York.

DEBYE, P. (1929), *Polar Molecules,* Dover, New York.

DE DOMINICIS, C. (1962), *J. Math. Phys. 3,* 183.

DE GROOT, S. R., and MAZUR, P. (1962), *Non-Equilibrium Thermodynamics,* North Holland Publishing Co., Amsterdam. Section XI.2.

DELEEUW, S. W., PERRAM, J. W., and SMITH, E. R. (1980), *Proc. R. Soc. Lond. A373,* 27, 57.

DEUTCH, J. M. (1977), *Faraday Symp. Chem. Soc. (Lond.) 11,* 26.

DEUTCH, J. M. (1980), *J. Chem. Phys. 73,* 4700.

DEUTCH, J. M., and OPPENHEIM, I. (1968), *Adv. Magn. Reson. 3,* 43.

DORFMAN, J. R., and VAN BEIJEREN, H. (1977), in *Modern Theoretical Chemistry* (B. J. Berne, ed.), Vol. 6: *Statistical Mechanics,* Part B, p. 65, Plenum Press, New York.

DYMOND, J. H., and SMITH, E. B. (1969), *The Virial Coefficients of Gases,* Clarendon Press, Oxford.

EBELING, W. (1964), *Z. Phys. Chem. 225,* 15.

EBELING, W. (1965), *Ann. Phys.* [7] *16,* 147.

EBELING, W., and GRIGO, M. (1980), *Ann. Phys. (Leipzig) 37,* 21.

EDMONDS, A. R. (1960), *Angular Momentum in Quantum Mechanics,* Princeton University Press, Princeton, N.J.

EGELSTAFF, P. A. (1967), *An Introduction to the Theory of the Liquid State,* pp. 66, 94, Academic Press, New York.

EINSTEIN, A. (1905), *Ann. Phys. 17,* 549.

EISENBERG, D., and KAUZMANN, W. (1969), *The Structure and Properties of Water,* pp. 151–155, Oxford University Press, New York.

ERDELYI, A. (1954), *Tables of Integral Transforms,* Vol. 1, McGraw-Hill, New York.

ERMAK, D. L., and MCCAMMON, J. A. (1978), *J. Chem. Phys. 69,* 1352.

ERNST, M. H., DORFMAN, J. R., HOEGY, W. R., and VAN LEEUWEN, J. M. J. (1969), *Physica 45,* 127.

EVANS, D. (1983) *Phys. Rev. A28,* 1016.

FELDERHOF, B. U. (1978), *J. Phys. A11,* 929.

FEYNMAN, R. P. (1951), *Phys. Rev. 84,* 108.

FLYGARE, W. H. (1978), *Molecular Structure and Dynamics,* Chapter 7, Prentice-Hall, Englewood Cliffs, N.J.

FOWLER, R., and GUGGENHEIM, E. A. (1952), *Statistical Thermodynamics,* Cambridge University Press, Cambridge.

FREDRIKZE, H. (1983), *Mol. Phys. 48,* 903.

FREED, J. H. (1968), *J. Chem. Phys. 49,* 376.

FREED, J. H. (1972), in *Electron Spin Relaxation in Liquids* (L. T. MUUS and P. W. ATKINS, eds.), p. 165, Plenum Press, New York.

FRIEDMAN, H. L. (1962), *Ionic Solution Theory,* Wiley, New York.

FRIEDMAN, H. L. (1975), *J. Sol. Chem. 1,* 387, 413, 419.

FRIEDMAN, H. L. (1981), *Annu. Rev. Phys. Chem. 32,* 179.

FRIEDMAN, H. L. (1982), *J. Chem. Phys. 76,* 1092.

FRIEDMAN, H. L., and DALE, W. D. T. (1977), in *Modern Theoretical Chemistry* (B. J. Berne, ed.), Vol. 5: *Statistical Mechanics,* Part A, p. 85, Plenum Press, New York.

FRIEDMAN, H. L., and RAMANATHAN, P. S. (1970), *J. Phys. Chem. 74,* 3756.

FRIEDMAN, H. L., KRISHNAN, C. V., and JOLICOEUR, C. (1973), *Ann. N.Y. Acad. Sci. 204,* 79. (Appendix.)

FUTRELLE, R. P., and MCGINTY, D. J. (1971), *Chem. Phys. Lett. 12,* 285.

GOLDSTEIN, H. (1950), *Classical Mechanics,* Addison-Wesley, Reading, Mass.

GORDON, R. G., and KIM, Y. S. (1972), *J. Chem. Phys. 56,* 3122.

GRAY, C. W. (1975), *Statist. Mech. 2,* 300 (Chemical Society, London, Specialist Periodical Reports).

GRAY, C. G., and GUBBINS, K. E. (1982), *Theory of Molecular Fluids,* Oxford University Press, London.

GROTE, R. F., and HYNES, J. T. (1982), *J. Chem. Phys. 77,* 3736.

GRUNDKE, E. W., and HENDERSON, D. (1972), *Mol. Phys. 24,* 269.

GUGGENHEIM, E. A. (1960), *Trans. Faraday Soc. 56,* 1159.

HALL, D. G. (1971), *J. Chem. Soc. Faraday Trans. 67,* 2516.

HANNA, M. W. (1981), *Quantum Mechanics in Chemistry,* 3rd ed., Benjamin-Cummings, Menlo Park, Calif.

HANSEN, J. P. (1970), *Phys. Rev. A2,* 221.

HANSEN, J. P., and MCDONALD, I. (1976), *Theory of Simple Liquids,* Academic Press, New York.

HELFAND, E. (1960), *Phys. Rev. 119,* 1.

HENDERSON, R. L., and GRAY, C. G. (1978), *Can. J. Phys. 56,* 571.

HEWISH, N. A., ENDERBY, J. E., and HOWELLS, W. S. (1982), *Phys. Rev. Lett. 48,* 756.

HILL, T. L. (1956), *Statistical Mechanics,* McGraw-Hill, New York.

HILL, T. L. (1960), *Statistical Thermodynamics,* Addison-Wesley, Reading, Mass.

HIRATA, F., ROSSKY, P. J., and PETTIT, B. M. (1983), *J. Chem. Phys. 78,* 4133.

HIRSCHFELDER, J. O., CURTIS, C. F., and BIRD, R. B. (1954), *The Molecular Theory of Gases and Liquids,* Wiley, New York.

HOOVER, W. G. (1983), *Annu. Reviews Phys. Chem. 34,* 103.

HØYE, J. S., and STELL, G. (1976), *J. Chem. Phys. 65,* 18.

HWANG, L. P., ANDERSON, C. F., and FRIEDMAN, H. L. (1975), *J. Chem. Phys. 62,* 2098.

HYNES, J. T., and DEUTCH, J. M. (1975), in *Physical Chemistry* (D. Henderson, ed.), Vol. XIB, Academic Press, New York.

JACKSON, E. A. (1968), *Equilibrium Statistical Mechanics,* Prentice-Hall, Englewood Cliffs, N.J.

JEPSEN, D. W., and FRIEDMAN, H. L. (1963), *J. Chem. Phys. 38,* 846.

JHON, M. S., and DAHLER, J. S. (1981) *J. Chem. Phys. 74,* 2477.

JOHNSON, E., and HAZOUME, R. P. (1979), *J. Chem. Phys. 70,* 1599.

JONES, R. B. and BURFIELD, G. S. (1982) *Physica 111A,* 562, 577.

JORGENSEN, W. L. (1981), *J. Am. Chem. Soc. 103,* 335, 341.

KADENOFF, L. P., and MARTIN, P. C. (1963), *Ann. Phys. 24,* 419.

KARPLUS, M., and PORTER, R. N. (1970), *Atoms and Molecules,* Sections 5.1–5.5, W. A. Benjamin, New York.

KEYES, T. (1977), in *Modern Theoretical Chemistry* (B. J. Berne, ed.), Vol. 6: *Statistical Mechanics,* Part B, p. 259, Plenum Press, New York.

KIRKWOOD, J. G. (1933), *Phys. Rev. 44,* 31; *45,* 116.

KIRKWOOD, J. G. (1939), *J. Chem. Phys. 7,* 911.

KIRKWOOD, J. G., and BUFF, F. P. (1951), *J. Chem. Phys. 19,* 774.

KIRKWOOD, J. G., DUNLOP, P. J., GOSTING, L. J., and KEGELES, K. (1960), *J. Chem. Phys. 33,* 1505.

KRAMERS, H. A. (1940), *Physica 7,* 284.

KUBO, R. (1962), *J. Phys. Soc. Jpn. 17,* 1100.

KUBO, R. (1966), *Rep. Prog. Phys. 29,* 255.

KUBO, R. (1969), *Adv. Chem. Phys. 15,* 101.

KUBO, R. (1973), in *Cooperative Phenomena* (H. Haken and M. Wagner, eds.), p. 140, Springer-Verlag, Berlin.

KUSHICK, J., and BERNE, B. J. (1977), in *Modern Theoretical Chemistry* (B. J. Byrne, ed.) Vol. 6: *Statistical Mechanics,* Part B, p. 41, Plenum Press, New York.

LADANYI, B., and CHANDLER, D. (1975), *J. Chem. Phys. 62,* 4308.

LADO, F. (1964), *Phys. Rev. A135,* 1013.

LADO, F. (1981), *J. Chem. Phys. 75,* 5461.

LADO, F., FOILES, S. M., and ASHCROFT, N. W. (1983), *Phys. Rev. A28,* 2374.

LANDAU, L. D., and LIFSCHITZ, E. M. (1957), *Mechanics,* Addison-Wesley, Reading, Mass.

LANDAU, L. D., and LIFSCHITZ, E. M. (1958), *Statistical Physics,* Addison-Wesley, Reading, Mass.

LANDAU, L. D., and LIFSCHITZ, E. M. (1959), *Fluid Mechanics,* Addison-Wesley, Reading, Mass.

LANTELME, F., TURQ, P., and SCHOFIELD, P. (1979), *J. Chem. Phys. 71,* 2507.

LARSEN, B. (1979), *Theory of Ionic Fluids. Statistical Mechanics of Charged Hard Spheres,* Dr. Tech. Thesis, University of Trondheim, Trondheim, Norway. See also *J. Chem. Phys. 68,* 5411 (1978).

LEBOWITZ, J. L. (1968), *Annu. Rev. Phys. Chem. 19,* 389.

LEBOWITZ, J. L., PERCUS, J. K., and VERLET, L. (1967), *Phys. Rev. 153,* 250.

LEVESQUE, D., and ASHURST, W. T. (1974), *Phys. Rev. Lett. 33,* 277.

LEVESQUE, D., and VERLET, L. (1970), *Phys. Rev. A2,* 2514.

LEVESQUE, D., and WEIS, J. J., and PATEY, G. N. (1980), *J. Chem. Phys. 72,* 1887.

LIGHTHILL, M. J. (1959), *Fourier Analysis and Generalized Functions,* Cambridge University Press, Cambridge.

LOVESLEY, S. W. (1973), *J. Phys. C6,* 399.

MADDEN, P. A., and KIVELSON, D. E. (1984), *Adv. Chem. Phys. 56,* 467.

MADDEN, W. G. (1981), *J. Chem. Phys. 75,* 1984.

MADDEN, W. G., and RICE, S. A. (1980), *J. Chem. Phys. 72,* 4208.

MARTIN, P. C. (1968), *Measurement and Correlation Functions,* Gordon and Breach, New York.

MATHEWS, J., and WALKER, R. L. (1965), *Mathematical Methods of Physics,* W. A. Benjamin, New York.

MAYER, J. E. (1942), *J. Chem. Phys. 10,* 629.

MAYER, J. E. (1950), *J. Chem. Phys. 18,* 1476.

MAZENKO, G. F. (1972), *Phys. Rev. A6,* 2545.

MAZENKO, G. F., and YIP, S. (1977), in *Modern Theoretical Chemistry* (B. J. Berne, ed.), Vol. 6: *Statistical Mechanics,* Part B, p. 181, Plenum Press, New York.

MCDONALD, I. R., and SINGER, K. (1973), *Chem. Br. 9,* 54.

MCMILLAN, W. G., and MAYER, J. E. (1945), *J. Chem. Phys. 13,* 276.

MCQUARRIE, D. A. (1974), *Statistical Mechanics,* Harper & Row, New York.

MEHROTRA, P. K., MEZEI, M., and BEVERIDGE, D. L. (1983), *J. Chem. Phys. 78,* 3156.

METROPOLIS, N., METROPOLIS, A. W., ROSENBLUTH, M. N., TELLER, A. H., and TELLER, E. (1953), *J. Chem. Phys. 21,* 1087.

MILLER, D. G. (1966), *J. Phys. Chem. 70,* 2639.

MITCHELL, D. J., MCQUARRIE, D. A., SZABO, A., and GROENEVELDT, J. (1977). *J. Statist. Phys. 17,* 15.

MONROE, J. L., and FRIEDMAN, H. L. (1977), *J. Chem. Phys. 66,* 955.

MORI, H. (1965), *Prog. Theor. Phys. (Japan) 33,* 423.

MORI, H., OPPENHEIM, I., and ROSS, J. (1962), in *Studies in Statistical Mechanics* (J. DeBoer and G. E. Uhlenbeck, eds.), Vol. 1, p. 217, North-Holland, New York.

MORITA, T., and HIROIKA, K. (1961), *Prog. Theor. Phys. 25,* 537.

NARTEN, A. H., JOHNSON, E., and HABENSCHUSS, A. (1980), *J. Chem. Phys. 73,* 1248.

NARTEN, A. H., THIESSEN, W. E., and BLUM, L. (1982), *Science 217,* 1033.

NEILSON, G. W., and ENDERBY, J. E. (1983) *Proc. Roy. Soc. Lond.* A *390,* 353.

NIENHUIS, G., and DEUTCH, J. M. (1972), *J. Chem. Phys. 56,* 5511.

NIJBOER, B. R. A., and RAHMAN, A. (1966), *Physica 32,* 415.

OUTHWAITE, C. W. (1975), *Statist. Mech. 2,* 188 (Chemical Society, London, Specialist Periodical Reports).

PATEY, G. N. (1977), *Mol. Phys. 34,* 427.

PATEY, G. N. (1978), *Mol. Phys. 35,* 1413.

PERCUS, J. K. (1964), in *The Equilibrium Theory of Classical Fluids* (H. L. Frisch and J. L. Lebowitz, eds.), W. A. Benjamin, New York.

PETTIT, B. M., and ROSSKY, P. J. (1982), *J. Chem. Phys. 77,* 1451.

RAHMAN, A. (1964), *Phys. Rev. A136,* 405.

RAHMAN, A., and STILLINGER, F. H. (1971), *J. Chem. Phys. 55,* 3336.

RAHMAN, A., STILLINGER, F. H., and LEMBERG, H. L. (1975), *J. Chem. Phys. 63,* 5223.

RASAIAH, J. C. (1972), *J. Chem. Phys. 56,* 3071.

RASAIAH, J. C., CARD, D. N., and VALLEAU, J. P. (1972), *J. Chem. Phys. 56,* 248.

ROSSKY, P. J., and DALE, W. D. T. (1980), *J. Chem. Phys. 73,* 2457.

ROSSKY, P. J., and FRIEDMAN, H. L. (1980), *J. Chem. Phys. 72,* 5694.

ROTENBERG, M., BIVINS, R., METROPOLIS, N., and WOOTEN, J. K. (1960), *The 3-j and 6-j Symbols,* MIT Press, Cambridge, Mass.

ROTHSCHILD, W. G. (1970), *J. Chem. Phys. 53,* 990.

RUSHBROOKE, G. S., STELL, G., and HØYE, J. S. (1973), *Mol. Phys. 26,* 1199.

RYCKAERT, J. P., and CICCOTTI, G. (1983), *J. Chem. Phys. 78,* 7368.

RYCKAERT, J. P., CICCOTTI, G., and BERENDSEN, H. J. C. (1977), *J. Comp. Phys. 23,* 327.

SANDSTROM, D. R., and LYTLE, F. W. (1979), *Annu. Rev. Phys. Chem. 30,* 215.

SCHOFIELD, P. (1975), *Statist. Mech. 2,* 1 (Chemical Society, London, Specialist Periodical Reports).

SCHRÖDINGER, E. (1960), *Statistical Thermodynamics,* Cambridge University Press, Cambridge.

SCOLES, G. (1980), *Annu. Rev. Phys. Chem. 31,* 81.

SKINNER, J. L., and WOLYNES, P. G. (1978), *J. Chem. Phys. 69,* 2143.

SKÖLD, K., ROWE, J. M., OSTROWSKI, G., and RANDOLPH, P. D. (1972), *Phys. Rev. A6,* 1107.

SMITH, W. R. (1974), *Can. J. Phys. 52,* 2022.

SMITH, W. R., and HENDERSON, D. (1970), *Mol. Phys. 19,* 411.

STEELE, W. A. (1969), in *Transport Phenomena in Fluids,* arcel Dekker, New York, H. J. M. Hanley, ed. 230–249.

STEELE, W. A. (1976), *Adv. Chem. Phys. 34,* 2.

STELL, G. (1964), in *The Equilibrium Theory of Classical Fluids* (H. L. Frisch and J. L. Lebowitz, eds.), W. A. Benjamin, New York.

STELL, G. (1967), in *Graph Theory and Theoretical Physics* (F. Harary, ed.), Academic Press, New York.

STELL, G. (1975), in *Phase Transitions and Critical Phenomena* (C. Domb, ed.), Vol. 5B, Academic Press, London.

STELL, G. (1977), in *Modern Theoretical Chemistry* (B. J. Berne, ed.), Vol. 5: *Statistical Mechanics,* Part A, p. 47, Plenum Press, New York.

STELL, G., HØYE, J. S., and PATEY, G. N. (1981), *Adv. Chem. Phys. 48,* 183.

STILLINGER, F. H. (1975), *Adv. Chem. Phys. 31,* 1.

STILLINGER, F. H. and LOVETT, R. (1968), *J. Chem. Phys. 49,* 1991.

STILLINGER, F. H. and RAHMAN, A. (1974), *J. Chem. Phys. 60,* 1545.

SUNG, W., and DAHLER, J. S. (1983), *J. Chem. Phys 78,* 6264.

SWOPE, W. C., ANDERSEN, H. C., BERENS, P. H., and WILSON, K. R. (1982), *J. Chem. Phys. 76,* 637.

SZABO, A., SCHULTEN, K., and SCHULTEN, Z. (1980), *J. Chem. Phys. 72,* 4350.

SZABO, A., SCHULTEN, K., and SCHULTEN, Z. (1981), *J. Chem. Phys. 74,* 4426.

THIELE, E. (1963), *J. Chem. Phys. 39,* 474.

VALLEAU, J. P., COHEN, L. K., and CARD, D. N. (1980), *J. Chem. Phys. 72,* 5942.

VANHOVE, L. (1954), *Phys. Rev. 95,* 249.

VAN KAMPEN, N. G. (1974), *Physica 74,* 215, 239.

VERLET, L., (1967), *Phys. Rev. A159,* 98.

VERLET, L., (1968), *Phys. Rev. A165,* 201.

VERLET, L., and WEIS, J. J. (1972), *Mol. Phys. 24,* 1013.

WAISMAN, E., and LEBOWITZ, J. L. (1972), *J. Chem. Phys. 56,* 3093.

WANG, M. C., and UHLENBECK, G. E. (1945), *Rev. Mod. Phys. 17,* 323.

WATTS, R. O. (1973), *Statist. Mech. 1,* 1 (Chemical Society, London, Specialist Periodical Reports).

WATTS, R. O., and MCGEE, L. J. (1976), *Liquid State Chemical Physics,* Wiley, New York.

WAX, N. (1954), *Noise and Stochastic Processes* (N. Wax, ed.), Dover, New York.

WEEKS, J. D., CHANDLER, D., and ANDERSEN, H. C. (1971), *J. Chem. Phys. 54,* 5237.

WERTHEIM, M. S. (1963), *Phys. Rev. Lett. 10,* 321.

WERTHEIM, M. S. (1964), *J. Math. Phys. 5,* 643.

WERTHEIM, M. S. (1971), *J. Chem. Phys. 55,* 4291.

WERTHEIM, M. S. (1973), *Mol. Phys. 26,* 1425.

WERTHEIM, M. S. (1979), *Annu. Rev. Phys. Chem. 30,* 471.

WERTHEIM, M. S. (1980), *J. Chem. Phys. 73,* 1398.

WIDOM, B. (1963), *J. Chem. Phys. 39,* 2808.

WIGNER, E. P. (1959), "*Group Theory,*" Academic Press.

WOLYNES, P. G. (1976), *Phys. Rev. A13,* 1235.

WOLYNES, P. G. (1978) *J. Chem. Phys. 68,* 473.

WOLYNES, P. G., and DEUTCH, J. M. (1977), *J. Chem. Phys. 67,* 733.

WOOD, D. W. (1979), in *Water, A Comprehensive Treatise* (F. Franks, ed.), Vol. 6, Plenum Press, New York.

WOOD, R. H., LILLEY, T. H., and THOMPSON, P. T. (1978), *J. Chem. Soc. Far. Soc. Trans. I 74,* 1301.

YARNELL, J. L., KATZ, M. J., WENZEL, R. G., and KOENIG, S. H. (1973), *Phys. Rev. A7,* 2130.

ZWANZIG, R. (1965), *Annu. Rev. Phys. Chem. 16,* 67.

Index

Activity, 18
  absolute, 9
  solute, 203, 209
Adjoint matrix or operator, 276, 286, 296
A^{ex}, excess free energy, 111
  virial expansion, 129 ff
Analytical solutions. *See* Exact results
Argon
  dynamic structure factor, 234
  mean square displacements, 248
  memory function for velocity correlations, 293
  pair correlation function, 100
  $S(k)$, various methods, 107
  van Hove self-correlations, 230
Atom–atom correlation functions, 171 ff
Axioms, 5

BBGKY hierarchy, 307
BD, Brownian dynamics simulation, 274 ff
Bjerrum's approximation, 151
Blip function theory, 186 ff, 197, 198
Blum–Torruella invariant expansion, 158 ff, 324
Boltzon, 40, 44 ff, 48
  mixture, 60
Born–Oppenheimer level, 85, 157
Boson, 38, 40, 60
Brownian motion, 244
  of oscillator, 256
  of rotator, 251 ff
Brownon, 244

Canonical ensemble
  distribution function, 6 ff
  equations, 13
  thermodynamics, 12 ff
Carnahan–Starling equation, 133, 148
Central limit theorem, 20
Chemical bond, 84
Chemical reaction, 55 ff
  equilibrium constant, 57, 61
  equilibrium constant, mass effects, 70
  models, 259
  rate constant for various models, 264
  rate constant theory, 258 ff
  role of heat bath, 263
  transition state theory, 262
Classical limit, 63 ff
  for Einstein crystal model, 33
  independent indistinguishable particles, 43 ff
Cluster expansion, 111 ff
Cluster function, 118
Cluster integral
  irreducible, 131
  multicomponent system, 131
  reducible, 135
Cluster perturbation theory, 189 ff
  optimized, 192 ff
Coherence, 243
Compressibility, isothermal
  definition, 9
  fluctuation formula, 28
  as functional of $g(r)$, 89
  for mixture, 28
Configurational free energy, 111
Configuration integral, 18, 111
Continuity equation, 240
Convolution, 138
  for Fourier–Laplace transforms, 238
  for molecular functions, 323
Correlation function, spatial, 77
  K-particle, 203
  pair, 82
  three-particle, 90
Correlation function, time, 219 ff
Correlation matrix, 291, 296 ff
Correlation time, 221
  T_1 and T_2, 257
Coulomb potential, 135
Crystal
  Debye model, 35, 59
  Einstein model, 31, 60
  heat capacity, 31
Current–current correlation function
  distinct and self parts, 233
Current field
  longitudinal and transverse parts, 232
  particle current field, 232

de Broglie
  thermal wavelength, 18, 47, 49, 68
Debye–Hückel
  approximation, linearized, 140 ff
  EXP approximation, 142, 156
  extended theory, 142
  limiting law, 156
  shielding factor, kappa, 141
Density
  density–density correlation function, 82, 229

Density (*cont.*)
field, 78, 79, 104, 228
local. *See* Density, field
operator, 78
Dielectric constant, thermodynamic, 167 ff, 181
for ISM, 173, 174
Diffraction, 104 ff
neutron, 81, 106
x-ray, 106
Diffusion coefficient
and diffusive displacement, 248
distinct, 234, 241
practical or fundamental, 234
reference frame, 234
self, 233, 240, 247, 269 ff, 283
self, Einstein expression for, 239
in terms of random force correlations, 292
Diffusion tensor, 277
Dipolar hard spheres
dielectric constant, 167
free energy, 166
MSA, 162 ff
Dirac delta function, 25, 66
orientational, 252
three-dimensional, 105
Direct correlation function, 137 ff
as a functional derivative, 136
as renormalized potential, 311
site–site, 175 ff
Dynamical variable, 2, 19, 216, 219
electric current, 225
electric polarization, 226
fluctuating part, 220
response to perturbation, 224

Einstein frequency, 292
Electrical conduction, 226 ff
Electrical potential, 140
Ensemble, 3
canonical, 4 ff, 75
distribution function, 5
equivalence of various, 24
grand, 4, 5, 39, 75, 88, 90
microcanonical, 4, 5, 36, 99
moments, ensemble average, 19
nonequilibrium, 217, 224, 246
replica systems in phase space, 217
Entropy
microcanonical ensemble equation, 36
of mixing, 60, 72
Ergodic theory, 6
Exact results
MSA for dipolar hard spheres, 162 ff
MSA for ionic system, 150
PY for hard spheres, 146
Excess free energy A^{ex}
definition, 72
virial expansion, 111 ff
EXP approximation, 142, 199
Exponential operator
definition, 26
perturbation expansion, 73

Fermion, 38, 40, 60
distribution function, 41
Fermi level, 42
Fick's law of diffusion, 267
Fluctuation–dissipation theorem, 225
second, 265, 317
Fluctuations, 19 ff
composition, 22 ff, 201
density, 21, 89
density, and κ_T, 28
energy, 20, 21
Fokker–Planck equation, 250, 266
propagator, 250
Fourier transform, 104
applied to ring integral, 322
of a convolution, 320
of Debye or Yukawa potential, 321
Fourier–Laplace transform, 220
of Gaussian function, 320
inverse transform, 319
for orientation-dependent functions, 323 ff
of spherical function, 320
in three dimensions, 319
Friction coefficient, 239
Functional, 75
Functional derivative, 114 ff
Lemma one, 123 ff
Functional expansion of A^{ex}, 199
Fused salt model
for LiBr, velocity self-correlations, 271
for LiCl, pair correlations, 81
primitive model, pair correlations, 153

Gaussian function, 20, 25, 47
Fourier transform, 320
integral over, 72
Gaussian–Markov process, 243, 245
Generalized hydrodynamics, 299 ff
Generalized Langevin equation, 265, 289 ff
characteristic frequencies, 290
of the phase-space density, 309
Gibbs–Bogoliubov inequality, 185 ff
Gibbs–Duhem equation for ionic system, 156

Gibbs equation, 10
Grand ensemble, 14 ff
  equations, 18, 19, 27
  $g(r)$, 82
  thermodynamics, 16 ff
Graph
  bond, 119
  bond and face, 122
  bridge, 139
  chain, 197, 320
  circle, 119, 122
  connected, 122
  definition, 121
  expansion for PY approximation, 146
  hair, 126 ff, 130
  hypervertex, 199
  irreducible, 122
  lemmas, 123 ff
  product graph, 123
  ring, 198, 322
  simple, 122
  symmetry number, 120, 131
  topological reduction, 121, 123, 188 ff
  value, 120, 122
Green's function, 218, 249

Hamiltonian
  equations of motion, 216
  function, 67, 71, 93
  level of, 85
  operator, 26, 29
  operator eigenfunction, 26, 64
  potential term, pairwise additivity, 85
  spin and lattice parts, 300
Hankel transform, 160, 325
Hard-sphere fluid
  dynamical structure factor, 314
  equation of state, 147, 148
  pair correlation function, 80, 87, 149
  pair potential, 83
  PY approximation for, 146
  as reference system, 183 ff
  $S(k)$, 107
  T-fluctuations, 108
  various approximations, 148
  virial coefficients, 113, 133
  virial expansion, 132
Heat bath, 6
Heat capacity, 26
Heisenberg equation of motion, 217, 218
Hermitian, anti-, 219
Hilbert space, 286, 287
HNC, hypernetted chain approximation, 150 ff
  for fused salt, 153
  LHNC for dipolar hard spheres, 196
  with reference system, RHNC, 195

Ideal gas, 27
  quantal, 37 ff
Independent subsystems, 29 ff
  distinguishable, 30 ff
  distinguishable particles, 59
  indistinguishable particles, 38 ff
  indistinguishable particles, classical limit, 43 ff
  limitations, 58
Inelastic scattering
  of electromagnetic radiation, 105, 281
  of neutrons, 278 ff
Inhomogeneous system
  due to fixed particle, 78 ff
  due to K fixed particles, 203
Integral equation
  extended MSA, 196
  HNC, 150 ff
  LHNC, 196
  MSA, 148 ff, 162 ff
  numerical solution, 152 ff
  perturbation theory, 195
  PY, 144 ff
  RHNC, 195
Interaction representation, 301
Intermolecular force, 83 ff
Invariant expansion, 158 ff
  convolution table, 162
Invariant pairwise function, 158, 323
Ionic systems
  Debye–Hückel approximation, 140 ff
  dielectric constant, ϵ_ρ, 168 ff
  HNC approximation, 151
  ionic solutions, 202
  MSA, 149
  Stillinger–Lovett moment conditions, 143
ISA, interaction site approximation, 178 ff
ISM, interaction site model, 171, 174 ff

Kac inverse range parameter, 198
Kinetic theory, 306 ff
  BBGKY hierarchy, 307
  from generalized Langevin equation, 308
  linearized Boltzmann equation, 313
  one-particle phase space distribution, 310
  particles with hard cores, 312 ff
  renormalized Enskog equation, 313
  Vlasov equation, 307
  Zwanzig–Vlasov equation, 311
Kirkwood–Buff theory, 201 ff
Kramers equation, 249

Kramers–Kronig relations, 299
Kubo theory, 223 ff

Lambda expansion, 184
Langevin's equation, 243 ff
 frictional force, 245
 generalized, 285 ff, 289
 generalized, continued fraction form, 293 ff
 mean square displacement, 248
 random force, 245
 simulation, 274
Lennard–Jones model. *See* 6–12 model
Linear response theory, 223 ff
 applications, 236
 for local fields, 235
 response function, 239
Line shape function, 238, 255
 Lorentzian, 256
 Lorentzian, and sum rules, 267
Liouville
 adjoint operator, 276
 equation, 218, 307
 operator, 216 ff
Local density, 78, 104
Long-range potential, 135
Long-time tails, 273

Magnetic resonance, 257
 Larmor precession, 258
Markov
 chain, 102
 process, 245
Mass
 in first quantum correction, 70
 reduced, 49
Mayer f function, 118
 in blip function theory, 186
 in cluster perturbation theory, 189
 in graph theory, 119
 with internal coordinates, 132
 with nonpairwise interactions, 207
 in solution theory, 204
MC, Monte Carlo simulation, 101
 $g(r)$, 100
 move algorithm, 102
 pass, 103
McMillan–Mayer level, 86
McMillan–Mayer solution theory, 203 ff, 209 ff
MD, molecular dynamics simulation, 96 ff
 $g(r)$, 100
 rate processes, 269
 Verlet algorithm, summed form, 109
Mean first passage time, 260 ff
Mean spherical approximation. *See* MSA
Memory function, 265, 289
 model, with Einstein frequency, 292
 model equation, 291
 moment expansion, 305
Metal
 crystalline, characteristic temperature, 34
 free electron model, 43
Methane, reorientation, 222
Methylene chloride, reorientation, 255
Mode
 -coupling theory, 297, 299, 314
 electronic, 54
 hydrodynamic, 296
 nuclear, 55
 rotation, 49 ff
 translation, 46
 vibration, 53
Molecular correlation functions, 157 ff
 coordinate system for, 160
 invariant expansion, 158 ff, 324
 Ornstein–Zernicke equation for, 158
Molecular parameters, 51
Moment of inertia, 49, 52
Moments of $h(r)$
 second, 142, 174
 zeroth, 142, 202
Moments of the power spectrum, 220, 316
Momentum operator, 65
Mori continued fraction, 293 ff
 various truncation schemes, 295
MSA, mean spherical approximation, 148 ff
 for dipolar hard spheres, 162 ff
 for ionic solutions, 151
 primitive model, 149
 with reference system, 196

Neumann expansion, 327
Neutron scattering length, 279
 coherent, 280
$NiCl_2$ solution in water
 cation–cation correlation, 206
 inelastic neutron scattering data, 282
Nitrogen, atom–atom pair correlation function, 179
Noisy oscillator, 256 ff
Norm, 243
Notation
 activity, z, 18, 75
 Boltzmann's constant k_B, 13
 complex number: Re F, Im F, 225
 de Broglie thermal wavelength, Λ, 18
 Dirac delta function, $\delta(y)$, 66
 direct correlation function, $c(1\ 2)$, 138 ff

exponential, e^Z, 26
for generalized Langevin equation, 290
isothermal compressibility, κ_T, 9
Liouville operator $H^\times$, 216
matrix algebra, 23
order in limit, $O(x)$, 70
packing fraction, η, 133
pair correlation function, $h(1\ 2)$, 137
Planck's constant, h, 18, 31, 63
potential of average force, $w(1\ 2)$, 138
reduced concentration of dipoles, y, 165
species of components, s, 10
spherical Bessel function, $j_m(x)$, 326
thermodynamic variables, 9
trace, tr, 65
unit step function, $\theta(x)$, 142
Wigner D function, $D^l_{mn}(A \leftarrow B)$, 325
Wigner $3j$ coefficient, (:::), 326
y correlation function, $y(1\ 2)$, 138

Occupation number, 39, 40
Ornstein–Zernicke equation, 137 ff, 149, 150
for effective direct correlation function, 208
for ion–dipole mixture, 168
matrix form for many components, 155
for molecular correlation functions, 158, 161
for multicomponent system, 155
with orientation-dependent correlations, 323
for site–site correlations, 174 ff
under invariant expansion, 163
ORPA, optimized random phase approximation, 193
and MSA, 199
ORPA + B_2, 193
Osmotic pressure, 202

Pair correlation function
canonical ensemble, 82
as chemical equilibrium constant, 87
core condition, 86, 142
Debye–Hückel approximation, 141
Fourier transform, 104
as functional derivative of A^{ex}, 118
as functional derivative of Ξ, 116
grand ensemble, 82
matrix form for many components, 155
moments (of $h(r)$), 142
from simulation, 99
site–site correlation function, 171 ff
solute–solute pair, 205
theory, 137 ff
and thermodynamic coefficients, 88, 89
virial expansion, 132

Pair potential model
Born–Mayer, 84
charged hard spheres, 84
charged soft spheres, 199
charged soft spheres, velocity correlations, 271
dipolar hard spheres, 162
exp-six, 84
hard sphere, 83
ion–dipole mixture, 168
ISM, 175
Lennard–Jones. *See* 6–12 model
molecule–molecule case, 157 ff
Morse, 84
primitive model, 84
6–12, 81, 83, 84
square well, 134
Pairwise additivity
N-particle potential, 85
for solution models, 207
Particles with internal structure, 44, 45, 68, 157
Partition function
canonical ensemble, 8
Einstein model crystal, 32
electronic, 54
grand, 15, 39
linear rotator, 50
nonlinear rotator, 52
one mode, 56
one particle, 30, 36, 46
as a phase integral, 67, 68
quantum correction, 69
simple harmonic oscillator, 31
spins on a lattice, 36
as a trace, 64
vibrational mode, 53
Periodic boundary conditions, 94
Perturbation theory
equilibrium, 183 ff
as functional expansion, 199
under linear response, 300 ff
Phase space, 67, 72
density operator, 308
Dirac delta function, 218
distribution function, 72
time-dependent distribution function, 306
Poisson bracket, 216
Potential of average force
pairwise, 86
two-point and three-point, 91
Projection operator, 286, 287, 315
Propagator, 218, 249
PY, Percus–Yevick approximation, 144 ff
compression factor of hard-sphere fluid, 148
graphical expansion of, 145
$g(r)$, ISM for nitrogen, 179

PY, Percus–Yevick approximation (*cont.*)
  $g(r)$ for hard-sphere fluid, 149
  $g(r)$ for hard-sphere mixture, 80
  $S(k)$ of hard-sphere fluid, 107

Quantum state, 3, 5

Radial distribution function, 82
Random force, 245 ff, 265, 274 ff, 289, 296 ff
Random number, 102
Random process, 265
Renormalization, 188, 311, 313
Replica system of ensemble, 3, 4
Restricted primitive model, 84
  electrolyte, osmotic coefficient, 151
  fused salt, spatial correlations, 153
RISM, reference interaction site model, 178 ff
  site–site correlation function, 181
Rotational constant, 51, 57
Rotational invariant, 160, 180
  convolution, 162

Scattering, 105, 278 ff. *See also* Diffraction; Inelastic scattering
Second virial coefficient
  chemical model for, 113
  as functional of pair potential, 112
  in solution theory, 202, 210
Setchenow coefficient, 213
Simple harmonic oscillator
  in Einstein crystal model, 31, 60
  equilibrium ensemble, 30
  under Langevin's equation, 256
Simulation, 93 ff
  BD, 274 ff
  boundary conditions, 94, 95
  control chart, 98
  MC, 81, 87, 101 ff
  MD, 81, 96 ff, 269 ff
6–12 model
  long-time tail, 317
  memory function, 293
  pair correlation function, 100
  parameters, 73
  second virial coefficient, 113
  van Hove distinct correlations, 231
  velocity self-correlation, 222
Smoluchowski equation, 251
  model for chemical reaction, 261 ff
  operator, 251, 277
  operator, adjoint, 276, 283
  propagator, 251
  simulation, 278
  for spherical rotator, 252
Solution theory
  Adelman, 207 ff
  chemical potential of solute, 211
  Henry's law, 209
  Kirkwood–Buff, 24, 28, 201 ff
  McMillan–Mayer, 203 ff, 209 ff
  osmotic pressure, 202, 204, 205, 211
  second virial coefficient of p_{osm}, 210, 212
  solubility interaction, 213
  solute activity, 203
  solute–solute correlations, 205
  use of Born–Oppenheimer-level results, 206
Solvent-averaged potential, 205, 206, 208
  K-point, 203
  pair, 204, 210 ff
  pairwise additivity, 207
Spherical Bessel function, 163, 323, 326, 327
Spherical harmonic, 252, 254, 267
  generalized, 160, 180, 323, 325
Spin state, 51
Square-well model, 113
Stationarity, 219
Statistical thermodynamics, 45 ff
Stillinger–Lovett moment conditions, 143
Stirling approximation, 44
Structure factor
  of Ar liquid, 107
  dynamical, 234, 235, 241, 266
  partial, 105, 202
Sum rule, 220, 240
Susceptibility
  Kramers–Kronig relations, 299
  for local fields, 235, 299
  as a response function, 225
Symmetry number, 50, 52, 120

Temperature, 6
  negative, 37
Temperature, characteristic
  Debye, 33, 34
  for rotation, 50, 52
  for vibration, 53
Thermodynamic functions
  as functionals of pair correlations, 88
  as functionals of U_N, 72, 76
  notation, 9
  for rotator, 53
  for simple harmonic oscillator, 31, 54
Thermodynamic limit, 21
Thermodynamics, 9 ff
  canonical ensemble, 11
  grand ensemble, 16

Time average, 2, 3
from MD simulation, 99, 270
Time correlation function, 219 ff, 288
cumulant expansion, 303
current–current, 232
distinct, 221
under Langevin equation, 246
memory function expansion, 305
mode–mode, 297
moment expansion, 301
self, 221, 222
as the susceptibility, 225
Time ordering, 301 ff
Time reversal symmetry, 221
Topological reduction, 188 ff
of chain sum, 197

van Hove correlation function, 226 ff
intermediate, 229, 297
intermediate, from Langevin equation, 266
partial dynamical structure factor, 234
self, 266
self and distinct, 228

Variance, 243
Virial
coefficient, 111. *See also* Second virial coefficient
equation, 77
expansion for hard spheres, 132 ff
expansion of A^{ex}, 111, 129 ff
expansion of p, 111
expansion of p_{osm}, 211

Water
correlation functions, 81
mean square displacements, 248
model, 157, 158, 174, 175
rotational correlation times, 254
velocity self-correlation, 222
Wave function
noninteracting particles, 66
orthonormal set, 26, 66
WCA theory, 193
Wigner rotation functions, 253, 325
Wigner 3j coefficient, 324, 326

Zwanzig–Vlasov equation, 311